W9-COB-037

High-performa
Chromatography

Edition

CHRISTIE
Lipid Metabolism in Ruminant Animals

Journals of related interest

Progress in Lipid Research

Food & Chemical Toxicology

Journal of Pharmaceutical and Biomedical Analysis

High-performance Liquid Chromatography and Lipids

A PRACTICAL GUIDE

by

WILLIAM W. CHRISTIE

The Hannah Research Institute, Ayr, Scotland

PERGAMON PRESS

OXFORD · NEW YORK · BEIJING · FRANKFURT
SÃO PAULO · SYDNEY · TOKYO · TORONTO

U.K.	Pergamon Press, Headington Hill Hall, Oxford OX3 0BW, England
U.S.A.	Pergamon Press, Maxwell House, Fairview Park, Elmsford, New York 10523, U.S.A.
PEOPLE'S REPUBLIC OF CHINA	Pergamon Press, Room 4037, Qianmen Hotel, Beijing, People's Republic of China
FEDERAL REPUBLIC OF GERMANY	Pergamon Press, Hammerweg 6, D-6242 Kronberg, Federal Republic of Germany
BRAZIL	Pergamon Editora, Rua Eça de Queiros, 346, CEP 04011, Paraiso, São Paulo, Brazil
AUSTRALIA	Pergamon Press Australia, P.O. Box 544, Potts Points, N.S.W. 2011, Australia
JAPAN	Pergamon Press, 8th Floor, Matsuoka Central Building, 1-7-1 Nishishinjuku, Shinjuku-ku, Tokyo 160, Japan
CANADA	Pergamon Press Canada, Suite No. 271, 253 College Street, Toronto, Ontario, Canada M5T 1R5

First edition 1987

Library of Congress Cataloging in Publication Data

Christie, William W.
High-performance liquid chromatography & lipids.
Bibliography: p.
1. Lipids—Analysis. 2. High Performance liquid chromatography. I. Title. II. Title: High-performance chromatography and lipids.
QP751.C488 1987 547.7'7046 87-2395

British Library Cataloguing in Publication Data

Christie, William W.
High-performance liquid chromatography & lipids.
1. Lipids—Analysis 2. High performance liquid chromatography
I. Title
547.7'7046 QP751

ISBN 0-08-034212-4

Phototypesetting by Thomson Press (India) Limited, New Delhi
Printed in Great Britain by Richard Clay Ltd., Bungay, Suffolk

Preface

Lipid analysts have been slow to take up high-performance liquid chromatography (HPLC), largely because of the lack of an all-purpose detection system suited to lipids. Nonetheless, with perseverance and ingenuity, some considerable advances have been made, and a substantial body of published work now exists. The subject is not suited to "recipe" treatment, as the wide variation in the equipment available to analysts, especially the detectors, means that some developmental work may be necessary to reproduce published separations. The practice of HPLC is thus a skilled occupation, in which the analyst can take some pride. About three years ago, I decided to attempt to replace all of the thin-layer chromatography methods in my laboratory with HPLC equivalents, and I have very nearly accomplished this. In addition to improving efficiency, this has resulted in a much cleaner working environment. There are still many gaps in our knowledge that remain to be filled, and many published procedures could be improved. Hopefully, this book will stimulate further endeavours. In the Preface to *Lipid Analysis*, I stated that I hoped the book would "remain on the laboratory bench, not on the library shelf". That comment is equally apposite here.

I am grateful to the Director and Council of the Hannah Research Institute and to the Department of Agriculture and Fisheries for Scotland for permission to write this book. John H. Shand read and criticized each of the chapters as they were written, and his assistance is gratefully acknowledged. My final thanks must go to my wife, Norma, and to our two sons, Paul and Matthew, who have had to put up with computers and papers strewn around our home for 18 months.

WILLIAM W. CHRISTIE

Contents

CHAPTER 1

An Introduction to Lipids and High-performance Liquid Chromatography

A. Why Separate Lipids by High-performance Liquid Chromatography?

The study of those compounds, that are included under the diffuse generic term of "lipids", has assumed considerable importance in recent years with the recognition that they are involved in many vital biological processes in animals, plants and microorganisms. Disturbances in lipid metabolism are known to accompany a variety of disease states, and the role of lipids in heart disease especially remains an unresolved controversy. A valuable perspective on the part played by fats in nutrition and disease has been provided by Gurr [263]. Such aspects of lipid metabolism have caught the public eye, but relatively few non-scientists recognize the very many essential functions of lipids. These have recently been reviewed by Hadley [266].

It has long been known, for example, that lipids serve as a major storage form for energy in animal and plant tissues, and that they are responsible for maintaining the structural integrity of cells as the principal components of the membranes. The dietary essential fatty acids, linoleic and linolenic acids, and the longer-chain components derived from them may be required to confer some distinctive physical properties upon the membranes, in addition to serving as the precursors of the prostanoids [328]. Lipids in the membranes of nervous tissue function in the transmission of electrical signals. Recently, it has become apparent that certain lipids such as phosphatidylinositol and its metabolites are vital cellular messengers, while the "platelet-activating factor" is capable of regulating various cellular processes when present at very low concentrations in tissues. Lipids are also required for insulation, integumental waterproofing, detergency, lubrication and other interfacial phenomena. In marine mammals, lipids are used to adjust buoyancy, and they function in echo-location as the major constituents of the sonar lens. Many insect pheromones are derived from lipids.

Oils and fats are of great economic importance as agricultural products, and they are major items of international commerce. They are ingredients of all foods. Fat amounts to approximately 40% of the dietary calories in

Western Europe and North America, and its composition is obviously vital to good nutrition. In addition, fats contribute greatly to the palatability, taste and structure of foods. Fatty substances are of considerable industrial significance as the precursors for such materials as detergents and lubricants.

Methods for the analysis of lipids are therefore of great importance for many research, clinical and quality control applications. High-performance liquid chromatography or "HPLC" was at first rather neglected for such purposes, and is only now coming rapidly into use. Certainly, it has the potential to become one of the major tools in the hands of those analysts who wish to determine lipid compositions and structure.

Lipid analysts were at the forefront in the development of gas–liquid chromatography (GLC) and thin-layer chromatography (TLC), and these techniques provided the springboard for the explosive growth of knowledge in the fields of lipid chemistry and biochemistry that has occurred over the last 25 years. It may seem surprising, therefore, that lipid analysts have not been quick to take advantage of HPLC. Indeed, only a few years ago, the author wrote of the technique in relation to lipids that "the cynical observer will note that many more papers are appearing describing model separations than those describing analyses of real samples" [133]. In the last year or two, the situation has been changing rapidly, and many separations of lipids by means of HPLC have been described that cannot be rivalled by other methods. The resolving power of HPLC has never been in doubt. The main barrier to progress and a wider use of the technique has always been the fact that lipids lack chromophores which facilitate spectrophotometric detection. Fortunately, new "universal" detectors have been developed, and are now available commercially, that have the potential to overcome many of the remaining problems. In addition, some lipid analysts have shown great ingenuity in adapting their methodology to circumvent the deficiencies of detection systems.

It could now be claimed that there is no separation, which has been accomplished by TLC, which could not now be repeated with greater convenience and speed by HPLC, if detector limitations could be overcome. Almost certainly, therefore, lipid analysts will make a greatly increased use of HPLC in the years to come, probably mainly at the expense of TLC, but also to a certain extent of GLC. HPLC will be used most for the analysis of lipids of high molecular weight or of high polarity, or for lipids containing labile functional groups. It will also be used for the isolation of such compounds, both on a small and a larger scale, for further analysis by other procedures. GLC is likely to remain the preferred method for the analysis of stable non-polar lipids or their derivatives with a molecular weight of up to about 700, as high resolution is attainable at lower capital and running costs than by means of HPLC. Preparative-scale GLC, a technique which has always been considered for use only as a last resort (because of poor recoveries and doubtful reproducibility among other reasons), will fall into disuse as reversed-phase HPLC takes its place.

TLC has served lipid analysts well over the years, especially for lipid class separations, and the required equipment is inexpensive and versatile in that a wide range of adsorbents can be employed and complexing agents, such as silver nitrate, can easily be incorporated into the layers. It is particularly useful for a rapid qualitative examination of a sample, for developing methodology, and for routine screening of large numbers of samples for the occurrence of certain compounds. A wide variety of specific spray reagents can be used to identify functional groups, such as amine, phosphate or carbohydrate moieties, in lipids. On the other hand, TLC does have a number of disadvantages. Reversed-phase TLC is a valid technique, for example, but it is seldom used by lipid analysts as it lacks reproducibility and can be messy. Quantification of components separated by TLC is rarely straightforward, and dyes used to detect or identify specific lipids can interfere with subsequent analytical steps. In preparative-scale applications, small amounts of extraneous impurities, including silica gel, dyes and binding or complexing agents, can be inadvertently introduced into samples. Lipid analytical laboratories soon acquire a patina of grey dust on benches and equipment that appears to confound the most assiduous of cleaners. While no ill effects of this dust on the lungs of laboratory personnel appear to have been recorded, there will always be lingering doubts as to safety.

HPLC equipment is undoubtedly costly, especially when it is designed with a capacity for gradient elution, and it can be tedious and wasteful to set it up for the analysis of a single sample say. It does, however, offer a number of advantages over other techniques. A wide range of column packing materials are available for specific applications, and the columns can be used very many times. Most separations of interest can be achieved at ambient temperature under anaerobic conditions, so HPLC is particularly well suited to compounds with reactive functional groups. In addition, resolution tends to fall off only slowly with increasing sample size, analysis times can be short, retention times of compounds under set conditions are reproducible, there is a sound theoretical base to the technique, and the nature of the equipment implies that it is capable of a high degree of automation, especially with respect to the quantification of separated components. Occasionally it is possible to transfer elution conditions known to be suited to a TLC separation to HPLC, but generally some developmental work is necessary.

It is hoped that this book will serve as a valuable practical guide to lipid analysts who are unfamiliar with HPLC, and as a comprehensive source of reference and perhaps inspiration to those who already have experience of the technique. The author's own book [133] is recommended to those who seek a wider knowledge of the practice of lipid analysis in general, although much information on ancillary chromatographic techniques is also included below. Short books by Runser [712] and by Hamilton and Sewell [274] are recommended to those who seek introductory practical guides to HPLC. Further comprehensive texts are discussed below. Short reviews of HPLC and other techniques in the analysis of lipids, or of classes of lipids, have appeared [13, 15, 137, 199, 276, 377, 455, 503, 696, 827].

B. A Summary

It is necessary to have some knowledge of the theory of HPLC and of the various separatory modes possible to make best use of the technique, even if the analyst is unwilling to pursue the mathematics and physics of the processes involved. Information on this subject is included in Chapter 2. In addition, the novice analyst or the newcomer to HPLC is confronted by a bewildering array of column types, pumps, detectors and fittings of various sorts, and this chapter treats these aspects systematically in relation to lipid analytical problems. The choice or availability of a specific detector may determine whether a given separation can be attempted. Chapter 2 should assist in formulating the needs of the analyst before he approaches particular problems in the analysis of lipids. Various precautions are described to preserve column life, and to ensure reproducibility in quantification, and details of these obviously cannot be repeated in each of the subsequent chapters.

Chapter 3 summarizes the structures, chemistry and compositions of lipids in animals, plants and microorganisms. By referring to selected analyses, some of the important compositional features of organs, tissues and membranes are illustrated. A knowledge of the lipid types, which may be encountered in an analysis, might assist in determining the best procedure to adopt.

The first step in the analysis of lipids generally involves the preparation of a lipid fraction, relatively free of non-lipid contaminants, by means of solvent extraction of the tissue. Methods of achieving this are described in Chapter 4. Unwanted degradation of lipids can occur during the storage and handling of tissues and lipid samples, and autoxidation of unsaturated fatty acids can be especially troublesome. Methods of avoiding these difficulties during extraction are described that are applicable to all stages of analytical procedures, and they are again relevant to each of the subsequent chapters. Some preliminary fractionation of lipid samples into simple lipid, glycolipid and phospholipid groups may then be desirable to facilitate their analysis or the isolation of single lipid classes on a preparative scale.

In Chapter 5, methods are described for the isolation and analysis of simple lipid classes. The availability of a particular type of detector is often a crucial factor, and elution conditions have been established that are suited to those in common use. Generally, adsorption chromatography with silica gel has been favoured as the separation mode, but nitrile-bonded phases hold promise. Specific HPLC methods have been described for the determination of such biologically-important lipids as cholesterol and free fatty acids. If it is intended that both the individual simple and complex lipids in a sample be analysed in one step, it is essential to employ a detector of the "universal" or "mass-selective" type, and a few procedures suited to this goal have been published and are described here. Finally in this chapter, alternatives to HPLC methods are summarized briefly.

Most lipid analysts have tended to approach the problem of the analysis of phospholipids by first isolating them as a group (see Chapter 4), before proceeding to separate the individual phospholipid classes. Chapter 6 contains descriptions of methods which utilize either isocratic or gradient elution for the latter purpose. Isocratic elution procedures require less sophisticated equipment and are of particular value in small-scale preparative applications. More different components can usually be separated from a single sample when gradient elution techniques are employed. As with many other problems in HPLC analyses, the choice of method may be governed by the nature of the equipment, especially the detector, available to the analyst. Again, to give a wider perspective to the discussion, alternatives to HPLC methods are briefly described, and in the last section a method for phosphorus determination is given.

HPLC would probably be the preferred method nowadays for the isolation (as opposed to the analysis) of specific fatty acids (Chapter 7) after these have been converted to various derivatives, appropriate to the particular detection systems available. Methods of preparing the more widely-used derivatives are described. Methyl esters have been used for many years for fatty acid analysis by means of GLC and TLC, and their chromatographic properties are well understood, but UV-absorbing, e.g. phenacyl esters, or fluorescent derivatives can be detected and quantified with very great sensitivity by HPLC. Columns containing silver ions have been prepared to separate esters according to degree of unsaturation, although it is probably true to say that most analysts would like to see further improvements to these before they are adopted more widely. In contrast, HPLC separation in the reversed-phase mode is now capable of very high resolution with excellent reproducibility; the main disadvantage is that with complex fatty acid mixtures the order of elution does not follow a pattern that can be described in a simple manner. The separation of fatty acids with labile functional groups, such as hydroperoxides, has been greatly aided by the advent of HPLC procedures. On the other hand, GLC with capillary columns of fused silica affords very high resolution, relatively simple elution patterns, and ease of quantification; it will probably remain the method of choice for analytical purposes for the common range of fatty acids in plant and animal tissues for some time to come. A range of complementary techniques are available to identify fatty acids separated by chromatographic procedures, and these are indicated briefly in this chapter also.

Each lipid class in a tissue exists as a complex mixture of related compounds in which the composition of each molecule varies. Methods for the isolation or analysis of such "molecular species" from glycerolipids are described in Chapter 8. While there have been a few attempts to adapt silver ion chromatography and HPLC towards this end, these have met with comparatively little success. However, HPLC in the reversed-phase mode has been increasingly adapted to the purpose, to the extent that it is now probably the most useful of all the methods available to lipid analysts. Methods using a variety of detection systems have been described for the fractionation of intact triacylglycerols, phospholipids and glycoglycerolipids. In addition, resolution can sometimes be improved and alternative analytical techniques can be employed with the last two groups, if they are converted to non-polar diacylglycerol derivatives prior to the analysis.

The sphingolipids tend to occur at low levels only in tissues, but have many vital functions. One of the first successful applications of HPLC to lipids was in the analysis of such compounds, and this together with methods developed subsequently are described in Chapter 9. There have been two approaches to the separation of glycosphingolipid classes by means of HPLC. Methods for isolation of compounds in the native form are important as this, for example, enables their antigenicity to be tested directly. On the other hand, they can be separated and quantified more easily and with much greater sensitivity by conversion to the perbenzoylated or related derivatives. Similar approaches have been utilized for separations of molecular species of glycosphingolipids and of sphingomyelin, i.e. by HPLC in the reversed-phase mode. Until recently, GLC would have been favoured for the analysis of the long chain bases in sphingolipids, but now HPLC affords improved resolution and sensitivity.

In this final chapter, some miscellaneous separations are described that do not fit logically elsewhere in this book. For example, lipoprotein classes can now be adequately resolved by HPLC in the gel-permeation mode. Although the required columns are expensive, the overall cost is much less than for ultracentrifugation, which had hitherto been the standard analytical and/or preparative technique. Molecular species of sterol esters and other simple non-glycerolipids are readily isolated or analysed by means of HPLC in the reversed-phase mode. Lastly, such lipids as acyl–coenzyme A esters or acyl carnitines, which have very great biological importance but which are present at trace levels only in tissues, have been separated and analysed by means of HPLC.

C. Abbreviations

The following abbreviations are used at various points in the subsequent text.

BHT 2,6-di-*tert*-butyl-*p*-cresol
C cholesterol (free or unesterified)

CDH	ceramide dihexoside
CE	cholesterol esters
CER	cerebrosides
CMH	ceramide monohexoside
DEAE	diethylaminoethyl
DG	diacylglycerol
DGDG	digalactosyldiacylglycerol
DPG	diphosphatidylgylcerol (cardiolipin)
ECL	equivalent chain-length
EDTA	ethylenediaminetetraacetic acid
FFA	fatty acids (free or unesterified)
GLC	gas–liquid chromatography
HDL	high density lipoproteins
HPLC	high-performance liquid chromatography
LDL	low density lipoproteins
LPC	lysophosphatidylcholine
LPE	lysophosphatidylethanolamine
MG	monoacylglycerol
MGDG	monogalactosyldiacylglycerol
MS	mass spectrometry
NMR	nuclear magnetic resonance
ODS	octadecylsilyl (phase)
PC	phosphatidylcholine
PDME	phosphatidyldimethylethanolamine
PE	phosphatidylethanolamine
PG	phosphatidylglycerol
PI	phosphatidylinositol
PMME	phosphatidylmonomethylethanolamine
PS	phosphatidylserine
RI	refractive index
SPH	sphingomyelin
SQDG	sulphoquinovosyldiacylglycerol
TG	triacylglycerol
TLC	thin-layer chromatography
UV	ultraviolet
VLDL	very low density lipoproteins

CHAPTER 2

High-performance Liquid Chromatography: Theoretical Considerations and Equipment

A. Definitions

The term "*high-performance liquid chromatography*" or "HPLC" is now understood to denote a form of chromatography in which a mobile liquid phase is forced under controlled high pressure through a relatively narrow bore column containing a stationary phase, which can be a solid surface or a "liquid" phase, the latter bonded chemically to an inert support. The technique has also been termed "high-*pressure* liquid chromatography" or "high-*speed* liquid chromatography".

Modern technology has been applied to each stage of the separatory process, from the design of the pumping and detection systems to the manufacture of the column packing materials and fittings, in order to improve the quality and reproducibility of the separations attainable and the accuracy of the quantification. Of the equipment required, the first essential is a solvent delivery system, which will comprise solvent reservoirs and filters with perhaps some means of degassing the solvents, and most important of all a pulse-free pump (or pumps), capable of delivering solvents at precise specified flow-rates under pressures of up to 6000 psi; some means of generating a gradient in the composition of the mobile phase is desirable for many applications. Next, an injection system is needed to ensure that a sample can be introduced on to the column without releasing the pressure. The column itself is a polished stainless steel (or sometimes glass) tube with the appropriate end couplings, and is packed with a micro-particulate adsorbent or other phase; most laboratories will soon acquire a range of columns for use in different analyses. Finally, some form of detection system is required so that the progress of an analysis can be monitored, and each of the components of the sample can be quantified.

B. The Theory of HPLC

It is not intended that a comprehensive treatment of the theory of liquid chromatography be supplied here. That is available in a number of texts of which the author has found those cited to be of particular value [274, 505, 621, 783]. Indeed, many experienced chromatographers have ignored the mathematics and the physical chemistry of the subject with no apparent detrimental effects on their work and prefer an empirical approach to the subject. While there is something to be said for this, there are some theoretical concepts which are of great practical value, especially to the novice confronted by a mass of advertising literature extolling the virtues of particular pieces of equipment in terms of their separatory "efficiencies". The mathematics of the separation processes need not be especially daunting, and Meyer [527] has shown in a concise manner how some formulae of great practical value can be derived from relatively simple parameters.

The process of liquid chromatography involves a partition of the components of a mixture between two phases—a mobile liquid phase and a stationary liquid or solid phase. A dynamic equilibrium state is set up between the phases and this can be characterized by an equilibrium coefficient or capacity factor (k'), defined as the ratio of the amount or concentration of a given component in the stationary phase (C_s) to that in the mobile phase (C_m), i.e.

$$k' = C_s/C_m.$$

This parameter can perhaps be seen to have more immediate relevance when it is defined in terms of the retention times (or volumes) of a solute, i.e.

$$k' = (t - t_0)/t_0$$

where $t =$ the retention time of a solute, and $t_0 =$ the time required for the solvent (or an unretained solute) to move from one end of the column to the other.

As migration of the component only takes place when it is in the mobile phase, the rate of movement is inversely proportional to the capacity factor. Different compounds exhibit different relative distributions and must migrate at different rates, so separation is possible. When the resolution is insufficient, it is necessary to alter the capacity factors by varying the selectivities towards either the stationary or mobile phases by changes in the elution conditions.

If there were no diffusion, each compound would migrate as a sharp band. In practice, components do diffuse and emerge from a column in the form of peaks, ideally with a Gaussian shape (see Fig. 2.1). The efficiency of a column can be calculated from the dimensions of the peaks by using the concept of "plate heights", derived initially from distillation theory. If it is assumed that a peak indeed has an ideal Gaussian shape, the number of theoretical plates (N) in a column is determined by the retention time of the component (t_r), i.e. the time from sample injection until the peak for a component reaches

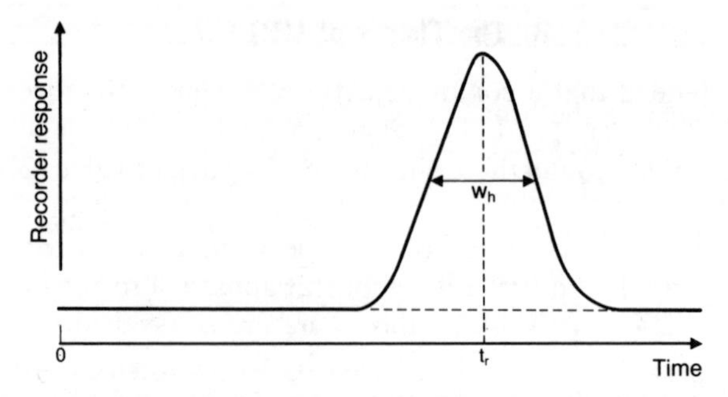

FIG 2.1. A chromatographic peak with an ideal Gaussian shape. Calculation of column efficiency.

its maximum height, and the width of the peak at half that height (w_h) by using the relationship

$$N = 5.54 \times (t_r/w_h)^2.$$

It follows that the wider a peak is at a given retention time, the lower is the column efficiency. An understanding of the factors which cause peak broadening, even without recourse to mathematics, can therefore assist the chromatographer in optimizing the separation conditions for a particular analysis. Commercial suppliers of HPLC equipment usually claim that their columns have a guaranteed efficiency, which may be expressed in terms of numbers of theoretical plates per column or per metre of column length, and often they will support this claim by providing a specimen chromatographic chart. It should be recognized that such separations were almost certainly obtained with a well-characterized synthetic sample under conditions which had been carefully optimized. While these plate numbers may be of some value in comparing allegedly similar columns from different suppliers, the analyst should not be too disappointed if he is unable to obtain similar efficiencies with his own, possibly more complex, samples under different elution conditions.

A number of factors contribute to broadening of chromatographic bands. Eddy diffusion occurs because of irregularities in the size and shape of the particles of packing material, and in turn irregularities are found in the size of the channels between particles. Solvents will move more rapidly through a wide channel than through a narrow one, and any sample molecules in a narrow channel could be delayed. Manufacturers have made great improvements in the production of particles of uniform size and shape to minimize such effects; spherical particles tend to pack more uniformly than those with irregular shapes. If a column is not properly consolidated during packing, its efficiency will be less than the optimum. In general, the smaller the diameter of the particles, the higher the efficiency of the column that should be attainable. Although it is technically more difficult to pack particles as small as 3 micrometres in diameter, most reputable suppliers have mastered the problems.

Further effects on the broadening of peaks are known as "mass transfer" phenomena and can be considered in terms of both the stationary and mobile phases. With a liquid stationary phase, the stationary-phase mass-transfer effect is a consequence largely of the depth of the liquid phase. Molecules which enter the liquid phase can diffuse further down into the layer and in consequence spend a longer time there than molecules which remain closer to the surface. When these molecules eventually enter the mobile phase, they will have been overtaken by the bulk of the material and peak broadening again is the result. A similar phenomenon is seen in adsorption chromatography except that in this instance, molecules are retarded because some of the sites on the adsorptive surface are more active than others and hold on to molecules for longer. Again, manufacturers are aware of the problems and have been steadily improving their packing materials to minimize the effects. Mobile-phase mass-transfer causes broadening of peaks for two main reasons. Firstly, not all molecules in a particular flow stream will move at the same rate, since those near the channel walls will travel more slowly than those in the centre of the stream. Secondly, because of irregularities in the shapes of the particles, there will always be some regions where the mobile phase is static. It may be necessary for molecules to spend time diffusing through such dead volumes, which will be variable in size, before they can enter the moving stream again.

Simple diffusion can have an effect on the broadening of bands. Longitudinal diffusion, i.e. in the direction of flow, brings about a symmetrical band broadening, although this is only troublesome when molecules have a long residence time on the column, as when very slow flow-rates are used. Radial diffusion, which occurs in all directions, causes band spreading indirectly at the column walls as molecules in this region will travel more slowly than those in mid-stream. To counter this, manufacturers polish the interior surface of their columns carefully. Equations have been derived to model all these phenomena and can be used, if need be, to ascertain the reason for diminished resolution in particular columns. With an ideal Gaussian distribution, sample size should have no effect on peak width, as there is then a linear relationship between the concentrations of compounds in the stationary and mobile phases. More often, however, there is a non-linear distribution isotherm and sample size can have a noticeable effect on the width, shape and retention time of a peak. This may be manifested by the appearance of skewed peaks, which exhibit "tailing" or "fronting". Similarly, badly shaped peaks may be observed if the solvent flow-rate is either too high or too low.

Many factors, which are external to the column, have an effect on peak broadening. The influence of temperature on a separation is proportional to the enthalpy of the transfer from the mobile to the stationary phase, and tends to be small (much less in fact than in GLC) but significant. Most HPLC separations are run at ambient temperatures, but thermostatically controlled ovens are available for some purposes. For example, higher temperatures may increase the solubility of compounds which are otherwise only sparingly soluble in the liquid phase, and so will facilitate analysis. Increased temperatures, will generally reduce solvent viscosities and this can also improve resolution. In practice, it is usually easier and more effective to use gradient elution techniques than temperature-programming in HPLC to achieve the same results.

Other deleterious external effects on column efficiency are observed as a consequence of the fact that the tubing and fittings between the injection valve and the detector have a finite volume—the "extra-column" volume, not to be confused with the "dead" or "unswept" volume. Resolution can be improved by good design of the equipment, and by keeping the length (certainly less than 30 cm in total) and the diameter (ideally 0.25 mm or 0.01 in) of the connecting tubing to the minimum. This is particularly important in HPLC with narrow-bore or capillary columns. The volume of the flow cell in a spectrophotometric detector for HPLC use should be no more than 10 μl. In addition, the volume of the injector loop, installed in the instrument, should be no larger than is necessary.

C. Modes of Separation and Column Packing Materials in HPLC

1. Introduction

There are six main modes of separation in HPLC that appear to be of interest to lipid analysts, i.e. adsorption, reversed-phase liquid partition, normal-phase liquid partition, ion-exchange, gel-permeation and chiral-phase chromatography. Complexation and ion-pair chromatography might perhaps be considered as additional forms, although these are always employed in conjunction with one of the principal modes. Separation in all of these modes arise from molecular interactions between the solute and the mobile and stationary phases, in which various types of forces are involved. Ionic forces arise when positively or negatively charged species are present; for example, a cation in the mobile phase will be strongly attracted to an anionic sample molecule. Polar forces are the result of uneven charge distributions (dipoles) over molecules, when electron-withdrawing or electron-donating substituent groups are present. In liquid–solid chromatography, molecules with such dipoles will initially be held strongly by the adsorbent, but will be released as the polarity of the mobile phase is increased, as this will compete for the adsorptive sites. Lastly, there are dispersive forces, which are weak impermanent electrical charges, and they may come in to play to a limited extent, usually during reversed-phase chromatography. With a complex sample, some or all of these forces may have some importance.

The newcomer to the technique will find that there is a large number of manufacturers, each with a comprehensive catalogue of packing materials, intended for use in all of the above separation modes. The particles will be described as spherical or pellicular; the former are preferred for most high resolution applications, while the latter tend now to be reserved for guard columns (as they are very easy to pack). They will be offered with diameters of 3, 5 or 10 micrometres for analytical purposes, or greater for preparative applications. Columns are available commercially in a variety of lengths and diameters pre-packed with such materials, and most workers probably buy them in this form. Finally, all have distinctive trade or proprietary names, such as "Spherisorb", "Zorbax", "Partisil", "Nucleosil", "LiChrosorb", "Bondapak", and so forth. Hamilton and Sewell [274] have cautioned that it should not be assumed that apparently comparable materials from different manufacturers will give identical separations. This is especially true of bonded phases, where differing production methods can lead to differences in the loading of the phase on the inert support, and on the stability to acidic or basic eluents of the chemical bond between the phase and the support. Some knowledge of the chemical nature of these packing materials and of the separatory processes involved may be of assistance in making choices for particular analytical applications, and in solving problems which arise during an analysis.

2. *Adsorption chromatography*

Silica gel has been widely used by lipid analysts as an adsorbent in low-pressure column chromatography and in TLC, and now has innumerable uses in HPLC, but especially for classes of lipids separated according to the number and nature of the various polar functional groups (ester bonds, phosphate, hydroxyl, and amine groups, for example) in the molecules. It is a porous solid and the surface area is inversely related to the size of the pores in the particles, 6 nm being the standard pore size for most purposes.

The adsorptive properties of silica gel are due to hydroxyl groups, which are attached to the surface and can be free or hydrogen-bonded. In addition, there is a water of hydration, which exists first in a strongly-bound layer and then in one or more loosely-bound layers on the surface. The loosely-bound water can have a marked effect on the reproducibility of separations, especially of non-polar lipids, as it is readily removed inadvertently or otherwise by elution with dry solvents. To ensure reproducibility in retention times and resolution, it is better to arrange that only the strongly-bound water layer is permitted to remain. This can be achieved by devising solvent elution schemes to remove as much as is required to produce the desired adsorptivity in the column, prior to an analysis and between gradient runs. Some examples are described in later chapters.

Different brands of silica gel can vary greatly in their properties, and indeed some variation between batches of the same brand can be observed. Some manufacturers produce synthetic silica gels with a lower content of extraneous metal ions than others, in order that the surface should have a more uniform activity. During prolonged use, the adsorptivity of a column can diminish as polar impurities accumulate at active sites; the activity can then often be restored by pumping polar solvents containing ionic species, followed by solvents of lower polarity, through the column. Furthermore profound damage to the surface can also occur, and while methods of regeneration have been described [712], they are likely to be of little value with modern microparticulate silica gels. Solvents containing water at a pH of below 2 or above 7.5 will slowly dissolve the silica gel at the surface; at a pH of 8.5 or more, the process can be quite rapid and the resolution will quickly deteriorate. If the pH of the eluent has to be fixed outside of the optimum range to achieve some specific separation, it is possible to insert a pre-column containing silica gel between the pump and the sample-introduction valve to saturate the eluent with silica gel before it enters the main column, so prolonging the working life of the latter. The solubility of silica gel also increases at elevated temperatures.

Many different solvents or solvent combinations have been used in adsorption chromatography, and the interactions of these with the silica gel surface have an effect on the nature of the separations. For example, some non-polar solvents such as heptane can be adsorbed onto the surface forming a layer, and in some circumstances they may interact directly with a solute [745]. Separations akin to those in reversed-phase HPLC may then be seen. A polar solvent can also form a monolayer, with which solvent molecules interact rather than with the silanol groups; in effect, the surface is partially deactivated. This is observed with heptane containing, say, a small amount of a polar moderator solvent (e.g. 0.5% ethyl acetate). With a careful choice of moderator, it is possible to produce surfaces of controlled activity [747].

One specialized type of adsorption chromatography of particular importance in lipid analysis is a form of *complexation* chromatography, i.e. *silver ion* or *argentation* chromatography. Silica gel impregnated with silver nitrate has been used in many forms of adsorption chromatography, including HPLC, to separate lipids according to the number and configurations of the double bonds in the acyl or alkyl moieties [133, 258, 548]. The principle of the method is that silver ions interact reversibly with double bonds (*cis* more strongly than *trans*) to form polar complexes; the greater the number of double bonds in a molecule, the stronger the complex formation and the longer it is retained.

Although alumina is also available as an adsorbent in HPLC, it reacts chemically with some lipids and solvents and is not recommended for analyses of lipids.

3. *Reversed-phase liquid partition chromatography*

The term "reversed-phase" implies that the stationary phase is a non-polar liquid and the mobile phase is a more polar solvent. HPLC in the reversed-phase mode has been much used by lipid analysts for the separation of molecular species of lipids within a single lipid class, i.e. they are separated principally according to the sum of the chain-lengths of the fatty acyl or alkyl moieties, together with an appreciable dependence on the number and configuration of any double bonds. As described above, separation depends on differences in the equilibrium distribution coefficients of the various components between the two phases. By far the most widely used stationary phase consists of octadecylsilyl ("C_{18}" or "ODS") groups, linked to a silanol surface by covalent bonds, i.e.

$$\text{(surface Si)—O—Si—}(CH_2)_{17}CH_3.$$

The surface may also be "end-capped" by a further silylation procedure to remove any residual active sites. Different manufacturers prepare the inert support and effect the chemical linkage by different means, so that the amount of liquid phase bound to the surface will vary. Commonly, the liquid phase comprises 8–10% by weight of the particles. In so-called "ODS2" columns, it can be much higher. Other stationary phases of this type include bonded octanyl (C_8), ethyl (C_2) and phenyl groups.

With such columns, there should be negligible liquid "bleed" with most of the solvents in common use. Basic elution conditions ($pH > 8$) will cause hydrolysis of the silanol bond of ODS phases, and strong acids ($pH < 2$) will have a similar effect. Solvents containing water or methanol may cause slow hydrolysis of C_8 and C_2 phases.

Newer reversed-phase materials, which are being offered commercially, consist of a polystyrene/divinylbenzene matrix, and are somewhat more polar than ODS phases. They can tolerate a much wider range of pH values (1 to 13) in the eluents.

4. *Normal-phase liquid partition chromatography*

A number of liquid stationary phases of higher polarity are available, consisting of a substituent group, such as diol, nitrile, nitro, methylcyano or phenylcyano, bonded chemically via a short alkyl bridge or spacer to a silanol support. Only a few of these appear to have been used to a significant extent in the analysis of lipids by means of HPLC. The nature of the bond between the support and the organic group is important in that it determines the stability of the phase under various elution conditions. Three different types of bond can be involved, i.e.

silicate ester	(surface Si)—O—R
silicone	(surface Si)—O—SiR_3
silicon–carbon	(surface Si)—R

Of these, the silicate esters are much the least stable, and they can only be used with non-aqueous solvents. Partial polymerization and cross-linking of the silanols is used to further stabilize the phases. The manufacturers' literature can often be an invaluable guide to the properties of such phases, the use of which should certainly be explored further.

The nature of the separations can be similar to that obtained with silica gel, especially when eluted with non-polar solvents. Indeed, the term "normal-phase chromatography" is sometimes used to encompass both adsorption and normal-phase partition chromatography, with some justification as bound water is probably a major factor influencing the separations obtained with both. Bonded phases tend to give much more reproducible separations with less tailing of peaks, and they also equilibrate much more rapidly with the mobile phase in gradient applications.

5. *Ion-exchange chromatography*

The cellulose-based ion-exchange media, diethylaminoethyl (DEAE)- and triethylaminoethyl (TEAE)-cellulose (anion exchangers) and carboxymethyl (CM)-cellulose (cation exchanger), have been much used by lipid analysts in low-pressure column applications for the separation of polar complex lipids [133]. Although silica gel-based bonded phases containing the appropriate functional groups are available for HPLC, they do not appear to have been applied to lipid separations. On the other hand, bonded phases (silica gel-based) with a variety of different amine groups (anion exchangers) or with sulphonic acid (a strong cation exchanger) have been applied to phospholipid analyses (see Chapter 6). In essence, the process of ion exchange can be considered as a competition between the solute ions and counterions present in the mobile phase for fixed sites of opposite charge on a support. The quality and extent of a given separation can be manipulated by varying the nature and concentration of the counterion or by varying the pH of the mobile phase.

When organic solvents are present in the mobile phase, a number of secondary effects come into play, especially adsorption and partition phenomena, and these contribute greatly to the selectivity of the separation. They may indeed be the primary effects when lipids are separated on ion-exchange chromatographic columns. Some ion-exchange media, in which the inert support is a resin, are not compatible with organic solvents.

Ion-exchange media have also been used to bind silver ions for use in complexation chromatography.

6. Gel-permeation chromatography

This technique, which is also known as "exclusion chromatography" or "gel-filtration chromatography", differs from the others above in that the separations achieved are based mainly on the size (and to a limited extent only on the shape) of the solute molecules. The stationary phase is generally an inert porous silica or polymer matrix, such as a dextran or polystyrene, in which the sizes of the pores are carefully controlled by the manufacturer, within limits that are predetermined for particular applications. During chromatography, small molecules may diffuse out of the mobile phase into the pores so their progress is retarded relative to that of larger molecules. The smaller the molecule, the further it will be able to penetrate into the pores and the more it will be retained. The largest molecules obviously elute first. The mobile phases in gel-permeation chromatography are chosen mainly on the basis of their capacity to solubilize the compounds of interest, although again not all resin-based stationary phases can be used with organic solvents. The theory of the separation process has been reviewed [370].

With a gel of a given type, the retention time of a compound of known molecular weight can be predicted with reasonable accuracy; conversely, the molecular weights of unknown compounds can be predicted from retention times determined experimentally. Components tend to elute as sharp bands, so are often detected relatively easily. Elution times are usually short, and solvent gradients are rarely required. The resolution attainable with the technique is limited to compounds that differ in molecular weight by at least 10%.

The technique has been little used with lipids *per se*, but it has proved to be of value for the separation of polymerized lipids and of lipoprotein complexes and apolipoproteins.

7. Chiral-phase chromatography

A number of stationary phases consisting of various chiral molecules, which are bonded chemically to a matrix of silica gel, have been described, and they have been used in HPLC columns for the separation of enantiomeric compounds, avoiding any need for the preparation of diastereoisomeric derivatives. The forces involved are highly complex, and depend on the natures of the bound molecules and of the solutes. The technique has been reviewed [34, 172, 656]. To date, only a few applications to lipids have been described, but many more will undoubtedly be published as the potential of the procedure becomes apparent to analysts.

8. Ion-pair chromatography

This technique is a form of liquid partition chromatography, usually but by no means always carried out in the reversed-phase mode, in which a counterion with both a lipophilic and ionic character is added to the mobile phase to form a steady-state distribution with the stationary phase. The counterion can then form an ion pair with an ionizable sample molecule and alter the manner of its elution, often greatly improving separations. Although the technique would seem to be of value for the separation of molecular species of phospholipids on ODS columns, there do not appear to be any published reports of such separations.

D. Detectors for HPLC

1. Detectors and quantification

When a separation by means of HPLC is underway, it is necessary to have some sensitive means of detecting the solutes in the mobile phase as they elute, so that defined peaks can be collected for analysis by other means, or so that they can be quantified. Detectors functioning according to many different principles are available, and these must be connected to a recording device so that the progress of the separation can be monitored visually, and ideally they should also be linked to an electronic integrator or computer. It may seem strange that the equipment involved in the last step of an analysis should be the first item of hardware for HPLC to be considered in this chapter, but the lack of any specific detection system for lipids in general has meant that the approach of the analyst to any separation problem is governed largely by the nature of the detector(s) available to him. It determines which solvents can be used, whether they can contain inorganic ions, whether gradient elution is appropriate, and possibly which mode of chromatography and which stationary phases are suitable. For example, if the only detector to which the analyst has access is a differential refractometer, it will be necessary for him to use an eluent of constant composition, i.e. *isocratic* elution, and incidentally to accept a compromise in the quality of the separation that might be achieved in many instances. As few analysts have an unlimited choice of equipment, it has been necessary in many of the chapters below to consider specific analytical problems in terms of the different detectors, which have been or might be used.

All detectors produce an amplified electronic signal, the strength of which bears some relationship to the amount of material eluting at that time. There can be little doubt that electronic digital integration is the most accurate and reproducible, not to mention the most rapid and convenient, means of quantifying chromatographic peaks. Of course, it is essential to ensure that the various instrumental parameters on the integrator, especially those defining the sampling rate, are appropriate to the elution volumes of peaks at various times during the analysis. If this is not done carefully, according to the manufacturer's instruction manual, it is possible to negate the advantages of the technique. It is not always easy to convince a novice analyst that a computer print-out of his results can be fallible.

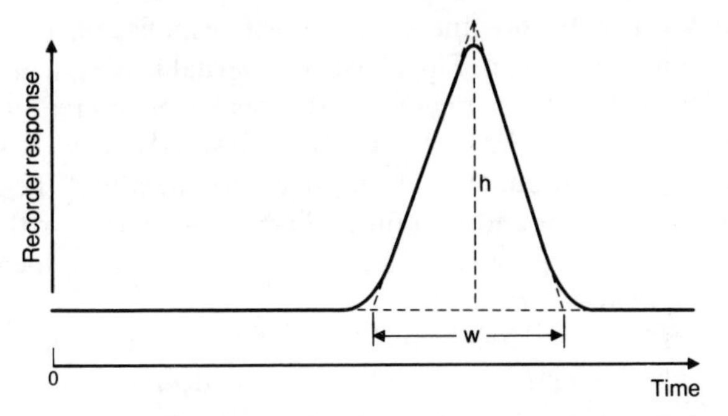

FIG 2.2. Measurement of peak area by means of triangulation.

When an electronic integrator is not available, there are manual methods which can be used. Many of the methods were developed originally for use with GLC, and their efficacy is proven. Planimetry and "cutting and weighing" (of the recorder chart paper) can give good results in the hands of a skilled technician, but are tedious and time-consuming. Acceptable results can even be obtained with unsymmetrical peaks. The best method for well-shaped peaks consists in using triangulation (Fig. 2.2). Tangents are drawn to the sides of the peak to intersect each other and the base-line; the height of the peak (h) is measured from the base-line to the point of intersection, while the width of the peak (w) is the distance between the two points of intersection with the base-line. The area (A) of the peak is then given by

$$A = h \times w/2.$$

It is obviously desirable to use a reasonably fast chart speed to reduce the error in the measurement of peak width.

When the areas of peaks have been measured by the most accurate means available, it is then necessary to convert the raw data to a form which gives a direct measure of the amount of material eluting. Each type of detector responds to lipids in a different way and must be calibrated accordingly. Idiosyncracies of particular detectors are discussed below. Calibration can be accomplished either by use of an external or an internal standard to determine response factors. The former is easier to use and can give more accurate results in HPLC than by other chromatographic procedures if a valve injection system is utilized, as this permits the repeated introduction of samples of identical volume. A plot of peak area against concentration, over the range likely to be encountered in natural samples, is then constructed for every compound of interest. If the plot is linear and passes through the origin, a simple arithmetic factor can be invoked to convert detector response to concentration. With a calibration method using an internal standard, the response of the detector to each component of interest is determined relative to a standard compound, added to the sample at a known concentration, but not present in it naturally; the nature of the response of the detector to this compound must be known. The choice of internal standard is important, as it must elute in the same region as the peaks of interest and yet be completely resolved from them. If an internal standard is used both to determine the relative amounts of different compounds and their absolute concentrations, it should be added at the earliest possible point in the analysis, so that it is carried through any sample work-up procedure, including any extraction and derivatization steps.

Most lipid classes are heterogeneous in that they contain aliphatic moieties with a range of chain-lengths, and with a variable degree of unsaturation. If the detector chosen is sensitive to such structural features, the standards used in calibration must closely resemble the same compounds in the samples to be analysed.

When HPLC is used in routine analytical applications, it is important to set up a proper system of quality control in order to ensure that the detector system functions correctly, and that it is not subject to gradual deterioration or excessive random variation. To do this, it is necessary to establish regular checks on procedures by testing them with defined primary standards. The results should then be evaluated objectively by statistical methods. Such systematic checks may also indicate whether reagents are deteriorating with age, or, for example, whether a faulty batch of solvent has been received. It might help in picking up unplanned changes in methodology that might have been introduced by unskilled technical support staff. Of course, these comments could equally be applied to most other aspects of lipid analysis. Quality control in the lipid laboratory has been reviewed by Naito and David [566].

2. *Ultraviolet spectrophotometric detectors*

Spectrophotometric detectors in the UV-visible range are probably by far the most widely used detectors for HPLC in general, so they are relatively inexpensive and they tend to be one of the first to be acquired by analysts. Their properties have been reviewed [44]. It is possible to use a standard UV spectrophotometer with a flow-cell as a detector, but the extra-column volume may be larger than is advisable and can cause some loss of resolution. Purpose-built HPLC detectors are therefore recommended, and these should ideally have a cell volume of about 8 μl (1 mm in diameter and 10 mm long). Two main types are available, i.e. those containing filters to offer a range of fixed wavelengths, and those affording continuously variable wavelengths, from 190 to 380 nm (with possible extension to 600 nm). Only the latter are of much value to lipid analysts. By far the best response is given with compounds containing conjugated double bond systems and aromatic rings, but such substituent groups are only rarely found in natural lipids unfortunately. UV detectors in general can give great selectivity and sometimes sensitivity in the analysis of specific compounds, and they are relatively insensitive to changes in temperature or flow-rate. They can frequently be used in gradient elution applications, although base-line drift can occasionally cause problems. In use, a detector cell can get dirty, although it may not be obvious to the eye, but the problem can often be remedied by pumping a range of pure solvents of first increasing and then decreasing polarity through it.

Quantitative analysis by peak height or peak area measurement using a UV photometric detector can be performed with a high degree of precision, by both gradient and isocratic methods, if the mobile phase composition and flow-rate are carefully controlled and standardized.

A few lipids only have functional groups with high extinction coefficients in the UV range. Some seed oils contain fatty acids with conjugated double bond systems, and these are frequently present also in lipids subjected to chemical or enzymatic hydroperoxidation. If a broad definition of "lipid" is accepted (see Chapter 3), carotenes and tocopherols can also be analysed with UV detection. In addition, it has proved possible to convert lipids to compounds which absorb strongly in the UV range. For example, fatty acids have been converted to aromatic esters for analysis, glycolipids have been benzoylated, and diacylglycerols derived from phospholipids have been esterified with aromatic acid derivatives. These and other examples are discussed in some detail in the chapters below. With derivatives of this kind, when a specific wavelength can be selected, good quantification is often possible.

Most lipids exhibit a weak absorbance in the range 200 to 210 nm, that is the result of the presence of isolated double bonds predominantly, although other functional groups, such as carbonyl, carboxyl, phosphate, amino and quaternary ammonium, have some effect [237, 300, 380]. It is sometimes termed "end absorption". There are a number of disadvantages to using UV detection at such wavelengths, however. Many of the solvents of proven value in the chromatography of lipids, and chloroform especially, absorb strongly between 200 and 210 nm so cannot be used. Those solvents which are transparent in this region, e.g. hexane, isopropanol, acetonitrile, methanol and water, must be of very high purity since traces of extraneous materials with high extinction coefficients, such as antioxidants or plasticizers, could seriously disturb the base-line and give high backgrounds. Similar impurities and natural substances, such as peroxidized lipids or tocopherols, in samples for analysis can appear as large peaks, obscuring the components of interest. Because small differences in the degree of unsaturation of each component can make a big difference to the response, quantification is not at all easy, and comparatively saturated lipids might be overlooked. Direct quantification has been attempted by determining the apparent extinction coefficient for each component in a standard mixture very similar to those to be analysed, but if the samples are likely to be variable in composition, it is necessary to collect the peaks for estimation by appropriate micro-methods. Most analysts have followed the latter approach. In spite of these problems, some excellent results have been obtained with this detection method, especially since gradient elution is possible.

3. *Fluorescence detectors*

Only a few rare lipids exhibit fluorescence naturally, but some analysts have made use of the high sensitivity (up to a hundred fold greater than absorption detectors) and selectivity of the technique by preparing fluorescent derivatives for chromatography, e.g. anthrylmethyl derivatives of fatty acids (see Chapter 7). Such detectors also have a wide dynamic range. As the response can be greatly modified by the nature of the mobile phase, careful calibration is necessary when gradient elution conditions are employed. The use of fluorescence detection in HPLC has been thoroughly reviewed [488].

4. *Infrared spectrophotometric detectors*

IR detectors have found some specialized use in the analysis of non-polar lipids, with which there is a specific absorbance for the carbonyl function between 1650 and 1860 cm^{-1} (or at about 5.75 μm). The detector is not sensitive to variations in temperature or solvent flowrate. One major problem is that most solvents tend to absorb to some extent at least in the spectral regions of interest, so causing high background values. The effect is diminished in cells with a short path length (1 to 2 mm), but the sensitivity is then also reduced. Gradient elution can be used, but marked base-line drift is usually seen. Acetonitrile, chloroform, methylene chloride and tetrahydrofuran appear to be the most useful solvents for the purpose. Because of the selective nature of the wavelengths used, good quantification can often be achieved.

5. *Differential refractometers*

There are three different types of refractive index (RI) detectors in common use, i.e. deflection, Fresnel, and interference refractometers. All are universal in their scope and can be used with any solute, the refractive index of which is different from that of the mobile phase. They function by monitoring continuously the difference in refractive index between the eluent and the pure mobile phase. An RI detector is probably the second most common type to be found in laboratories, and has been used especially for preparative-scale chromatography and for gel-filtration. By their nature, they can only be used under isocratic elution conditions; indeed volatile solvents are best avoided to minimize changes in solvent composition during analysis. If a static reference cell is used, it may be necessary to flush it out with fresh mobile phase at some point during a day's use. In addition, samples should be dissolved in fresh mobile phase for introduction into the system. RI detectors are very sensitive to changes in temperature, bubbles of gas in the solvents, the flow-rate of the solvent, leaks in the system, back pressure and pulsations of the pump. Many commercial detectors have some means of controlling the temperature, varying from water-circulation to a more sophisticated thermostatted system. The detector should be sited well away from sun-lit windows and from draughts, such as are found near doors or fume cupboards. One further disadvantage is that sample peaks can be both positive and negative. With simple lower-cost RI detectors in routine use, it is doubtful whether concentrations of much less than 1 ppm of solute in the mobile phase could be detected. With accurate control of temperature and other chromatography parameters in the best commercial instruments available, a 10 to 100 fold improvement in this sensitivity is sometimes possible.

Comparatively little use has been made of RI detectors for the direct quantitative analysis of lipids. They are best regarded as sturdy work horses for the isolation of particular lipid components for analysis by other methods.

6. Mass detectors

A new optical detection system that is beginning to make an impression has been variously termed a "mass detector", "evaporative analyser" or "light-scattering detector". With this instrument, the solvent emerging from the end of the column is evaporated in a stream of air or nitrogen in a heating chamber (see Fig. 2.3); the solute does not evaporate, but is nebulized and passes in the form of minute droplets through a light beam, which is reflected and refracted [118]. The amount of scattered light is measured and bears a relationship to the amount of material in the eluent. A commercial detector, based on this principle, is available at a cost comparable to that of other optical detectors from Applied Chromatography Systems (ACS) Ltd (Macclesfield, Cheshire, UK). There are no special wavelength requirements for the light source, and in the commercial instrument, it is a projector lamp. A prototype experimental detector has been described, that is perhaps slightly more sensitive, and uses a laser light source [793, 795]. These detectors can be considered to be universal in their applicability, in that they will respond to any solute that does not evaporate before passing through the light beam. They give excellent results under gradient elution conditions, and are simple and rugged in use. The sensitivity is comparable to that of a refractive index detector, but the mass detector is not affected by changes in ambient temperature or small variations in the flow-rate of the eluent. Indeed once the instrument has warmed up and is running, there is little base-line drift during continuous operation even with abrupt changes in solvent composition.

As with all detectors, there are a number of disadvantages. A source of

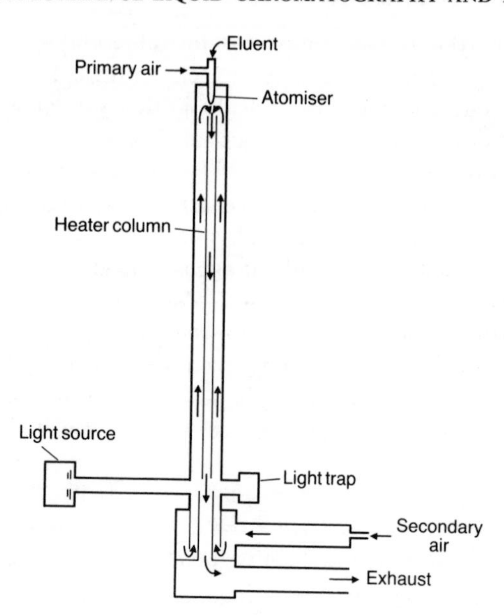

FIG 2.3. A schematic diagram of the ACS mass detector. (Reproduced by kind permission of Applied Chromatography Systems Ltd (Macclesfield, UK).)

dry, filtered compressed air, that is capable of delivering 5 l/min, is required and in practice, this means that an air compressor must be used; a standard cylinder of air or nitrogen is emptied in about 4 hr at this rate. The effluent containing the evaporated solvent must be conducted to the outside of the laboratory or into a fume cupboard. There are limitations on the range of solvents that are suitable, as they must be sufficiently volatile to evaporate in the heating chamber; the author [136] has used isopropanol containing 25% water without difficulty, but formic and acetic acids at a level of about 1% caused "spikes" on the base-line. Small but significant amounts of salts or ionic materials can be incorporated into the eluents, without adversely affecting base-line stability, when the detector is operated at optimum sensitivity levels [139]. In addition, there is a report that 0.1 M aqueous ammonium acetate was usable as an eluent at a cost of a 10-fold reduction in detector sensitivity [556]. The detector is destructive in that the sample is lost, but it is possible to insert a stream splitter between the end of the column and the detector to divert much of the sample to a collection device.

Doubts also have been expressed as to detector linearity, and investigators working with different instruments have described the response variously as linear, sigmoidal or exponential. In the most comprehensive practical investigations, use was made of the commercial ACS detector [556, 612]. It was found that the detector response increased sigmoidally with increasing sample size, in a manner which could be predicted by changes in the particle size distribution in the aerosol, i.e. at low solute concentrations, the droplets scatter light to a proportionately lesser extent. To maximize the response, it is necessary to adjust the flow-rate of the nebulizer gas and the temperature of the evaporator chamber to the optimum. Small variations from the optimum temperature were found to have comparatively little effect on response, and at set gas pressures, the response was relatively constant. The same response was obtained to a given solute when the eluting solvent was toluene, 2-butanone or tetrahydrofuran, and it was concluded that solvent viscosity was of relatively little importance. Anomalous results were obtained with dichloromethane, apparently because of its much greater density. Finally, the response was dependent on the refractive index of the sample. In summary, good quantitative results could be obtained if the instrumental parameters were adjusted carefully, and if the calibration conditions were rigidly set to be the same as in the analysis of real samples. Similar conclusions were reached in more empirical studies of lipid class separations and quantification, in which the mass detector was used [136, 698]. It was noted especially [698] that the detector was sensitive to small changes in the evaporator gas pressure at the nebulizer inlet. In a custom-built detector incorporating a laser light source, the response was found to be related to the solute mass raised to the power of 1.35 [644, 793, 795].

The mass detector has been used to a limited extent only with lipids to date, but it will undoubtedly find many more applications. In particular, the author has found it to be of great value when developing methods.

7. *Transport-flame ionization detectors*

This type of detector was first introduced by Pye Unicam Ltd and, like the mass detector, it is also of universal applicability. In the first model, the eluent was fed continuously onto a moving wire (cleaned in a furnace) and thence to an evaporator oven to remove the solvent before being passed into a flame ionization chamber for detection; in a subsequent more-sensitive model, the solute was pyrolysed to carbon dioxide which was reduced to methane before entering the detector. These instruments were soon discontinued, apparently because they lacked sensitivity and were unreliable, although second-hand equipment was widely sought after for some years.

Privett and colleagues constructed and described a transport-flame ionization detector of improved design, in which the eluent was entrained on a helical wire, so that more of the sample was combusted, thereby increasing the sensitivity of detection [797]. Subsequently, the wire was replaced by a stainless-steel belt of perforated structure that allowed most or all of the eluent to be collected [680]. After evaporation of the solvent, the solute was converted to hydrocarbons in a stainless-steel reactor of sandwich construction and was combusted for detection in a hydrogen flame. A schematic diagram is shown in Fig. 2.4. It has been shown to have a good linear response

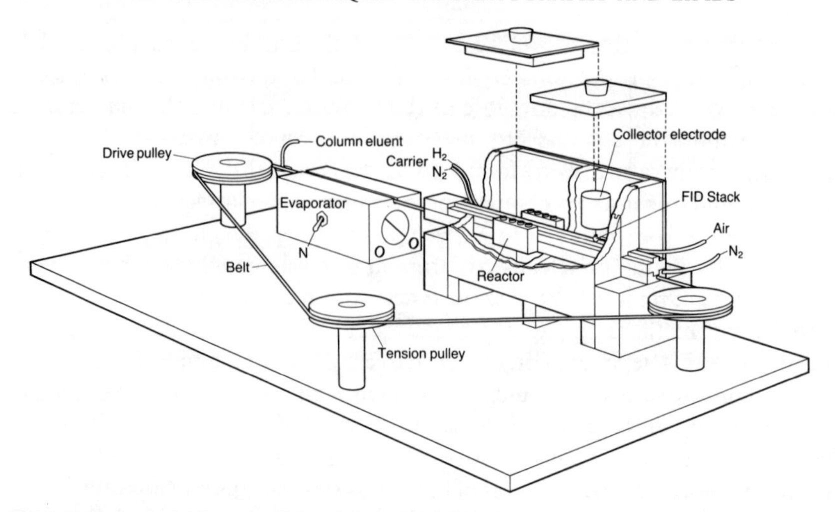

FIG 2.4. A schematic diagram of the transport-flame ionization detector designed by Privett and Erdahl [680]. (Reproduced by kind permission of the authors and of *Analytical Biochemistry*.)

to sample size, and the author was impressed on seeing it in operation in 1985. While it is a matter for regret that this detector has not yet been taken up by a commercial manufacturer, it has been made available to a few other research groups and more reports of its use will undoubtedly appear.

A new commercial detector, operating on a similar principle, is now manufactured by Tracor instruments (Austin, Texas, USA; UK distributor—Kemtronix Ltd, Compton, Berks), and its construction and some applications have been described [191]. It consists of a fibrous quartz belt at the periphery of a rotating disc, and enclosed in a heated, ventilated housing containing the solvent applicator and a dual-flame detector (Fig. 2.5). In operation, the total column effluent is applied to the rotating belt, the volatile solvent is vaporized and removed by a vacuum pump, before the involatile solutes are carried into the two flames of the detector, where they are combusted; the total ion current is transmitted to an amplifier and then to a recorder and integrator. Any residual sample is removed in a pyrolyser unit before the belt picks up fresh eluent. Unfortunately, the major disadvantage of the detector is likely to be cost, currently 4 to 5 times greater than many optical detection systems. The Tracor detector is claimed to give satisfactory results with microgram amounts of solute. As traces of inorganic ions greatly disturb the base-line, it is necessary to use solvents of very high purity. To this date only a few applications to lipids have been described, but the detector is likely to be used extensively in the future by lipid analysts, and further publications are awaited with interest.

Most investigators, working with differing models, have found that the response of detectors of this kind bears a good linear relationship to the amount of solute, although some correction factors may have to be

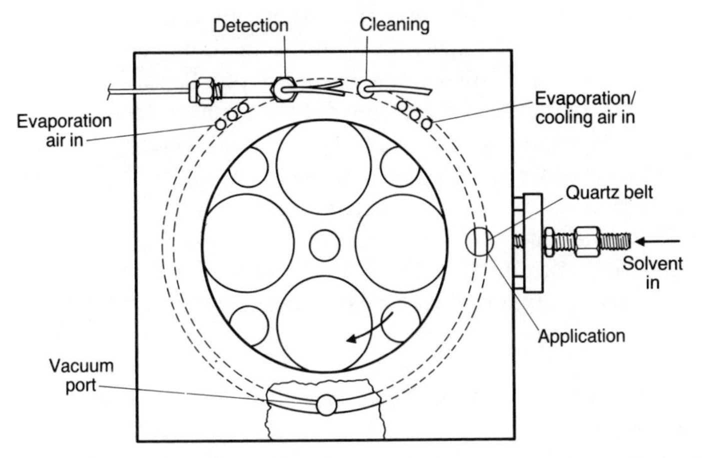

FIG 2.5. A schematic diagram of the Tracor transport-flame ionization detector. (Reproduced by kind permission of Tracor Instruments Inc. (Austin, Texas, USA).)

introduced to compensate for non-combustible moieties in lipids. The solvent limitations are similar to those for the mass detector above, i.e. solvents must be reasonably volatile and should not contain inorganic ions. Although the solute is destroyed during the detection process, it would presumably be possible to introduce a stream-splitter to divert a proportion of the eluent to a collection system.

If a reliable and relatively-inexpensive detector of this type were to be made available commercially, it would fulfill a long-felt need. Indeed, it would not be an overstatement to suggest that it would revolutionize the practice of lipid analysis.

8. Mass spectrometry detectors

Mass spectrometry (MS) is a powerful analytical tool which can supply qualitative and quantitative data, not readily obtained by other means. For example, it can provide the molecular weight, the empirical formula and often the complete structure of an unknown compound. For such determinations, it has been necessary to volatilize a sample in the ion source of the instrument before ionizing it by electron-impact or chemical-ionization techniques, but recently methods have been developed for producing ions from materials in the condensed phase. Combined GLC-MS has been available for many years, but it has proved more difficult to marry HPLC with MS, because of problems in removing the solvent prior to ionization (the vacuum systems available can cope with only about 0.01 ml/min). However, many of the problems have now been overcome and several different HPLC-MS interfaces can be purchased from commercial sources. The construction and properties of these have been reviewed [402, 583, 681, 852]. Increasing numbers of reports of the use of such systems in the analysis of lipids are appearing, especially for the identification of molecular species within specific lipid classes.

One of the first and still among the most popular methods of sample introduction to be developed was the field-desorption technique, in which the sample is deposited on a wire emitter in liquid solution, dried and then inserted into the mass spectrometer in order to obtain the spectrum. The technique is thus a discontinuous one, but can give excellent results.

A "thermospray" method has been devised in which a supersonic jet of vapour carrying entrained particles or droplets of solute is produced by controlled rapid heating of the capillary tube connecting the HPLC equipment and the mass spectrometer. The droplets travel at high speed through the ion source, where they continue to vaporize and are ionized by conventional chemical-ionization methods. The technique has been applied to some relatively involatile samples. In general, direct inlet interfacing methods are best suited to microbore or capillary column HPLC applications (see Section I below).

The third approach to interfacing an HPLC eluent with a mass spectrometer makes use of a mechanical transport device, in which the sample is deposited on a moving wire or belt and the solvent is evaporated before the sample reaches the ion source. The method of deposition of the sample on the belt is important to ensure even evaporation, and a spray deposition technique is now favoured. In essence the vaporization–ionization principle is field desorption, but a wide variety of supplementary excitation methods have been applied to generate ions characteristic of the intact molecules. Almost any HPLC mobile phase can be used, including aqueous buffers, at almost any practical flow-rate. This method appears to have the widest potential in that it also permits spectra to be obtained by electron-impact, chemical and desorption ionization.

With these methods, most structural information is obtained with relatively volatile samples; with compounds of higher molecular weight, it may be possible to obtain molecular weights only (although this information is not negligible). Very little quantitative work has been done by HPLC-MS as yet, but the technique is undergoing rapid development. At the moment, the equipment is probably much too costly to be considered for routine use, but such is the nature of technological advance that this may not always be so.

9. Radioactivity detectors

The sensitivity of methods for detecting radioisotopes are in general much better than many chemical and physical procedures, and the use of lipids labelled with the beta-emitters ^{14}C, ^{3}H and ^{32}P has revolutionized the study of lipid metabolism. In fact, the ease and accuracy of quantifying radioactive lipids is such that they are often used in the development of new methods, especially to test recoveries. Lipids are sometimes converted to isotopically-labelled derivatives for quantification purposes [133].

There are two basic approaches to interfacing HPLC and liquid-scintillation counting, i.e. discontinuous and continuous methods. The former consists simply in collecting discrete fractions of the eluent, drying them down in a stream of nitrogen and re-dissolving the eluted material in a toluene-based scintillant for counting purposes. If the elution solvent does not cause appreciable quenching, it can be mixed directly with the scintillant solution. A procedure of this kind can be used with many mobile phases and all the isotopes of interest in lipid research, and it can be very sensitive if lengthy counting times can be accepted. On the other hand, a considerable time can elapse between completing the chromatographic step and obtaining any results, and the procedure is tedious and labour intensive.

Alternatively, it is possible to monitor the HPLC eluent continuously by methods which can be classified as either heterogeneous or homogeneous counting systems. In the former, the eluent from the HPLC column is passed through a flow cell, packed with a solid scintillator such as yttrium silicate or cerium-activated lithium glass, and positioned in a liquid-scintillation counter; solutes can subsequently be recovered for analysis by other methods. In homogeneous counting, the HPLC eluent is mixed with a "cocktail" of liquid scintillant before passing into the counter. The results are obtained immediately with these methods, and the progress of a separation can be monitored. There are limitations on the range of mobile phases that can be used; chloroform, for example, will cause strong quenching. However, the major disadvantage is that the residence time of a sample in the flow cell must be short to maximize resolution, so limiting the number of counts that can register. Depending on the mode of HPLC and the form of counting, the minimum limits of detection of ^{14}C and of ^{3}H can be as high as 800 dpm and 27,000 dpm respectively [688]. In the analysis of lipids, the most successful application appears to have been to the separation of phospholipids labelled with the high energy emitter—^{32}P. Conventional liquid-scintillation counters equipped with a flow cell can be used as HPLC detectors, but the best results are obtained with purpose-built counter-detectors, as these can be designed to eliminate or at least minimize chemiluminescence effects due to flow phenomena. In a recent paper [223], three commercially-available radioactivity detectors, linked to an HPLC system, were compared with respect to static efficiency, dynamic efficiency, background measurements, and day-to-day variability.

Analysts should not attempt to use radioactivity detection systems until they are familiar both with the potential hazards involved and with the precautions required to ensure the safety of the operator and to prevent contamination of the equipment.

10. Chemical detection systems

It is possible to monitor the composition of an HPLC eluent by incorporating a reaction chamber after the column, where solutes can be mixed with specific chemical or enzymatic reagents for detection and quantification. Among applications to lipids, that are described below, are chemical determination of phosphate in phospholipids, and enzymatic determination of triacylglycerols and cholesterol in lipoproteins.

11. Nuclear magnetic resonance (NMR) spectroscopy

Recently, it has been demonstrated that it is possible to interface an HPLC column to NMR spectroscopy [476]. Detection limits of 50 μg were demonstrated for an aggregate of eight scans. Although this work is still at a developmental stage, it would obviously be of great advantage to analysts if the capabilities of NMR spectroscopy in structural analysis could be harnessed in this way in a practical system.

E. Solvent Delivery Systems

One of the primary requirements for an HPLC system is a pump, capable of propelling the mobile phase through the micro-particulate stationary phase in a column under high pressures (up to 6000 psi), at predetermined flow-rates and in a pulse-free manner. Pumps must be manufactured from materials that are resistant to attack by any of the mobile phases likely to be encountered in analyses (as indeed should all parts of the system). In practice, this means that the construction should be of high quality stainless-steel or inert polymers, such as polytetrafluoroethylene (PTFE, Teflon™ or Fluon™), while ruby or sapphire may be utilized in the construction of the valves. Mobile phases containing strong mineral acids or chloride ions can react with stainless steel and are best avoided. The pump should have a low internal volume to facilitate changing of the mobile phase, and it must be capable of delivering solvents at flow-rates in the range 0.5 to 3 ml/min in general use, within 5% of the set value, so that any variations in sample retention times are not significant; slower flow-rates may be required with microbore columns and higher with preparative columns. In addition, the pump should not contribute to detector noise, either through stray electrical signals or through pulsations in the solvent flow.

The first pumps manufactured commercially to meet these criteria were constant-pressure or constant-volume syringe pumps. These were capable of sustaining high pressures and constant flow-rates, but tended to have high internal volumes that made solvent changing difficult. Most modern HPLC equipment will have reciprocating pumps with a chamber of low internal volume (0.1 to 0.2 ml). The mobile phase is drawn into the pump on the upstroke and expelled into the system on the down stroke, by means of check valves on either side of a piston or diaphragm. A pulsed flow is produced, but this can be evened out by a damping device. In use, pump seals have to be replaced at regular intervals, and the check valves (stainless steel balls in sapphire seats) may require attention. Reciprocating pumps have the advantages that they are compact and relatively inexpensive, they can have very large solvent reservoirs and they are simple to use and maintain. When utilized in the development of new methods, their low internal volume is of great utility as it is easy to change solvents at frequent intervals.

For many analytical purposes, isocratic elution procedures will suffice. The equipment required is then relatively inexpensive as only a single pump and a simple control station are needed. In addition in comparison to gradient elution techniques, a wider range of solvents and detectors can be used, less preparatory work may be necessary to set up the analysis, and no time and solvent are wasted re-equilibrating columns between analyses. Indeed, the savings in time and money can be such that lipid analysts have spent much time exploring chromatographic conditions that permit isocratic elution for particular routine analyses, even if this sometimes involves compromise on the resolution attainable.

The complexity of many lipid samples is such that a gradient in the composition of the mobile phase may be essential to obtain effective separations. In modern instruments, this is achieved essentially in two ways, i.e. either by high-pressure or by low-pressure mixing. In the former approach, the different solvents are propelled into a mixing chamber ahead of the injection system by means of separate pumps. The form of the gradient is determined and controlled by microprocessors (often a microcomputer), which adjust the flow-rates from each pump to give the desired compositions. It is of course very important that the equipment should function reproducibly, so that constant elution times are obtained for the compounds of interest. Often it is possible to choose a linear gradient or one with a convex or concave profile, or to change the composition in steps at preset times. The extra pump required is costly, but this method is favoured by many analysts because of its simplicity and versatility, and because the solvent mixing is particularly efficient. The accuracy of the gradient can diminish when one component is less than 10% of the other. Although some systems permit the introduction of a third pump so that a ternary solvent gradient can be developed, there are less costly yet equally efficient means of effecting this.

In gradient systems with mixing on the low-pressure side, a single reciprocating pump only is required to maintain a constant flow of solvent, and the inlet line is connected to two or more switching valves, which are in turn connected to the solvent reservoirs. The gradient programmer is again a microprocessor driven device and controls the opening and closing of the valves to produce the desired gradient. Very accurate timing is necessary and the valves must respond instantaneously or the gradient will be irregular and irreproducible. A well-designed mixing chamber (often unfortunately of greater volume than in a high-pressure system) also is required, as each of the solvents enters it as a separate slug in the single inlet line. With the best modern commercial systems, excellent results can be obtained. This approach to gradient formation is much less costly than high-pressure mixing when ternary solvent gradients are required.

In all gradient analyses, it is of course essential to ensure that the various solvents or solvent combinations selected are truly miscible. Problems can arise with aqueous mobile phases especially, and with solvents at extremes of the polarity spectrum.

F. Injection Systems

The introduction of the sample to the top of the column is an important part of the chromatographic process, as any unnecessary diffusion of the sample will cause band broadening and loss of resolution. Ideally the sample should be introduced in the minimum volume of solvent so that it reaches the column as a discreet band.

The choice of solvent in which the sample is stored and injected is also important. The most suitable one will often be the same as the mobile phase at the start of a chromatographic run, as this should have a negligible effect on the nature and quality of the separation and should not disturb the detector base-line unduly, when it passes through the column. However, it is not always possible to do this, for example when some of the components in the sample are only partially soluble in the starting eluent. It is then necessary to find an alternative solvent for sample introduction that will have the minimum effect on the various chromatographic parameters, and the nature of this will vary with the type of sample and the mode of separation employed. It is an aspect of HPLC analysis that it sometimes not adequately considered.

The simplest and most efficient (in chromatographic terms) method of sample introduction is on-column injection by means of a syringe through a septum. Unfortunately, septa cannot withstand pressures greater than about 1000 psi, and the life of the expensive syringes can be short. Stop-flow injection was introduced to reduce these problems, but it is inconvenient to have to stop and start the pump with each injection.

Most modern instruments are now supplied with off-column valve injection systems, which are robust and convenient in use. They are of two types, i.e. with internal and external loops. With the former, the sample loop consists simply of a groove cut into the rotor surface, into which the sample is injected, and from which the sample is introduced into the solvent stream when the rotor is turned. The size of the loop can only be varied by changing the complete valve. External loop valves, in which the loop is a length of calibrated capillary tubing mounted on the outside of the valve, are now generally preferred as loops of different volumes can be selected according to the scale of a given separation. With care, there should be very little band spreading. A flat-tipped syringe, designed for the purpose, must be used to avoid damage to the rotor seal.

External loop injection systems can be used to deliver precise volumes of sample onto the column. This can be done by partially filling the loop by using a microsyringe to determine the volume of sample; the volume injected should then be no greater than half that of the loop. When small amounts only of sample are available, this is the preferred method. Alternatively, the sample loop can be filled completely to obtain a precise volume. Because the locus of the sample–solvent interface remains diffuse in this circumstance, some solvent remaining close to the wall of the loop, it is necessary to introduce 2 to 3 times the volume of the loop of sample to achieve 95% of the maximum possible filling; for high precision, it is recommended that 5 to 10 loop volumes be used. In practice, the optimum volume may have to be determined experimentally for a particular loop. The latter method is preferred when large amounts of sample are available. With care, a precision of injection of 1% (coefficient of variation) should be attainable by either method.

It should be recognized that while loop injectors permit the introduction of controlled amounts of sample with high reproducibility, the absolute amount of the sample may not be known with accuracy, unless the loop is calibrated precisely. For example, most manufacturers state that the smaller commercial loops deliver about ± 17% of their nominal volume.

G. Columns and Fittings

1. Column construction

The column is now one of the least costly components in an HPLC system, although the recurrent cost can be high, but it is vital that it be assembled and packed correctly for maximum resolution to be attained. It is perfectly feasible for analysts to pack their own columns, with all but the smallest size particulate phases, provided that they are prepared to acquire the necessary equipment. However, it is apparent that most analysts find it more cost-effective to purchase columns in prepacked form. There are a number of small companies who specialize in manufacturing such columns at competitive prices, and often it is possible to have spent columns repacked at a discount.

The design and construction of columns are still evolving rapidly. Typically, columns are constructed of high quality, precision-bore stainless-steel, and are highly polished internally to minimize the effects of the wall on peak broadening. The packing is retained by porous stainless-steel frits (mesh size 2–5 μm) at either end, and these are held in place by means of steel connector fittings. From time to time, the frits may be blocked by particulate impurities and may have to be replaced, so they must be reasonably accessible. Column end fittings with both "male" and "female" unions are available, and the latter are generally preferred as they can be constructed to have a smaller dead volume, and they are less liable to sustain damaged threads or deformation of the ferrules if columns have to be removed and reconnected repeatedly. A number of manufacturers (e.g. Valco, Waters, Swagelok, Upchurch, SSI) supply fittings, which are superficially similar and perform the same functions with apparently equal efficiency, but which are *not* in fact interchangeable. It is important that the analyst should standardize upon one of these at the outset to avoid any possibility of confusion and of damage to columns. Recently "universal" fittings, containing ferrules manufactured from a stable polymeric material (Kel-F™), have become available and these can be used with a variety of different makes of column endings. The author has found them to be very useful in his own laboratory. All fittings are designed for use with stainless-steel tubing of 1.59 mm ($\frac{1}{16}$ in) o.d.; the internal diameter should be no greater than 0.25 to 0.50 mm (0.01 to 0.02 in). They should never be tightened further than is necessary to give a good seal.

At present, the most widely used chromatographic columns would appear to be 250 mm long by 4 to 5 mm i.d., containing 5 or 10 μm packing materials. Under optimum test conditions, they can generate from 10,000 to 20,000 theoretical plates. These columns can be used both for purely analytical purposes and for the isolation of particular components on a small scale (1 to 10 mg depending on the ease of separation and the chromatographic mode). Shorter versions of the columns (50 to 150 mm) are sometimes preferred for rapid separations. Wider bore columns (up to 22 mm) are usually selected for larger-scale preparative purposes, since the carrying capacity of the column is proportional to the square of the diameter; such columns may also be longer than the average and tend to be packed with stationary phases of greater diameter (10 to 40 μm) to avoid too great a pressure drop along the column. Similarly, gel-permeation columns are often larger than the average. The highly-efficient packing materials of 3 μm diameter are generally supplied in short columns (100 mm), again to avoid too great a pressure gradient, of 4 to 5 mm i.d. They offer comparable separation efficiencies to the longer standard columns (in terms of plate numbers), but the analysis can be much more rapid and much less solvent is required, simply because of the reduced dimensions. In addition, the 3 μm stationary phases equilibrate very rapidly with the mobile phase and, therefore, tolerate a greater range of flow-rates before appreciable loss of resolution occurs, so offering further improvements in analysis times. One manufacturer is marketing 80 × 6.2 mm columns with 3 μm stationary phases for use in rapid analyses.

Cartridge systems are a newer development in column technology. It is necessary to purchase a special holder, constructed with low dead volume fittings which will accept the cartridges. The latter can be constructed from inert materials such as glass-lined steel or even of glass alone, in order to minimize peak broadening due to wall effects. If a glass column should break during use, and the maximum permissible pressure is only half that with stainless-steel columns, there is little danger to the operator as the kinetic energy of a compressed liquid is low. A further advantage of a glass column is that contamination of the column, or channelling of the packing in use can easily be observed. With most cartridge systems, it is possible to connect two or more columns is series via low dead volume couplings to increase column efficiency. Much is claimed (in the manufacturers' brochures) for cartridge systems, but few applications to lipid analyses appear to have been published.

2. *Care of columns*

When a new column is used for the first time, or if one has been in storage for a lengthy period, it should be flushed with pure solvents at slower flow-rates than the intended analytical conditions for up to five column volumes. This preconditioning can lengthen column life, and improve base-line stability and the reproducibility of retention times.

Guard columns in front of the main column may prolong column life, by filtering out particulate matter and other impurities, and they are especially recommended with dirty samples, such as body fluids. They can be simply short versions (say 50 × 5mm) of the main column, packed with the same stationary phase, and connected via a short length of $\frac{1}{16}$ in tubing and standard column fittings. Although they have to be repacked periodically, this is less costly than repacking or replacing the main column. Recently, "direct-connect" guard columns have become available, that are screwed straight into the standard columns, minimizing the amount of extra-column volume. These can be packed by the analyst by a simple tap and fill procedure, requiring no special equipment, if a dry pellicular packing (30 to 45 μm diameter) of an analogous type to that in the main column is utilized. Some loss of resolution (2 to 10%) is inevitable when guard columns are used, but this is generally preferrable to frequent replacement of columns.

A partial substitute for a guard column is a filter, consisting of a stainless steel porous frit (2 μm pores) in a low dead volume holder, inserted between the valve and the column. These should not contribute significantly to band broadening, they are relatively cheap and the frits are easily replaced. Most particulate impurities will be trapped, although soluble contaminants will pass through to the column. Prefiltering of samples through Millipore™ filters also assists in extending column life. The author has used precolumn filters in preference to guard columns in his own laboratory with some success.

Most columns are sold with an arrow on them to indicate the direction of flow during packing, and it is usually recommended that the mobile phase should only be pumped through in the direction indicated. However, the author observed no loss of resolution when columns of silica gel were reversed to alleviate difficulties caused by blocked frits.

A "silica-saturation" column can lengthen the useful lives of columns containing silica gel or bonded phases based on a matrix of silica gel.

It is of the utmost importance that the column pressure be monitored at regular intervals during operation. A sudden pressure rise could mean a partial blockage of a filter, frit, guard column, valve or pressure tubing. Other causes of fluctuating pressures are air bubbles in the system or pump malfunction. If these are not attended to rapidly, damage to the column can result. Columns should be changed only after slowly decreasing the flow-rate down to zero, and leaving for 2 min or so.

At the end of a day's work, polar solvents, especially if they contain water and inorganic ions, should be flushed out and replaced with a relatively inert one, if need be in stages so there are no problems of miscibility. After use, columns should never be left to dry out, but should be filled with an inert dry solvent, and stored with the ends sealed at an even equable temperature. In addition, they should always be properly labelled (date of purchase, type of packing, storage solvent) and boxed. Columns should never be handled roughly.

H. Mobile Phases

1. Solvent selection

The choice of solvents for use in the mobile phase is dependent on the nature of the separation mode adopted. While it is susceptible to a theoretical treatment [782, 783], which can be of great value, most analysts rather short-sightedly tend to approach the problem empirically, learning by experience which solvent combinations are likely to be of value for their particular analytical problems, or by tinkering with eluents developed by others. For lipid analyses of most kinds, organic rather than aqueous eluents are almost invariably required. In formulating mixed solvents, it is essential that they be fully miscible with each other, otherwise droplets of a second phase may be trapped in the system to appear at an inopportune moment and ruin a separation. Miscibility problems can also arise in replacing the solvent utilized in the column for storage, with the one to be used in the analysis. If this problem arises, an intermediate solvent must first be used to flush out the column. For reproducible results when mixing solvents to prepare an eluent, it is important to measure out each of the solvents separately before mixing (and not to use a single measuring cylinder and make up each solvent to the appropriate mark).

It is essential that none of the constituents of the mobile phase should interact with the stationary phase and affect it adversely. For example, basic eluents are harmful to silica gel and many bonded phases (see Section C above). Ketones, aldehydes and peroxides can interact irreversibly with bonded-amine phases. Nor should the stationary phase affect the solvent. Basic alumina, for example, can catalyse a condensation reaction with acetone to form diacetone alcohol [274].

All solvents can contain impurities, some of which (e.g. antioxidants) are indeed introduced deliberately by manufacturers to improve stability. For HPLC use, the solvents should be the highest quality available, or they should be purified by distillation or other means by the analyst prior to use [133]. Special "HPLC grades" of many solvents are sold, and this implies that they contain low levels of UV-absorbing materials, which could otherwise give high background values with UV detectors. Unfortunately, there are no objective standards which suppliers must meet. With other detectors, the requirements can be less stringent. Solvents should be stored in the dark in a flameproof cabinet, and in a cool dry place. The inlet lines from solvent reservoirs should incorporate filters to remove any particulate matter, including dust and bacteria, in solvents. Indeed, it has been recommended that solvents should be filtered through inert micropore filters (*not* filter paper) before they are introduced into the reservoirs, in order to prevent premature clogging of the in-line filters [712].

The physical properties of some solvents, which are used frequently by lipid analysts, are listed in Table 2.1. (adapted from Snyder and Kirkland [783]). The polarity of a solvent is probably one of the first factors to consider, especially in adsorption chromatography. Where two solvents appear to be equally suited to a purpose, it is usually recommended that the one of lower viscosity be selected. Then the solvent should not be so volatile that it tends

TABLE 2.1

The physical properties of some organic solvents of special interest to lipid analysts [783]

Solvents	Refractive index (at 25°C)	Boiling point (°C)	Polarity factor[a]	Viscosity (cP, 25°C)
2,2,4-trimethyl-pentane	1.389	99	0.1	0.47
n-hexane	1.372	69	0.1	0.30
n-heptane	1.385	98	0.2	0.40
toluene	1.494	110	2.4	0.55
diethyl ether	1.350	35	2.8	0.24
tetrahydrofuran	1.405	66	4.0	0.46
methanol	1.326	65	5.1	0.54
ethanol	1.359	78	4.3	1.08
isopropanol	1.384	82	3.9	1.90
n-butanol	1.397	118	3.9	2.60
water	1.333	100	10.2	0.89
acetone	1.356	56	5.1	0.30
butan-2-one	1.376	80	4.7	0.38
ethyl acetate	1.370	77	4.4	0.43
acetonitrile	1.341	82	5.8	0.34
acetic acid	1.370	118	6.0	1.10
chloroform	1.443	61	4.1	0.53
dichloromethane	1.421	40	3.1	0.41
1,2-dichloroethane	1.442	83	3.5	0.78

[a] As defined by Snyder [782].

to evaporate at an appreciable rate in the reservoir; tetrahydrofuran is often utilized instead of diethyl ether, for example, for this reason, although they have slightly different selectivities. Certain solvents have spectroscopic properties, which are particularly suitable for some purposes, and hexane, isopropanol and water mixtures, for example, can be used with UV detection at 205 nm, where most other solvents are opaque. Freshly distilled de-aerated tetrahydrofuran can also be used at low wavelengths, but it soon oxidizes in air so that unacceptably high backgrounds result. With refractive index detection, better base-line stability can sometimes be obtained when the component solvents in an eluent have refractive indices that are reasonably close to each other.

Solvents such as hexane, chloroform, ethers, methanol, isopropanol and toluene have been used for decades in the isolation and analysis of lipids. While acetone is a poor solvent for phospholipids, it is an excellent one for glycolipids and this property is of value in some circumstances. Since the advent of HPLC, acetonitrile has found many uses in the analysis of lipids, especially in the reversed-phase mode, for separations according to the chain-lengths of the fatty acyl groups.

In order to define solvent selectivities in greater detail, Snyder [782, 783] grouped them according to their physical properties as shown in Table 2.2. Eight main selectivity groups were evident. Within each group, substitution of one solvent for another would not be expected to have much effect on the nature of a particular separation. For example, substitution of isopropanol for methanol in an eluent should not have a great effect on resolution, although it might facilitate the separation if a particular component of the solute was solubilized better. On the other hand, substitution of one solvent in an eluent by another from a different selectivity group would be expected to have some influence on the nature of a separation, although the magnitude and direction of any effect is not always easily predicted. Aliphatic hydrocarbons are not sufficiently polar to be subjected to Snyder's theoretical

TABLE 2.2
Solvent selectivity groups, according to Snyder [782, 783][a]

Group	Solvents
I	Aliphatic ethers, trialkyl amines
II	Aliphatic alcohols, cyclohexanone
III	Tetrahydrofuran, dimethylsulphoxide, amides, pyridine, methoxyethanol
IV	Acetic acid, benzyl alcohol, ethylene glycol, formamide
V	Dichloromethane, 1,2-dichloroethane
VIa	Aliphatic ketones and esters, polyethers, dioxane
b	Sulphones, nitriles
VII	Aromatic hydrocarbons, halogenated aromatic hydrocarbons, nitro compounds, aromatic ethers
VIII	Fluoroalkanols, water, chloroform

[a]Aliphatic hydrocarbons are not sufficiently polar to be subjected to the theoretical treatment.

treatment, and should perhaps be regarded as an additional group. Again, the nature of solvent selectivity effects is dependent on the mode of chromatography.

2. *De-gassing*

All solvents contain dissolved air, the solubility of which is increased at high pressures. With the sudden release of pressure at the end of the chromatographic column, bubbles can form that cause pressure fluctuations and interfere with detection. Dissolved air can also have deleterious effects on solutes, for example by causing autoxidation of double bonds in lipids. It is, therefore, desirable to remove all dissolved gases from solvents prior to an analysis.

Perhaps the commonest method of achieving this consists in refluxing solvents for a short time, and some commercial equipment has a facility for this built in. However, this method can be hazardous, additional cooling time is needed before the solvent is used, and the solvent must be purged continuously with an inert gas, otherwise air will rapidly be taken up again. Air can also be removed by subjecting solvents to a vacuum for a time, and this method is often preferred for aqueous eluents, although again they must be purged with an inert gas to ensure that air does not redissolve. A less effective method is to use ultrasonic vibrations.

The simplest and least hazardous method of removing air from solvents is to purge them with helium gas, which displaces any dissolved gases but which itself is virtually insoluble in solvents [43]. All that is necessary is to purge the solvents with helium via a porous stainless steel frit or filter at a flow-rate of about 200 ml/min for about 5 min, then to reduce the flow-rate to a trickle to prevent any redissolution of air. Some commercial HPLC equipment is sold with a facility for this incorporated into it.

3. *Hazards*

Most solvents exhibit some degree of toxicity if inhaled in large amounts. Benzene, in particular, is frequently mentioned in the older literature as a solvent in the analysis of lipids, but it is now known to be extremely toxic and is best avoided entirely; toluene has comparable chromatographic properties and is much less hazardous. Similarly, it was once thought that chloroform was relatively safe, but it is now known that there are real hazards and the levels permitted in the atmosphere of laboratories have been steadily reduced over the years. Acetonitrile also has toxic properties, and indeed analysts should not view any solvent with complacency. During chromatographic analysis, most vapour tends to enter the atmosphere on mixing solvents for use as mobile phases, on filtering, on degassing, or when the eluent leaves the column. When it is feasible, fume cupboards should always be used. Otherwise, care should be taken to prevent spillages, to keep storage vessels closed, and generally to minimize any exposure of the laboratory personnel to solvent vapours.

Nor should it be forgotten that many solvents, especially low molecular weight hydrocarbons, ethers and alcohols, are highly inflammable. All electrical equipment, not only that in the HPLC system itself, should be correctly wired and earthed to minimize the risk of sparks. No naked flames should be permitted in any part of a laboratory in which solvents are used.

I. Microbore Columns in HPLC

In recent years, there has been a trend in research laboratories towards columns of smaller diameters than the normal 0.46 to 0.5 cm, with the objectives of increasing efficiency and resolution, of decreasing the use of expensive and environmentally hazardous solvents, and of saving on analysis time. At the same time, the use of such columns affords scope for unconventional solvents and detection systems, for example deuterated solvents and NMR detection. Microbore columns seem to have been used to a limited extent only in general laboratories to date, and very few applications to lipids have appeared, perhaps because of detector limitations, but a wider utilization will certainly be seen in the future. The theory and applications of such columns have been comprehensively reviewed [448, 746].

Three basic types of microbore columns have been described, i.e. columns of 1 to 2 mm internal diameter packed in a conventional manner (sometimes termed "small bore"), open tubular (capillary) columns, and packed microcapillary columns. For most analysts, the small bore packed columns will be of greatest interest, if only because they are available commercially at relatively low cost and need not necessarily require special equipment for the chromatography stage. For example with care, columns of 2 mm i.d. can be used with conventional pumps and detectors (indeed some workers would no longer consider these as true microbore columns). Solvent flow-rates are typically 25% or less of those in normal columns. Such small bore columns have the same geometrical characteristics as the common packed columns. In open tubular microbore columns, the internal column diameter is very small (50 μm or less) to allow for the slow diffusion time in liquids, and the stationary phase can be a liquid deposited mechanically or bonded chemically to the wall of the tube, or a finely dispersed solid. Packed open tubular columns have somewhat larger internal diameters, and the particles of the solid support or adsorbent are distributed uniformly along the column, with some actually embedded in the wall. The solvent flow-rates in such columns can be as little as 1 μl/min. Experimental capillary columns affording 1,000,000 theoretical plates have been described [508]. The theory of open tubular columns in HPLC [448] is very different from that of packed columns and cannot be discussed in detail here.

For some time, it was doubted whether microbore packed columns could indeed offer improvements in resolution over those obtained in conventional columns. The probable reasons for this were that it was extremely difficult technically to pack the columns (and relatively few commercial suppliers even now have mastered the technique), the pumps available could not cope with the low flow-rates required, and it was relatively easy to inadvertently overload the columns. Instrument manufacturers are now responding to the challenge of microbore columns and have designed pumps to deliver solvents at very low flow-rates reproducibly, and with the minimum amount of extra-column volume in other parts of the equipment; detectors have also been miniaturized. With columns of 1 mm i.d. or less, such equipment is mandatory.

It is to be expected that microbore column technology will improve, that the instrumentation will become more widely available and that many more applications to lipids will be described in the years to come. In particular, the flow-rates are well suited to mass spectrometric detection systems, also a technology that is developing rapidly [98, 402].

CHAPTER 3

Lipids: Their Structures and Occurrence

A. Definitions

Lipid chemists and biochemists tend to have a firm understanding of what is meant by the term "lipid", although no satisfactory or widely-accepted definition exists. Most general textbooks describe lipids as a group of naturally-occurring compounds, which have in common a ready solubility in organic solvents such as chloroform, benzene, ethers and alcohols; such diverse compounds as fatty acids and their derivatives, steroids, carotenoids, terpenes, and bile acids are included. Otherwise, many of these compounds have little by way of structure or function to relate them. In fact, a definition of this kind is positively misleading, since many of the substances which are now widely regarded as lipids may be more soluble in water than in organic solvents.

The historical origins of the term "lipid" make interesting reading, and should assist with the definition [241]. The term has evolved via "lipine", "lipin", "lipoid" and "lipide" and was originally used in a more restricted sense. Thus, Bloor [85] in 1920 classified lipoids into three groups, *simple lipoids* (fats and waxes), *compound lipoids* (phospholipoids and glycolipoids) and *derived lipoids* (fatty acids, alcohols and sterols). It appears that only later and rather arbitrarily were compounds such as carotenoids and terpenes included in the term. A more specific definition than one based simply on solubility is necessary, and most workers in this field would happily restrict the use of "lipid" to fatty acids and their naturally-occurring derivatives (esters or amides) and to compounds related closely through biosynthetic pathways (e.g. prostanoids, aliphatic ethers or alcohols) or by their functions (e.g. cholesterol) to fatty acid derivatives. Accordingly, the author submits the following definition for consideration.

> "Lipids are fatty acids and their derivatives, and substances related biosynthetically or functionally to these compounds."

This definition is close to that of Bloor [85]. It treats cholesterol as a lipid, and could be stretched to include bile acids. It does not include other steroids, fat-soluble vitamins, carotenoids or terpenes, except in rare circumstances, but these are in any case covered by generic terms in their own right. Only compounds encompassed by this new definition are considered in this book.

Having defined "lipids" in this way, it is now necessary to define the "*fatty acids*". They are compounds synthesized in nature via condensation of malonyl coenzyme A units by a fatty acid synthase complex. In general, they contain even numbers of carbon atoms in straight chains (C_4 to C_{24}), although the synthases can also produce odd- and branched-chain fatty acids to some extent when supplied with the appropriate precursors; other substituent groups, including double bonds are normally incorporated into the aliphatic chains later by different enzyme systems.

The main lipid classes of plant and animal origin consist of fatty acids linked by an ester bond to the trihydric alcohol glycerol, or to other alcohols such as cholesterol, or by amide bonds to long-chain bases (sphingoids or sphingoid bases), or on occasion to other amines. In addition, they may contain alkyl moieties other than fatty acids, phosphoric acid, organic bases, carbohydrates and many more components, which can be released by various hydrolytic procedures. A further subdivision into two broad classes is convenient for chromatographers [133]. *Simple* lipids are those which on hydrolysis yield at most two types of primary products per mole; *complex* lipids yield three or more primary hydrolysis products per mole. The terms "neutral" and "polar" lipids respectively are also used frequently to define these groups, but are less precise and can be misleading. In practice, it is often necessary to subdivide the main groups further. For example, the complex lipids for many purposes are best considered in terms of either the *glycerophospholipids* (or simply as *phospholipids*), which contain a polar phosphorus moiety and a glycerol backbone, or the *glycolipids* (both glyceroglycolipids and glycosphingolipids), which contain a polar carbohydrate moiety, since these are more easily analysed as separate groups. Comprehensive review volumes have recently been published on the occurrence, chemistry and biochemistry of lipids [260, 844], phospholipids [299] and glycolipids [397, 883].

In order to analyse a lipid sample completely, it is necessary to separate it into simpler classes, according to the nature of the various constituent parts of the molecules, and these in turn may have to be identified and quantified. It may, therefore, be helpful in the next section to consider the kinds of lipids to be found in various tissues. The nomenclature recommended by recent IUPAC–IUB commissions has been followed here [358, 359].

B. The Principal Simple Lipids and Complex Glycerolipids of Animal and Plant Tissues

1. Triacylglycerols and related compounds

Triacylglycerols (commonly termed "triglycerides") consist of a glycerol moiety, each hydroxyl group of which is esterified to a fatty acid. In nature,

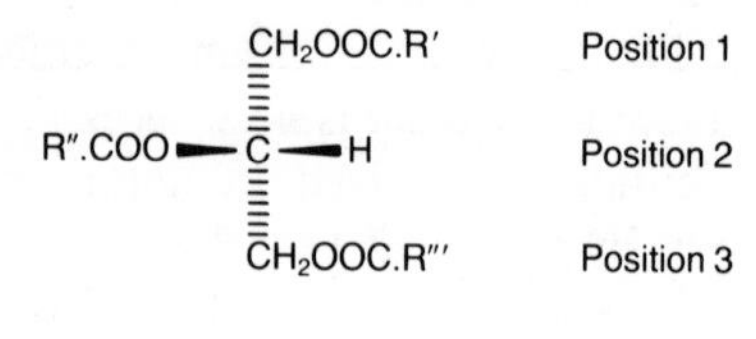

FIG. 3.1. The structure of a triacylglycerol.

these compounds are synthesized by enzyme systems, which determine that a centre of asymmetry is created about carbon-2 of the glycerol backbone, and they exist in different enantiomeric forms, i.e. with different fatty acids in each position. A "stereospecific numbering" system has been recommended to describe these forms [358, 359]. In a Fischer projection of a natural L-glycerol derivative (Fig. 3.1), the secondary hydroxyl group is shown to the left of C-2; the carbon atom above this then becomes C-1 and that below becomes C-3. The prefix "*sn*" is placed before the stem name of the compound. If the prefix is omitted, then either the stereochemistry is unknown or the compound is racemic. Smith [775] has reviewed glyceride chirality. Nearly all the commercially important fats and oils of animal and plant origin consist almost exclusively of this simple lipid class.

Diacylglycerols (less accurately termed "diglycerides") and monoacylglycerols (monoglycerides) contain two mole and one mole of fatty acids per mole of glycerol respectively, and are rarely present at greater than trace levels in fresh animal and plant tissues. 1,2-Diacyl-*sn*-glycerols, however, are important as intermediates in the biosynthesis of triacylglycerols and other lipids. In addition, it has become evident that they are important intracellular messengers, generated on hydrolysis of phosphatidylinositol and related compounds by a specific phospholipase C, and that they are involved in the regulation of vital processes in mammalian cells. 2-Monoacyl-*sn*-glycerols are formed as intermediates or end-products of the enzymatic hydrolysis of triacylglycerols.

Acyl migration occurs rapidly with such compounds, especially on heating, or in alcoholic solvents or when protonated, so special procedures are required for their isolation or analysis if the stereochemistry is to be retained.

2. *Other simple lipids*

Cholesterol is by far the most common member of a group of steroids with a tetracyclic ring system; it has a double bond in one of the rings and one free hydroxyl group (Fig. 3.2). It is found both in the free state, where it has a vital role in maintaining membrane fluidity, and in esterified form, i.e. as cholesterol esters. *Wax esters* in their most abundant form consist of fatty acids esterified to long-chain alcohols with similar aliphatic chains to the

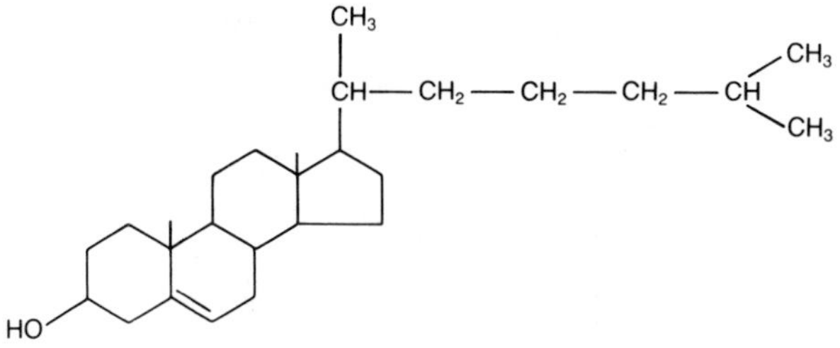

FIG. 3.2. The structure of cholesterol.

acids; these compounds are found in animal, plant and microbial tissues and have a variety of functions, such as acting as energy stores, waterproofing and even echo-location. In some tissues, such as skin and avian preen glands, the wax components can be much more complicated in their structures and compositions. For example, they can contain aliphatic diols, free alcohols, hydrocarbons (especially squalene), aldehydes and ketones. The composition and biochemistry of waxes in nature, and methods for their analysis, have been reviewed in a comprehensive monograph [436].

3. *Glycerophospholipids*

The structures of the common glycerophospholipid constituents of plant and animal tissues are shown in Fig. 3.3. *Phosphatidic acid* or 1,2-diacyl-*sn*-glycerol-3-phosphate is found naturally in trace amounts only in tissues, but it is important metabolically as a precursor of most other glycerolipids. It is strongly acidic and is usually isolated as a mixed salt. As it is somewhat water-soluble, it may be necessary to take special precautions during the extraction of tissues to ensure quantitative recovery.

Phosphatidylglycerol or 1,2-diacyl-*sn*-glycerol-3-phosphoryl-1′-*sn*-glycerol tends to be a trace constituent of tissues, although it does appear to have important functions in lung surfactant and in plant chloroplasts. *Diphosphatidylglycerol* (or cardiolipin) is related structurally to phosphatidylglycerol, and is an important constituent of mitochondrial lipids, especially in heart muscle; its occurrence and properties have been reviewed [353]. These lipids also are acidic.

Phosphatidylcholine or 1,2-diacyl-*sn*-glycerol-3-phosphorylcholine (commonly termed "lacithin") is usually the most abundant lipid in the membranes of animal tissues, and is often a major lipid component of plant membranes, and sometimes of microorganisms. Together with the other choline-containing phospholipid, sphingomyelin, it comprises much of the lipid in the external monolayer of the plasma membrane of animal cells. Lysophosphatidylcholine, which contains only one fatty acid moiety in each molecule, generally in position *sn*-1, is sometimes present in tissues also but

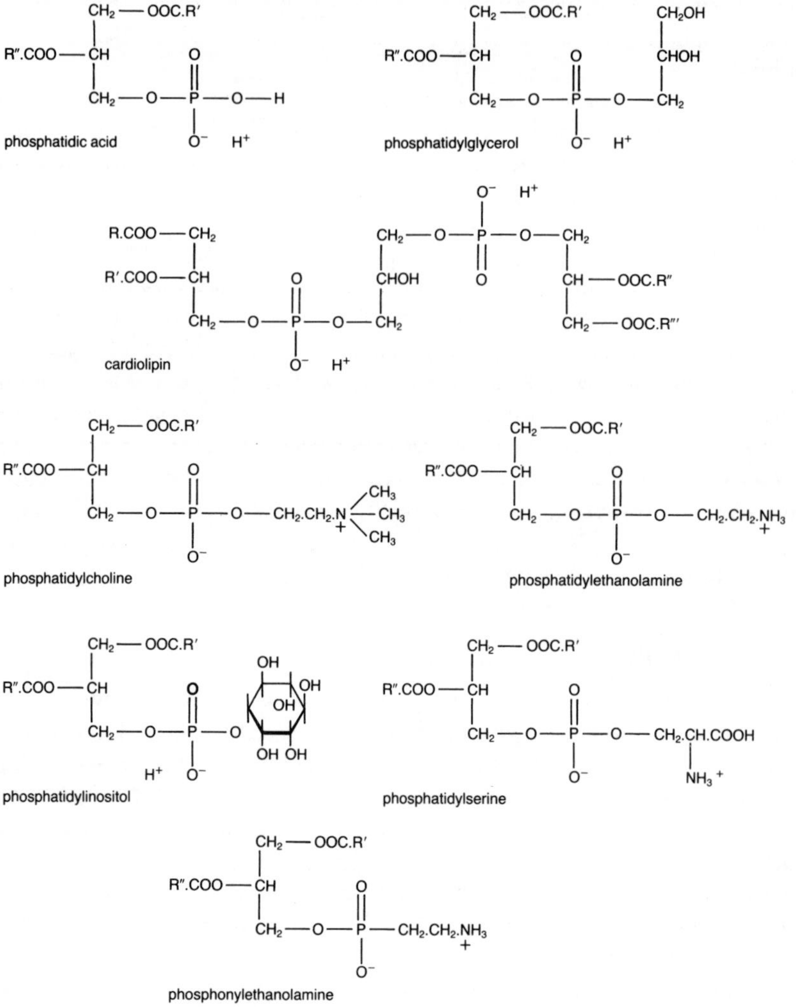

phosphatidic acid

phosphatidylglycerol

cardiolipin

phosphatidylcholine

phosphatidylethanolamine

phosphatidylinositol

phosphatidylserine

phosphonylethanolamine

Fig. 3.3. The structures of the principal glycerophospholipids.

as a minor component; it is more soluble in water than most other lipids and can be lost during extraction, unless precautions are taken.

Phosphatidylethanolamine (once trivially termed "cephalin") is frequently the second most abundant phospholipid class in animal and plant tissues, and can be the major lipid class in microorganisms. The amine group can be methylated enzymically, as part of an important cellular process, to yield as intermediates first phosphatidyl-*N*-monomethylethanolamine and then phosphatidyl-*N*,*N*-dimethylethanolamine; the eventual product is phosphatidylcholine. *N*-Acyl-phosphatidylethanolamine is a minor component of some plant tissues, and is also found in animal tissues under certain conditions. Lysophosphatidylethanolamine contains only one mole of fatty acid per mole of lipid.

Phosphatidylserine is a weakly acidic lipid, so is generally isolated from tissues in salt form. It is present in most tissues of animals and plants and is also found in microorganisms. Its biochemistry has been reviewed [53]. *N*-Acylphosphatidylserine has been detected in certain animal tissues.

Phosphatidylinositol, containing the optically-inactive form of inositol, myoinositol, is a common constituent of animal, plant and microbial lipids. Often in animal tissues, it is accompanied by small amounts of phosphatidylinositol 4-phosphate and phosphatidylinositol 4,5-bisphosphate (polyphosphoinositides). These compounds have a rapid rate of metabolism in animal cells, and with their diacylglycerol metabolites are important in regulating vital processes. The topic has recently been reviewed [326].

Phosphonolipids are lipids with a phosphonic acid derivative esterified to glycerol; the carbon–phosphorus bond is not easily hydrolysed by chemical reagents. The commonest glycerophosphonolipid is phosphonylethanolamine. Such lipids are found mainly in marine invertebrates and in protozoa, and their occurrence has been reviewed [334, 555].

4. *Glyceroglycolipids*

Plant tissues especially tend to contain appreciable amounts of lipids in which 1,2-diacyl-*sn*-glycerols are joined by a glycosidic linkage at position *sn*-3 to a carbohydrate moiety (see Fig. 3.4). Their structures and compositions have been reviewed [291, 292, 724]. The main components are the *mono-* and *digalactosyldiacylglycerols*, but related lipids have been found containing up to four galactose units, or in which one or more of these is replaced by glucose moieties. In addition, a 6-*O*-acyl-monogalactosyldiacylglycerol is occasionally a component of plant tissues. A further unique plant glycolipid is *sulphoquinovosyldiacylglycerol* or the "plant sulpholipid", and contains a sulphonic acid residue linked by a carbon–sulphur bond to the carbohydrate moiety of a monoglycosyldiacylglycerol; it is found exclusively in the chloroplasts.

Monogalactosyldiacylglycerols are known to be present in small amounts in brain and nervous tissue in some animal species, and a range of complex glyceroglycolipids have been isolated and characterized from intestinal tract and lung tissue. Such compounds would be destroyed by certain of the methods used in the isolation of glycosphingolipids, and may be more widespread than is thought. A complex sulpholipid, termed "seminolipid", of which the main component is 1-*O*-hexadecyl-2-*O*-hexadecanoyl-3-*O*-(3′-sulpho-*beta*-D-galactopyranosyl-*sn*-glycerol, is the principal glycolipid in testis and sperm [560, 812].

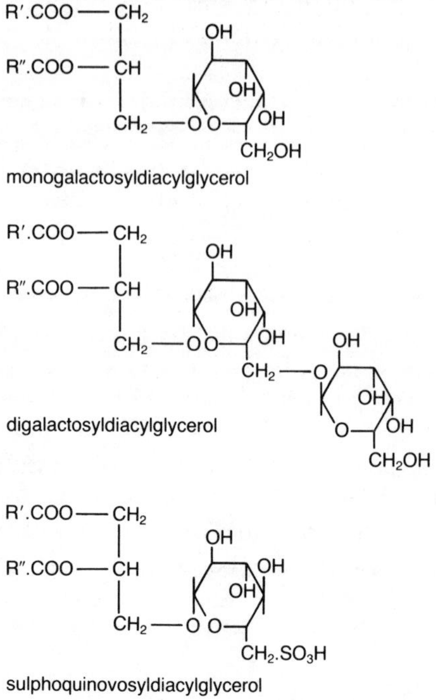

monogalactosyldiacylglycerol

digalactosyldiacylglycerol

sulphoquinovosyldiacylglycerol

FIG. 3.4. The structures of glycosyldiacylglycerols.

Glycolipids, unlike phospholipids, are soluble in acetone and this property can be used in isolating them by chromatographic means.

5. *The fatty acid components*

The common fatty acids of plant tissues are C_{16} and C_{18} straight-chain compounds with zero to three double bonds of a *cis* (or *Z*) configuration. Such fatty acids are also found in animal tissues, together with other even numbered components with a somewhat wider range of chain-lengths and up to six *cis* double bonds separated by methylene groups (methylene-interrupted). The most abundant saturated fatty acid is hexadecanoic acid

$$CH_3(CH_2)_{14}COOH$$

more often referred to as "palmitic" acid. It can also be designated a "16:0" fatty acid, the first numerals denoting the number of carbon atoms in the aliphatic chain and the second denoting the number of double bonds. Oleic acid or *cis*-9-octadecenoic acid, the most abundant monoenoic fatty acid in nature, is designated as an "18:1" acid, or more precisely as "18:1(n – 9)", to indicate that the last double bond is 9 carbon atoms from the terminal methyl group. The systematic and trivial names of the more abundant fatty

TABLE 3.1
The names and designations of the more common fatty acids

Systematic name	Trivial name	Shorthand designation
butanoic	butyric	4:0
hexanoic	caproic	6:0
octanoic	caprylic	8:0
decanoic	capric	10:0
dodecanoic	lauric	12:0
tetradecanoic	myristic	14:0
hexadecanoic	palmitic	16:0
octadecanoic	stearic	18:0
eicosanoic	arachidic	20:0
cis-9-hexadecenoic	palmitoleic	16:1(n – 7)
cis-9-octadecenoic	oleic	18:1(n – 9)
trans-9-octadecenoic	elaidic	
cis-11-octadecenoic	*cis*-vaccenic	18:1(n – 7)
9,12-octadecadienoic[a]	linoleic	18:2(n – 6)
9,12,15-octadecatrienoic	linolenic	18:3(n – 3)
6,9,12-octadecatrienoic	γ-linolenic	18:3(n – 6)
8,11,14-eicosatrienoic	homo-γ-linolenic	20:3(n – 6)
5,8,11,14-eicosatetraenoic	arachidonic	20:4(n – 6)
5,8,11,14,17-eicosapentaenoic		20:5(n – 3)
7,10,13,16,19-docosapentaenoic		22:5(n – 3)
4,7,10,13,16,19-docosahexaenoic		22:6(n – 3)

[a]When not specified, the double bond configuration is *cis*.

acids, together with their shorthand designations, are listed in Table 3.1. A comprehensive compilation has recently been published [701]. All the even-numbered saturated fatty acids from C_2 to C_{30} have been found in nature, but apart from an occasional exception, only the C_{14} to C_{18} homologues are likely to be encountered in appreciable concentrations in glycerolipids. Similarly, the most abundant *cis* monoenoic acids fall into the same range of chain-lengths, i.e. 16:1(n – 7) and 18:1(n – 9). Fatty acids with double bonds of the *trans* (or *E*) configuration are found occasionally in natural lipids, or are formed during food processing and so enter the food chain, but they tend to be minor components only of animal tissue lipids. Their suitability as food constituents is currently a controversial subject (reviewed by Kinsella *et al.* [424]).

The C_{18} polyunsaturated fatty acids, linoleic or *cis*-9, *cis*-12-octadecadienoic acid (18:2(n – 6)) and linolenic or *cis*-9, *cis*-12, *cis*-15-octadecatrienoic acid (18:3(n – 3)), are major components of most plant lipids, including many of the commercially important vegetable oils. They are essential fatty acids in that they cannot be synthesized in animal tissues (the topic has been thoroughly discussed in a recent review volume [328]). However, as it is virtually always present in food materials, linoleic acid tends to be relatively abundant in animal tissues. In turn, these acids are the biosynthetic precursors

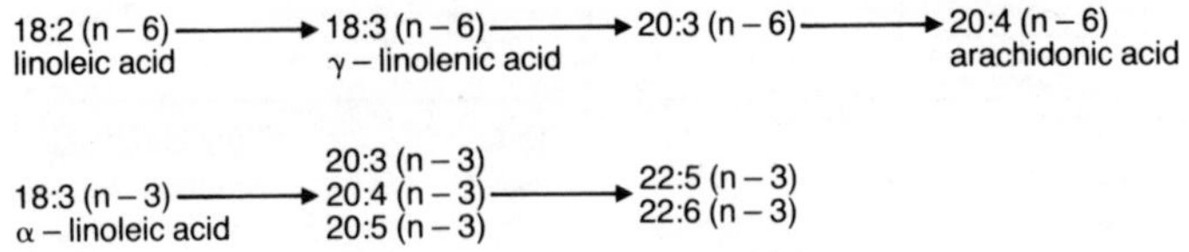

18:2 (n – 6) → 18:3 (n – 6) → 20:3 (n – 6) → 20:4 (n – 6)
linoleic acid — γ – linolenic acid — arachidonic acid

18:3 (n – 3) → 20:3 (n – 3), 20:4 (n – 3), 20:5 (n – 3) → 22:5 (n – 3), 22:6 (n – 3)
α – linoleic acid

FIG. 3.5. Biosynthesis of long-chain polyunsaturated fatty acids by chain-elongation and desaturation of linoleic and linolenic acids.

in animal systems of C_{20} and C_{22} polyunsaturated fatty acids, with 3 to 6 double bonds, via sequential desaturation and chain-elongation steps, as shown in Fig. 3.5 (animal desaturases can only insert a double bond on the carboxyl side of an existing double bond). Those fatty acids derived from linoleic acid, especially arachidonic acid (20:4(n – 6)), are important constituents of the membrane phospholipids in mammalian tissues, and are also the precursors of the prostaglandins and related compounds. In fish, linolenic acid is the more important essential fatty acid, and polyunsaturated fatty acids with an (n – 3) terminal double bond are found in greater abundance.

Fatty acids with many other substituent groups are found in plants (reviewed by Smith [774]) and in animals. These substituents include acetylenic bonds, conjugated double and triple bonds, allenic groups, cyclopropane, cyclopropene, cyclopentene and furan rings, and hydroxy-, epoxy- and keto-groups. Often the fatty acids containing such substituents have only been detected in trace amounts, or are known only from obscure seed oils. Branched-chain fatty acids are synthesized by many microorganisms (most often with an *iso*- or an *anteiso*-methyl branch), and occasionally can be produced in animal tissues, especially those of ruminants, and thereby enter the food chain. Phytanic acid, 3,7,11,15-tetramethylhexadecanoic acid, is a metabolite of phytol and is found in animal tissues, but generally at low levels only. 2-Hydroxy fatty acids are synthesized in animal tissues, and are often major constituents of the sphingolipids (see Section C below). In some circumstances, furanoid fatty acids can accumulate in the reproductive tissues of fish.

Certain of these substituent groups are not stable chemically, so that the fatty acids are not readily analysed by GLC, which has been the most widely used technique for fatty acid analysis in general; such unusual fatty acids can now often be analysed or isolated more easily and without degradation by means of HPLC.

6. *Ether lipids*

Many glycerolipids, including simple lipids, phospholipids and glycolipids, and especially those of animal and microbial origin, contain aliphatic residues linked either by an ether bond or a vinyl ether bond (Fig. 3.6) to position 1 of L-glycerol. The generic term "plasmalogen" is used for any lipid containing

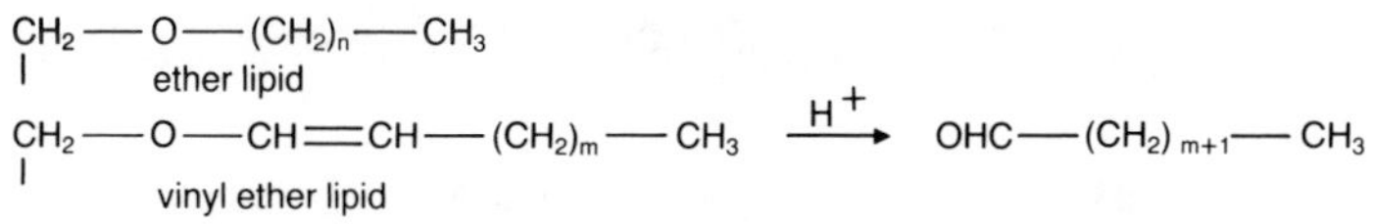

$$CH_2—O—(CH_2)_n—CH_3$$
ether lipid
$$CH_2—O—CH{=}CH—(CH_2)_m—CH_3 \xrightarrow{H^+} OHC—(CH_2)_{m+1}—CH_3$$
vinyl ether lipid

FIG. 3.6. Ether and vinyl ether bonds in lipids.

a vinyl ether bond. The occurrence, chemistry and biochemistry of such lipids have been reviewed in a comprehensive monograph [509]. 1-Alkyl-2,3-diacyl-*sn*-glycerols tend to be present in trace amounts only in tissues, but can be major constituents of certain fish oils. Related compounds containing a 1-alk-1′-enyl moiety (neutral plasmalogens) are occasionally present also. In the phospholipids of animals and microorganisms, analogues containing vinyl ether and ether bonds tend to be much more abundant than in the simple lipids, especially in the phosphatidylethanolamine fraction; in this instance, it has been suggested that they should be termed "plasmenylethanolamine" and "plasmanylethanolamine" respectively.

One ether-containing phospholipid, in particular, which is presently being studied intensively because it can exert profound biological effects at minute concentrations, is 1-alkyl-2-acetyl-*sn*-glycerophosphorylcholine or "platelet-activating factor". The chemistry and biochemistry of this compound have recently been reviewed [277].

On hydrolysis of glycerolipids containing an alkyl ether bond, 1-alkyl-glycerols are released that can be isolated for analysis. Similarly, when plasmalogens are hydrolysed under basic conditions, 1-alkenylglycerols are released; on acidic hydrolysis on the other hand, aldehydes are formed (see Fig. 3.6). With both groups of compounds, the aliphatic residues are generally 16 or 18 carbon atoms long and are saturated or may contain one additional double bond, that is remote from the ether linkage.

Phospholipid classes isolated by chromatographic means tend to be a mixture of the diacyl, alkylacyl and alkenylacyl forms. To indicate that this is so, they are sometimes termed the "diradyl" form of the appropriate phospholipid.

C. Sphingolipids

1. Long-chain bases

Long-chain bases (sphingoids or sphingoid bases) are the characteristic structural unit of the sphingolipids, the chemistry and biochemistry of which have been thoroughly reviewed [397, 883]. The bases are long-chain (12 to 22 carbon atoms) aliphatic amines, containing two or three hydroxyl groups, and often a distinctive *trans*- double bond in position 4 (see Fig. 3.7). The commonest or most abundant is sphingosine ((2S, 3R, 4E)-2-amino-4-octa-

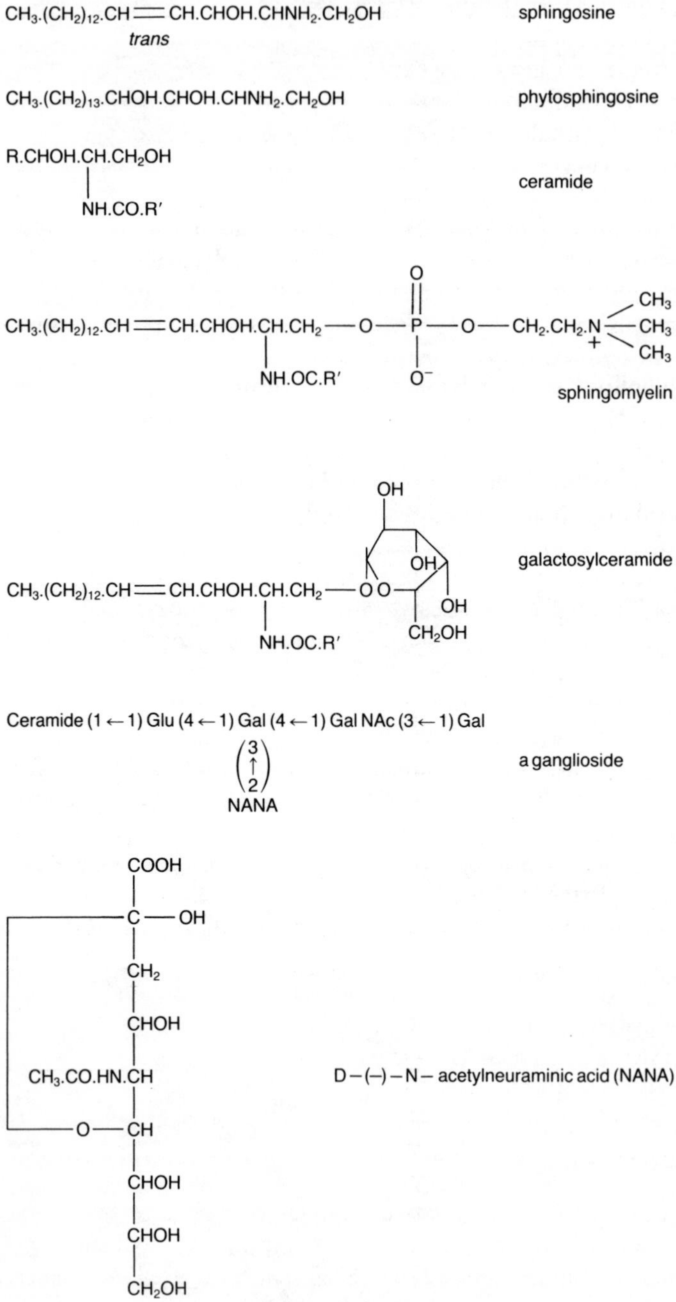

$CH_3.(CH_2)_{12}.CH{=}CH.CHOH.CHNH_2.CH_2OH$ sphingosine
trans

$CH_3.(CH_2)_{13}.CHOH.CHOH.CHNH_2.CH_2OH$ phytosphingosine

$R.CHOH.CH(NH.CO.R').CH_2OH$ ceramide

$CH_3.(CH_2)_{12}.CH{=}CH.CHOH.CH(NH.OC.R').CH_2{-}O{-}P(=O)(O^-){-}O{-}CH_2.CH_2.N^+(CH_3)_3$ sphingomyelin

$CH_3.(CH_2)_{12}.CH{=}CH.CHOH.CH(NH.OC.R').CH_2{-}O{-}$[galactose: OH, OH, OH, CH_2OH] galactosylceramide

Ceramide (1 ← 1) Glu (4 ← 1) Gal (4 ← 1) Gal NAc (3 ← 1) Gal
(3 ↑ 2) NANA (on Gal) a ganglioside

$COOH{-}C(OH){-}CH_2{-}CHOH{-}CH(NH.CO.CH_3){-}CH(O{-}){-}CHOH{-}CHOH{-}CH_2OH$ (ring: O linking C to CH) D-(−)-N-acetylneuraminic acid (NANA)

FIG. 3.7. The structures of sphingolipids and of their component parts.

decen-1,3-diol). More than 60 long-chain bases have been found in animals, plants and microorganisms [268, 403, 404], and many of these may occur in a single tissue, but always as part of a complex lipid as opposed to in the free form. The aliphatic chains can be saturated, monounsaturated and diunsaturated, with double bonds of either the *cis* or *trans* configuration, and they can also have methyl substituents. In addition, saturated and monoenoic straight- and branched-chain trihydroxy bases are found. The commonest long-chain base of plant origin, for example, is phytosphingosine ((2S, 3S, 4R)-2-amino-octadecanetriol). For shorthand purposes, a nomenclature similar to that for fatty acids can be used; the chain length and number of double bonds are denoted in the same manner with the prefix "d" or "t" to designate di- and tri-hydroxy bases respectively. Thus, sphingosine is d18:1 and phytosphingosine is t18:0.

2. *Ceramides*

Ceramides contain fatty acids linked to the amine group of a long-chain base by an amide bond (Fig. 3.7). Generally, they are present at low levels only in tissues, but are important as intermediates in the biosynthesis of the complex sphingolipids. The acyl groups of ceramides are long-chain (up to C_{26}, but occasionally longer) odd- and even-numbered saturated or monoenoic fatty acids and related 2-D-hydroxy fatty acids.

Tetraacetylsphingosines have been detected among the extracellular lipids of yeasts [792]. An unusual ceramide has been characterized in which the fatty acids linked to the sphingoid base consist of C_{30} and C_{32} *omega*-hydroxylated components, with predominantly the essential fatty acid, linoleic acid, esterified to the *omega*-hydroxyl group [873]; it is located in the epidermis and appears to have the important function of regulating the permeability of the membranes.

3. *Sphingomyelin and other sphingophospholipids*

Sphingomyelin consists of a ceramide unit linked at position 1 to phosphorylcholine, and it is found as a major component of the complex lipids of all animal tissues, but is not present in plants or microorganisms. It resembles phosphatidylcholine in many of its physical properties, and can apparently substitute in part for this in membranes. For example, it is a major constituent of the plasma membrane of cells. Sphingosine is usually the most abundant long-chain base constituent, together with sphinganine and C_{20} homologues.

Ceramide phosphorylethanolamine is a component of the lipids of insects and some freshwater invertebrates; the phosphonolipid analogue, ceramide 2-aminoethylphosphonic acid, has been found in sea anemones and protozoa.

4. Neutral glycosylceramides

The most widespread glycosphingolipids are the *monoglycosylceramides* (or cerebrosides), and they consist of the basic ceramide unit linked at position 1 by a glycosidic bond to glucose or galactose. They were first found in brain lipids, where the principal form is a monogalactosylceramide, but they are now known to be ubiquitous constituents of animal tissues. In addition, they are found in plants (monoglucosylceramides only), where the main long-chain base is phytosphingosine. *O*-Acyl-glycosylceramides have been detected in small amounts in some tissues.

Di-, tri- and tetra-glycosylceramides (oligoglycosylceramides) are usually present also in animal tissues. The most common diglycosylceramide is lactosylceramide, and it can be accompanied by related compounds containing further galactose or galactosamine residues, for example. Tri- and tetra-glycosylceramides with a terminal galactosamine residue are sometimes termed "globosides", while glycolipids containing fucose are known as "fucolipids". Oligoglycosylceramides with more than 20 carbohydrate residues have been isolated from animal tissues, those from intestinal cells having been studied with particular intensity. They appear to form part of the immune response system. Although certain of these lipids have been found on occasion to have distinctive long-chain base and fatty acid compositions, the complex glycosyl moiety is considered to be of primary importance for their immunological function and therefore has received most attention from investigators. Certain glycolipids can accumulate in persons suffering from rare disease syndromes, characterized by deficiencies in specific enzyme systems related to glycolipid metabolism.

Sulphate esters of galactosylceramide and lactosylceramide (often referred to as "sulphatides" or "lipid sulphates"), with the sulphate group linked to position 3 of the galactosyl moiety, are major components of brain lipids and are also found in trace amounts in other tissues; their chemistry and biochemistry have been reviewed [213].

Complex plant sphingolipids, phytoglycosphingolipids, containing glucosamine, glucuronic acid and mannose linked to the ceramide via phosphorylinositol, were isolated and characterized from seeds initially, but related compounds are also known to be present in other plant tissues and in fungi.

5. Gangliosides

Gangliosides are highly complex oligoglycosylceramides, which contain one or more sialic acid groups (*N*-acyl, especially acetyl, derivatives of neuraminic acid, abbreviated to "NANA"), in addition to glucose, galactose and galactosamine. They were first found in the ganglion cells of the central nervous system, hence the name, but are now known to be present in most animal tissues. The nature of the long-chain base and fatty acid components of each ganglioside can vary markedly between tissues and species and is related in some way to its function.

D. The Lipid Composition of Animal Tissues

1. Some general comments

Literally thousands of papers have appeared over the last 25 years detailing the structures and compositions of lipids from particular tissues and species, as determined by modern chromatographic methods, but there appears to have been very little effort to collate and critically compare this data in any systematic way, or to relate the compositions of lipids to their functions. Among other consequences of this, there remain anomalies and gaps in our knowledge. Comprehensive accounts of the lipids of the tissues of ruminant animals [132], tissue and membrane phospholipid compositions [333, 875] and triacylglycerol compositions [90, 490, 744] have appeared, however, and there are miscellaneous reviews of the compositions of specific lipid classes or tissues in the literature. For example, one lipid material to have been studied in minute detail is milk fat, because of its nutritional and commercial importance [132, 135, 552, 632]. It is unique in many ways and certainly is not typical of other tissue lipids, so is of little value for illustrative purposes in this section. It is only possible here to summarize briefly, some of the more significant features of lipid composition in animal tissue lipids, with some highly selective examples of relevant analyses. Where possible, data are presented for the liver, because it contains all of the important lipid classes, and because of the central importance of this organ in lipid metabolism.

One problem in comparing data from different sources is the method of presentation, which is often dependent on the nature of the analytical methods used. For example, it is easier technically to analyse the phospholipids and glycolipids, following isolation as distinct groups, separately from the simple lipids. Often there is no attempt subsequently to integrate the data. To add to the confusion, results of analyses of simple lipids are often reported in terms of weight % of each lipid class, since the data are acquired in this form, while the results for phospholipids are most frequently recorded as molar %, especially when phosphorus analysis is used as the means of quantification.

2. Lipid class compositions

In tissues, the structural lipid components, such as the phospholipids, tend to be rather constant in composition so meaningful comparisons between different organs can be often be made. On the other hand, the proportions of the simple lipids, especially the triacylglycerols, can vary greatly according to the dietary or physiological state of the animal, and this information is not always recorded in papers. As examples, some data on the lipid

TABLE 3.2
The composition of the lipid classes (weight % of the total) in rat heart, liver, erythrocytes and plasma [136]

Lipid class	Tissue			
	Heart	Liver	Erythrocytes	Plasma
CE[a]	tr	2	—	16
TG	4	7	—	49
C	4	5	30	6
DG	1	—	tr	tr
unknown	2	—	1	—
FFA	—	tr	—	2
DPG[b]	12	5	—	—
PE	33	20	21	—
PI	4	4	3	—
PS	—	—	3	—
PC	39	55	32	24
SPH	2	2	8	2
LPC	—	—	1	1

[a]See Chapter 1 for a list of abbreviations; tr = trace.
[b]Also contains cerebrosides and phosphatidylglycerol.

composition of selected rat tissues are shown in Table 3.2 [136]. Those lipids listed are by far the most abundant in these tissues, and they are those most often seen in other organs, although other lipids can assume importance in some circumstances. Lipids present at relatively low concentrations only, such as the glycosphingolipids, were not determined in these analyses, but they are of course of great metabolic importance. Nor were the diacyl, alkylacyl and alkenylacyl forms of the phospholipids distinguished. Cholesterol esters, triacylglycerols and free (unesterified) cholesterol tend to be the most abundant simple lipids; when substantial amounts of diacylglycerols and free fatty acids, especially the latter, are encountered in a sample, it is usually indicative of artefactual hydrolysis during storage or extraction of the tissues. The choline-containing constituents, i.e. phosphatidylcholine and sphingomyelin, but predominantly the former, tend almost always to be the most abundant phospholipids, amounting to 50 to 60% of the total, followed by phosphatidylethanolamine, and then by phosphatidylinositol and phosphatidylserine. The results quoted for rat liver lipids are therefore typical of this and many other organs. Diphosphatidylglycerol is a major component of mitochondrial lipids in particular and so is found in appreciable amounts in heart muscle; phosphatidylglycerol only appears to assume importance in lung surfactant. Again, when appreciable amounts of phosphatidic acid or lysophospholipids are encountered in samples, artefactual hydrolysis may have occurred.

In erythrocytes, all the lipids appear to be constituents of the membranes, and the results of the analysis listed here indicate that only cholesterol and the phospholipids are present. These data would be typical of many other species, and any departure from this general pattern is symptomatic of an underlying metabolic difference; for example in ruminant erythrocytes, there is virtually no phosphatidylcholine, which is replaced entirely by sphingomyelin (reviewed by Christie [132]).

The compositions of the plasma lipids are of particular importance as they supply fatty acids to all tissues, and are assumed to be in some form of compositional equilibrium with them. Of course, biopsy samples of plasma are easy to obtain. Those lipids reported as plasma lipid constituents here are typical; phosphatidylcholine is present as a higher proportion of the total phospholipids than in any other tissue, while cholesterol esters tend to be more abundant than in most organs (other than in steroidogenic tissues such as the adrenals). The proportion of triacylglycerols in the plasma in this instance was higher than normal, because an old obese rat was taken for the analysis. Lipids are transported in plasma in the form of complexes with proteins, which render them compatible with their aqueous environment. Lipoprotein fractions with different compositions and metabolic functions vary in density and can most conveniently be separated by means of ultracentrifugation, although HPLC methods are becoming increasingly important (see Chapter 10). It should be recognized that an analysis of the total lipids in plasma can give only part of the picture, and it may be necessary to determine the compositions of each of the lipoprotein fractions before definitive metabolic conclusions can be drawn.

All membranes (indeed each side of the bilayer) in a tissue can have distinctive compositions that are in some way related to their function. The results of some analyses of the phospholipids of the membranes of rat liver are recorded in Table 3.3. Naturally, the same phospholipids are present as

TABLE 3.3

The phospholipid composition of whole tissue and membrane preparations from rat liver (mol % lipid phosphorus)

		Membrane			
Lipid class	Whole tissue	Nuclei	Mitochondria	Microsomes	Plasma membrane
DPG[a]	5	—	15	2	—
PE	25	26	34	22	20
PI	7	4	7	8	7
PS	3	6	1	4	4
PC	51	57	41	59	43
SPH	4	6	2	4	23
LPC	1	—	1	2	2
Reference	[893]	[264]	[153]	[153]	[153]

[a] See Chapter 1 for a list of abbreviations.

is described above for the intact organ, but the relative proportions vary markedly. For example, diphosphatidylglycerol and sphingomyelin are much more abundant in mitochondria and plasma membrane respectively than in any of the other membranes. The data listed in Table 3.3 are incomplete, however, as cholesterol is an important constituent of membranes, modifying their fluidity, and the proportion relative to the phospholipids should be determined. In plasma membrane preparations from rat liver, the molar ratio of cholesterol to phospholipids is 0.76, while that in microsomes and mitochondria is 0.1 [153]. In addition, only a part of each phospholipid is the diacyl form and in most tissues, the phosphatidylethanolamine especially tends to contain a high proportion of plasmalogens. On the other hand, the plasmalogen form of phosphatidylcholine appears only to occur in high proportions in heart muscle and reproductive tissues. With a few significant exceptions, the lipid compositions of equivalent membranes from different tissues and even from different species are in general very similar, probably because these membranes have common functional requirements.

3. *Fatty acid compositions of glycerolipids*

The fatty acid composition of each lipid in a tissue is frequently distinctive and can vary markedly between species. It is obviously greatly, but not entirely, dependent on the nature of the diet of the animal concerned (although in many analytical studies, such details are not given). Kuksis [451] has systematically reviewed the fatty acid compositions of animal glycerolipids. Data for the fatty acid compositions of the depot fats, which are highly responsive to dietary influence, in several animal species are listed in Table 3.4. In all species, there are relatively high contents of 16:0 and 18:1 fatty

TABLE 3.4
The fatty acid compositions (mol % of the total) of the adipose tissue lipids of various animal species

	Species					
Fatty acid	Rat	Pig	Sheep	Horse	Herring	Seal
16:0	23	29	21	26	21	10
16:1	5	3	2	8	11	16
18:0	6	18	35	5	—	—
18:1	35	41	31	31	23	26
18:2	19	8	2	9	1	2
18:3	2	—	—	18		
20:1					10	14
20:5					9	7
22:1					10	7
22:6					5	8
Reference	[95]	[147]	[148]	[95]	[94]	[94]

acids. Species differences are most apparent in the relative concentrations of the saturated and polyunsaturated components. In the rat, pig and horse, for example, there are appreciable amounts of linoleic acid; this is a minor component in the sheep, because it is subjected to microbial biohydrogenation in the rumen, and in the herring and seal, because it is present at low levels only in their food chain. In addition, the sheep and horse may consume a very similar diet, but there are high levels of an 18:3 fatty acid, derived from the herbage, in the lipids of the horse and not in those of the sheep, again because of biohydrogenation in the rumen of the latter. Fish, such as the herring, tend to contain relatively high amounts of C_{20} and C_{22} fatty acids, derived from the microflora and microfauna which they consume. In turn, the composition of the depot fat of the seal reflects its diet of fish.

Each lipid class in a tissue also has a distinctive fatty acid composition, probably related to its function but in a way that is still only partly understood. Some data for the main glycerolipids and the cholesterol esters of rat liver are listed in Table 3.5. In essence, only C_{16} and C_{18} fatty acids are found in the triacylglycerols, although most of the glycerophospholipids contain substantial proportions of the longer-chain polyunsaturated components. For example, the phosphatidylcholine contains 50% of saturated fatty acids, while arachidonic acid constitutes 20% of the total. The phosphatidylethanolamine has a similar proportion of saturated fatty acids, but somewhat less linoleic acid and correspondingly more of the C_{20} and C_{22} polyunsaturated fatty acids. Characteristically, high proportions of stearic and arachidonic acids are present in the phosphatidylinositol, while the composition of the phosphatidylserine is similar except that the 22:6(n – 3) fatty acid substitutes for part of the arachidonic acid. The diphosphatidylglycerol differs markedly from all the other glycerophospholipids in that the single fatty acid, linoleic acid, comprises nearly 60% of the total. The composition of the cholesterol

TABLE 3.5

The fatty acid compositions (mole % of the total) of the main glycerolipids and the cholesterol esters of rat liver

	Lipid class						
Fatty acid	TG[a]	PC	PE	PI	PS	DPG[b]	CE
16:0	27	14	18	7	4	7	24
16:1	4	1				8	3
18:0	7	34	37	40	47	4	20
18:1	27	10	8	3	3	20	15
18:2	12	12	5	3	2	59	16
20:4(n – 6)	1	20	23	40	25	2	14
22:6(n – 3)	—	4	7	2	16	—	
Reference	[888]	[888]	[888]	[332]	[333]	[153]	[157]

[a]See Chapter 1 for a list of abbreviations.
[b]Of the mitochondrial fraction.

esters tends to resemble that of the phosphatidylcholine, but there is somewhat less stearic acid and more palmitic acid in the former. Similar general compositional trends are seen, although the absolute values may differ, in comparing the same lipids in other tissues and, with dietary influences superimposed, in comparing the corresponding lipids of other species.

4. Fatty acid distributions within glycerolipid molecules

In addition to each lipid class in a tissue having a distinctive fatty acid composition, each position of the glycerol moiety tends to have a unique fatty acid composition, that is determined during the biosynthesis of a lipid mainly by the specificities of various acyltransferases. Again, the positional distributions of fatty acids in particular lipid classes can vary markedly between tissues and species, and can have some metabolic importance. Data for triacylglycerols [90, 138, 490, 744] and for glycerophospholipids [333] have been thoroughly reviewed.

Some results for the principal glycerolipids of rat liver, taken from a single systematic analytical study [888], are listed in Table 3.6. In the triacylglycerols, the highest proportion of the 16:0 fatty acid is in position *sn*-1, but the other saturated fatty acid, 18:0, is distributed mainly between positions *sn*-1 and *sn*-3. The 16:1 fatty acid is found in equal abundance in all three positions, but the C_{18} unsaturated fatty acids are distributed to a large extent between positions *sn*-2 and *sn*-3. In the phosphatidylcholine, much of the saturated fatty acids are in position *sn*-1, although a significant relative proportion of the 16:0 fatty acid is also found in position *sn*-2. Some of the 18:1 fatty acid is in position *sn*-1, but virtually all of the remaining unsaturated fatty acids are in position *sn*-2. Similar relative distributions are seen in the

TABLE 3.6
Positional distributions of fatty acids in each position of the principal glycerolipids of rat liver (results are expressed as mol % of the total in each position) [888]

	TG[a]			PC		PE	
Fatty acid	*sn*-1	*sn*-2	*sn*-3	*sn*-1	*sn*-2	*sn*-1	*sn*-2
16:0	57	13	12	23	6	25	11
16:1	4	3	4	1	1	tr	tr
18:0	9	4	8	65	4	65	8
18:1	22	58	61	7	13	8	8
18:2	4	19	11	1	23	—	10
20:4(n – 6)	—	1	1	tr	39	—	46
22:6(n – 3)				—	7	—	13

[a]See Chapter 1 for a list of abbreviations; tr = trace.

phosphatidylethanolamine, except that a slightly greater proportion of the saturated fatty acids are present in position *sn*-2. 1,2-Diacyl-*sn*-glycerols are a common intermediate in the biosynthesis of each of these lipid classes, but there is little correspondence in the compositions of this part of the molecules, when the phospholipids are compared with the triacylglycerols. In addition, analyses of the molecular species distributions in the three classes of lipids indicated that there was little in common between the structures of the triacylglycerols and those of the phospholipids [888]. This suggests that there is great selectivity for particular molecular species of the intermediate diacylglycerols, utilized for the synthesis of each lipid class, or that some hydrolysis and resynthesis of lipids occurs to give the final compositions.

Much of the published work on the structures of animal triacylglycerols has been concerned with those of adipose tissue, which tends to contain most of the body stores of fat in the form of this single lipid class. The results quoted here for rat liver triacylglycerols are not dissimilar to those obtained for adipose tissue for many animal species. Brockerhoff [93] has shown that in most instances position *sn*-1 contains appreciable amounts of the saturated fatty acids, position *sn*-2 contains mainly unsaturated and any shorter-chain fatty acids, while position *sn*-3 consists predominantly of unsaturated and longer-chain fatty acids. The principal exception to this type of distribution was thought to be the triacylglycerols of the pig and related species, in which position *sn*-2 is occupied largely by palmitic acid (70% or more), but increasing numbers of tissues containing triacylglycerols with a structure of this kind are being revealed as research progresses. For example, milk fats from all the higher mammals studied have more than half of their total content of palmitic acid in position *sn*-2 (reviewed by Christie [135]), and the same is true of the triacylglycerols of lymph, plasma and adrenals in ruminant animals. Another interesting feature of milk fats, especially those of ruminants, is that all the short-chain fatty acids (4:0 and 6:0 specifically) are located exclusively in position *sn*-3. A number of excellent analytical studies of the molecular species distributions in animal triacylglycerols have been published, but as the results are not easily summarized, readers are referred to specialist texts for detailed information [90, 138, 490, 744].

The positional distributions of fatty acids in each of the two most abundant phospholipids of rat liver, as listed in Table 3.6, are typical for these lipids in many other tissues and species. In some tissues, the proportions of saturated fatty acids in position *sn*-2 can be somewhat higher, but polyunsaturated fatty acids never appear to occur in appreciable amounts in position *sn*-1. Perhaps the best known example of a glycerophospholipid containing a high proportion of saturated fatty acids is the phosphatidylcholine of lung surfactant, of which up to 60% can be the dipalmitoyl form. The distribution of fatty acids in phosphatidylinositol follows the common pattern, and a high proportion consists of the 1-stearoyl-2-arachidonoyl molecular species.

TABLE 3.7
The sphingolipid composition of different bovine kidney regions (results expressed as mg lipid per g dry tissue) [405]

Lipid class	Cortex	Medulla	Whole tissue
ceramides	1.0	0.7	—
glucosylceramides	0.6	0.6	0.8
galactosylceramides	0.2	1.3	0.4
di- and triglycosylceramides	0.3	0.7	0.4
sulphatides	0.1	0.9	—
sphingomyelins	17.8	9.8	14.4

5. *Sphingolipid compositions*

The sphingolipid components of many tissues and species have been analysed, but those of bovine kidney have been studied in very great detail, with respect both to the relative amounts in various regions of the organ and to the nature of the lipid and non-lipid portions of the molecules [405, 406]. A few of the results obtained are listed in Table 3.7 for illustrative purposes. Sphingomyelin is by far the most abundant sphingolipid but significant amounts of ceramides and mono-, di- and tri-glycosylceramides, and sulphatides are also present. The relative proportions of each vary appreciably at different sites in the kidney, presumably because of the differing functions of the membranes in each region.

The fatty acid composition of the sphingomyelin fraction of rat liver [214] is listed in Table 3.8. In comparison to the glycerophospholipids, a very

TABLE 3.8
The fatty acid composition (weight % of the total) in the sphingomyelin of rat liver [214], and the long-chain base composition (weight % of the total) of ceramide from human liver [403]

Rat liver sphingomyelin		Human liver ceramide	
fatty acid	%	long-chain base	%
16:0	22	d16:0	tr[a]
18:0	10	d17:0	tr
18:1	4	d16:1	5
18:2	2	d18:0	2
20:0	2	d17:1	3
22:0	14	d18:1(*i*)	2
23:0	9	d18:1	77
24:0	24	d18:2	8
24:1	13	d19:1(*ai*)	3

[a] Abbreviations: tr = trace, *i* = *iso*-methyl branch, *ai* = *anteiso*-methyl branch.

different range of fatty acids is seen. For example, virtually no long-chain polyunsaturated fatty acids are found in sphingomyelin, but C_{20} to C_{24} saturated and monoenoic fatty acids, with odd or even chain-lengths, are present in addition to the normal C_{16} and C_{18} components. A similar range of fatty acid constituents is seen in most other sphingolipids, and the bimodal chain-length distribution often causes individual sphingolipids separated by chromatographic means to appear as two adjacent bands. Also, glycosphingolipids frequently contain a similar range of fatty acids to that listed, but with a hydroxyl substituent in position 2.

The long-chain base composition of ceramide, which is probably similar to that of sphingomyelin, from human liver [403] is also listed in Table 3.8. Sphingosine accounts for 77% of the total, but a number of other bases are present in small amounts. Relatively simple base compositions of this kind are found in the other sphingolipids of tissues from most simple-stomached animals. In ruminants, on the other hand, the long-chain base compositions can be much more complex, and may include trihydroxy bases (reviewed by Christie [132]).

E. The Lipid Compositions of Plant Tissues

1. Lipid class compositions

The compositions of the lipids of plant tissues have been reviewed [291, 324, 520]. Some results are listed in Table 3.9. Those plant tissues that serve as major food materials have obviously received most study. Triacylglycerols tend to be the most abundant class of storage lipid in tissues that are rich in lipids, such as the commercially important oil seeds. On the other

TABLE 3.9

The lipid class compositions (weight % of the total lipids) of various plant tissues

Lipid class	Potato tuber	Apple fruit	Soybean seed	Clover[a] leaves	Rye[a] grass	Spinach chloroplast
MGDG[b]	6	1	tr	46	39	36
DGDG	16	5	tr	28	29	20
SQDG	1	1	tr	4	4	5
TG	15	5	88			
PC	26	23	4	7	10	7
PE	13	11	2	5	5	3
PI	6	6	2	1	2	2
PG	1	1	tr	6	7	7
others	15	42	5	3	4	
Reference	[229]	[230]	[290]	[708]	[291]	[882]

[a]Weight % of total acyl lipids.
[b]See Chapter 1 for a list of abbreviations; tr = trace.

hand, there are exceptions and jojoba oil, for example, consists mainly of wax esters. In storage tissues with low concentrations of lipid, such as the potato tuber or apple, the complex glycolipids and phospholipids predominate. In addition to those lipids listed, sterols, sterol esters, acylated sterolglycosides, phytoglycolipid, ceramide, glycosylceramide, phosphatidic acid, *N*-acylphosphatidylethanolamine, and phosphatidylserine, among others, may be found. It should be recognized that seeds and tubers do not have a homogeneous lipid distribution, the endosperm, germ, bran and other organelles each having a distinctive composition.

The glycosyldiacylglycerols, i.e. mono- and di-galactosyldiacylglycerols and sulphoquinovosyldiacylglycerol, are the most abundant lipid classes in leaf tissues. Glycerophospholipids, such as phosphatidylcholine, phosphatidylethanolamine, phosphatidylinositol and phosphatidylglycerol, are also present and other complex lipids are occasionally reported. Phosphatidylglycerol appears to be especially characteristic of photosynthetic tissue, and it can be the main glycerophospholipid in certain green algae. Triacylglycerols are virtually absent from leaves. As in animal tissues, each of the membranes or organelles in the leaf has a characteristic lipid composition. Spinach chloroplasts have received a great deal of attention, because they can be prepared relatively easily for biochemical experiments, and like the intact leaf they contain appreciable amounts of the glycolipids and a smaller proportion of glycerophospholipids. In contrast, as in animal tissues, the plasma membrane has a high content of phosphatidylcholine, while the mitochondria contain diphosphatidylglycerol.

A further region of plants with a distinctive composition is the epidermis or cuticle. The lipids here tend to be rich in waxes, and also contain cutin and suberin, which are complex polyesters of hydroxy fatty acids [437].

2. *Fatty acid compositions*

The fatty acid compositions of the seed oils of importance to commerce have been reviewed [758], and data for a few typical ones are listed in Table 3.10. Maize (corn), sunflower and safflower oils are of great nutritional value since they contain appreciable amounts of the essential fatty acid, linoleic acid. Excessive amounts of linolenic acid, as in soybean oil, lower the value of an oil because it is then more susceptible to rancidity problems because of autoxidation; it is therefore a common industrial practice to subject the oil to hydrogenation. In contrast, there are no such problems with olive oil with its high content of oleic acid. Palm oil contains a high proportion of the single molecular species, 1,3-dipalmitoyl-2-olein, and is of great value in the manufacture of the confectionary products. Cocoa butter is similar. Rapeseed is one of the few oil crops capable of being grown in more northerly climates. In its native form, it tends to have a high content of erucic acid (22:1(n − 9)), which may have some harmful properties to the consumer,

TABLE 3.10
The fatty acid compositions (weight % of the total) of some seed oils [758]

Fatty acid	Soybean	Maize	Safflower	Rapeseed[a]	Olive	Palm
16:0	11	11	6	3	12	42
18:0	4	tr	3	1	2	4
18:1	23	25	12	11	72	38
18:2	51	57	73	13	8	9
18:3	7	1	1	9	1	
C_{20}–C_{22}				55		

[a]Newer cultivars can contain much less erucic acid.

although this is still a matter for controversy. However, new cultivars with negligible levels of this component are now widely grown. Cotton seed oil resembles maize oil in its composition, but also contains small amounts of the cyclopropene fatty acid, "sterculic" or 9,10-methyleneoctadecenoic acid, which has well-established toxic properties and must be removed during refining. Palm kernel and coconut oils are noteworthy for a high content of saturated fatty acids of medium chain-length. In addition, there are many seed oils which may have limited or negligible commercial value, but contain fatty acids with unusual substituent groups and are of great interest to biochemists [774].

As in animal tissues, each of the lipids in a plant tissue can have a characteristic fatty acid composition and for illustrative purposes, some results for spinach leaf lipids are listed in Table 3.11 [25]. Glycosyldiacylglycerols tend to consist mainly of the unsaturated fatty acids, linoleic acid and especially linolenic acid; a hexadecatrienoic acid is often present also. On the other hand, the glycerophospholipids contain higher proportions of saturated fatty acids, generally palmitic acid, in addition to the unsaturated components. Phosphatidylglycerol is unique in that it contains a substantial amount of an unusual fatty acid, i.e. *trans*-3-hexadecenoic acid. The fatty

TABLE 3.11
The fatty acid compositions (weight % of the total) of the individual lipids of spinach leaves [25]

Fatty acid	MGDG[a]	DGDG	SQDG	PG	PC	PI	PE
16:0	tr	6	27	22	20	41	46
16:1(3*t*)	tr			35	tr		
16:3	30	3			tr		2
18:0		1		tr		1	1
18:1	1	4	6	2	11	6	2
18:2	1	3	39	5	30	25	7
18:3	67	84	28	36	40	27	43

[a]See Chapter 1 for a list of abbreviations; tr = trace.

acid compositions of plant tissues can vary with climatic and other cultivation conditions, and with the stage of development of the tissue, and major species differences occur. However, the results listed in Table 3.11 are not untypical.

3. Fatty acid distributions within glycerolipid molecules

Once more, the triacylglycerols of the commercial seed oils are those to have been subjected most frequently to more detailed structural analyses [90, 138, 490, 744]. In general, there tends to be little difference between the compositions of positions *sn*-1 and *sn*-3 of the glycerol moiety, but the saturated fatty acids are concentrated in the primary positions and the unsaturated are in greatest abundance in position *sn*-2. In some instances, there appears to be a higher proportion of longer-chain fatty acids (C_{20} to C_{24}) in position *sn*-3 than in position *sn*-1, and sometimes the more unusual fatty acids are concentrated in position *sn*-3. In many instances, good agreement has been obtained between the proportions of various molecular species of triacylglycerols, determined experimentally by chromatographic means, and those calculated from the positional distribution data using a "1, 3-random-2-random" model.

The positional distributions of fatty acids in many of the glycerophospholipids of plants seem to resemble those of animal tissues in that the saturated fatty acids are concentrated in position *sn*-1 and the unsaturated in position *sn*-2 [324]. However, the phosphatidylglycerol from spinach leaves is unusual in that the major molecular species contains linolenic acid in position *sn*-1 and *trans*-3-hexadecenoic acid in position *sn*-2 [298]. Relatively little information is available on the positional distributions of fatty acids in the glycosyldiacylglycerols, but there appears to be a tendency for slightly more of the linolenic acid to be in position *sn*-1 than in position *sn*-2 [324]. Although ether-containing lipids have been detected in higher plants, they are not common.

F. The Lipids of Microorganisms

1. Lipid class compositions

It is not easy to generalize about the lipids of microorganisms, as each family tends to have a distinct and characteristic lipid composition, and there are many lipid classes which are unique to particular groups. In addition, the fatty acid components are often very different from those of animal and plant tissues, and there are some which have as yet been found in certain rare species of microbes only. Microbial lipids have been reviewed elsewhere in greater detail than is possible here [247, 292, 480]. The nature and composition of microbial lipids have proved to be of some taxonomic value

[480]. Their study has also assisted towards an understanding of the molecular basis of evolution [578].

Simple glycerolipids such as triacylglycerols are often the most abundant lipids in fungi, but bacteria do not normally accumulate storage lipids of this type, although they are known to be present in a few species; diacylglycerols may be formed transiently as intermediates in the biosynthesis of glycerophospholipids. Sterols are found in yeasts, fungi and algae, but in essence are absent from bacterial membranes.

Most of the common glycerophospholipids of plants and animals, that are described above, are also found in microorganisms. For example, phosphatidylethanolamine is often the most abundant lipid class in many groups of microorganisms (including both Gram-negative and Gram-positive bacteria), and may be accompanied by phosphatidylserine, phosphatidylglycerol, and phosphatidylinositol. The *N*-mono- and *N*,*N*-di-methyl derivates of phosphatidylethanolamine are found in some bacterial genera. On the other hand, phosphatidylcholine is not a ubiquitous constituent of microbial lipids, rarely being found in bacteria, although it is often the major lipid class in eukaryotes. In bacteria, phosphatidylglycerol especially is widely distributed and is found in most genera; uniquely it is the only glycerophospholipid in the membranes of certain cyanobacteria. Phosphatidylglycerol is usually accompanied by diphosphatidylglycerol, which is often a major lipid component of bacteria, although it is only found as a constituent of mitochondrial membranes in eukaryotes. In addition, derivatives of phosphatidylglycerol (lipoamino acids) in which the glycerol moiety is esterified to an amino acid, such as lysine, ornithine, or alanine, are not uncommon in Gram-positive bacteria.

The plant glycosyldiacylglycerols are also found as major constituents of the membranes of algae and cyanobacteria, although glucose may replace a galactose unit in some species. Diglycosyldiacylglycerols, in which the carbohydrate moiety may be two glucose, two mannose or less often two galactose units, are found as minor components of the membranes of Gram-positive and occasionally of Gram-negative bacteria. Similarly, the plant sulpholipid, sulphoquinovosyldiacylglycerol, is present in small amounts in certain photosynthetic bacteria, and an analogous sulphonate of a digalactosyldiacylglycerol is a constituent of *Bacillus acidocaldarius*. Other bacterial sulpholipids tend to be sulphate esters of glycolipids.

Complex phosphorylceramides containing inositol are major components of yeast lipids, but sphingolipids are rarely found in bacteria. On the other hand, there always appear to be exceptions to any rule of this kind, and ceramidephosphorylethanolamine and ceramidephosphorylglycerol have been detected in some anaerobic bacteria. Free ceramides have been found in some species of *Bacteriodes*. Aminolipids containing alkylamides, but lacking the hydroxyl groups of the sphingoid bases, occur in some bacteria, while others contain sulphonolipids (capnoids), which appear to be related structurally to the long-chain bases. For example, capnine is 2-amino-3-hydroxy-15-methylhexadecane-1-sulphonic acid and is found in the free form or as the *N*-acyl derivative in some species of gliding bacteria.

Many bacterial species contain certain glycerophospholipids and glycolipids (including glycophospholipids), which are apparently found nowhere else in nature. The structures of some of these are shown in Fig. 3.8. Glycoglycerophospholipids are common constituents of bacterial membranes. For example, although phosphatidylinositol itself is rarely found, various mannoside derivatives do occur in some species. Glycosylated forms of phosphatidylglycerol are perhaps the most common of all (e.g. the structures (a), (d) and (f) in Fig. 3.8). To add to the complexity, the glyceryl moiety can also be esterified with an amino acid, the nature of which is dependent on the bacterial species, and the stereochemistry of the glyceryl group can vary between species

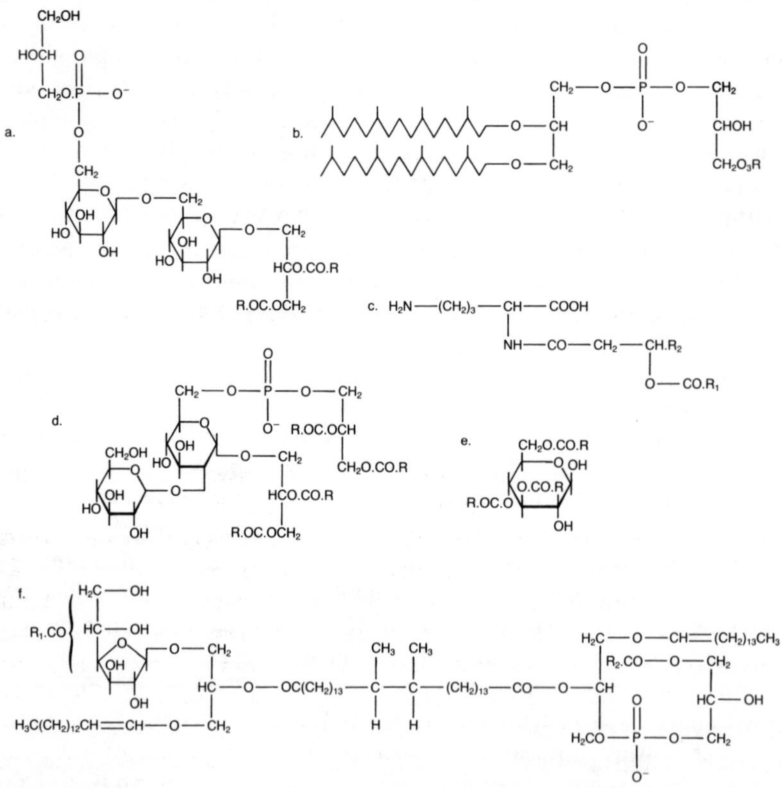

FIG. 3.8. Some distinctive complex lipids of microorganisms; (a) glycerylphosphodiglucosyldiacylglycerol, (b) phytanyl ether lipids from archaebacteria (R can be H or a complex phosphoryl moiety), (c) an ornithine lipid, (d) phosphatidyldiglucosyldiacylglycerol, (e) a triacylglucose, (f) phosphoglycolipid containing diabolic acid from rumen bacteria (R_1 = butyroyl, R_2 = butyroyl or palmitoyl).

in these lipids. Glucosaminylphosphatidylglycerol has been found in both Gram-negative and Gram-positive bacteria. Certain rumen bacteria contain lipids with complex glycerophosphoryl groups and glycosylglycerol groups, linked by a C_{32} dicarboxylic acid (diabolic acid), that may span the membrane bilayer. The lipoteichoic acids of the cell walls of Gram-positive bacteria are polymers of glycerol-1-phosphate and other complex organic groups linked to a diglycosyldiacylglycerol.

Relatively simple glycolipids, such as di- and tri-acylglucoses (structure (e) in Fig. 3.8), and an analogous rhamnolipid are present in some bacteria. Much more complex lipid polysaccharides, for example "Lipid A", occur in the cell envelope of Gram-negative bacteria [227]. Similarly, the cell walls of the Mycobacteria contain a wide range of lipid polymers, including trehalose derivatives, phosphate esters, peptidolipids and phenolic lipids, and these may be esterified to a bewildering array of unusual fatty acids (see below).

A further group of distinctive bacterial lipids are the acylornithines (e.g. structure (c) in Fig. 3.8). In these, the amine group of ornithine is attached via an amide bond to a fatty acid with a 2- or 3-hydroxyl group, which can in turn be esterified to a further fatty acid. The carboxyl group of the ornithine can also be linked to an aliphatic alcohol. The precise nature of the aliphatic groups and of the linkages can be characteristic of particular species.

The glycerophospholipids of bacteria exist not only in the diacyl forms but also as ether lipids, most often the plasmalogen forms. In addition, many different complex di- and tetra-ether phospholipids and glycolipids, in which the aliphatic moieties are phytanyl or related structures, are present in archaebacteria, thermoacidophiles and thermohalophiles (e.g. structure (b) in Fig. 3.8). They have been reviewed by Kates [407].

2. *Fatty acid compositions*

Polyunsaturated fatty acids of the kind found in plant lipids also occur in algae (green and brown), fungi and cyanobacteria, but are not present in other bacteria. In general, bacterial lipids tend to contain appreciable amounts of C_{14} to C_{18} straight-chain saturated and monoenoic fatty acids. The common C_{18} monoenoic acid is oleic acid, however, but *cis*-vaccenic acid (18:1(n − 7)). In addition, bacterial lipids can contain odd-chain, branched-chain (mainly *iso*- and *anteiso*-methyl, but 10-methyloctadecanoic or "tuberculostearic" acid is characteristic of some species), cyclopropane and 3-hydroxy fatty acids, which are only rarely synthesized by other organisms. It appears that the presence of an *iso*- or *anteiso*-methyl branch or of a cyclopropane ring in the fatty acids in a membrane increases its fluidity in an analogous manner to that of polyunsaturated fatty acids in the membranes of higher organisms. In comparing the detailed fatty acid compositions of bacteria, it is important to recognize that they can vary greatly with culture conditions and with stage of growth.

The Mycobacteria and certain related species contain a highly distinctive range of very long-chain *alpha*-branched *beta*-hydroxy fatty acids, known as the "mycolic" acids. Different species synthesize mycolic acids with quite characteristic structures. For example, Mycobacteria contain C_{60} to C_{90} acids with C_{20} to C_{24} *alpha*-branches, while the Nocardiae produce C_{38} to C_{60} acids with C_{10} to C_{16} branches. They may also contain additional carbonyl groups, methyl branches, cyclopropane rings and isolated double bonds. They occur in the bacterial cell walls in the free form, as wax esters and as components of complex lipids.

CHAPTER 4

Prelude to Analysis: Extraction, Storage and Preliminary Fractionation of Lipids

A. Introduction

Before any analysis of lipid samples can be commenced, it is necessary to extract the lipids from their tissue matrices and free them of any non-lipid contaminants. Ideally, this should be done immediately after removal of the tissue from the living organism, but if this is not possible, the tissue should be stored in such a way that it does not deteriorate significantly. During extraction, as in many other aspects of analysis, it is possible to inadvertently introduce contaminants or to bring about some unwanted change in the composition of the lipids. Autoxidation of double bonds in fatty acids, for example, is particularly troublesome and care must be taken at all steps in the analysis of lipids, not just during storage and extraction, to eliminate the problem. Similarly, with a little care, many other difficulties can be avoided. Having obtained a lipid extract, it may then be necessary or desirable to put it through some preliminary fractionation step to obtain simple lipids, phospholipids and glycolipids as distinct groups for further analysis. There are potential hazards to the analyst from all operations that involve large quantities of solvents, either because of their inflammability or toxicity. Familiarity should never be allowed to breed contempt. These preliminary aspects of analysis have been reviewed in other publications [133, 566, 915].

B. Some Practical Considerations

1. Storage of tissues and lipid samples

The removal of tissues from a living animal is a specialized task, which can only be performed under licence in most countries, and it is not easily reviewed here. Blood is probably the commonest biopsy tissue, and some of the precautions to be taken have been discussed elsewhere [566]. It is easier technically and legally to obtain tissues from a dead animal, and the required organs should be dissected out immediately after the animal has been killed by a method that minimizes trauma. As soon as they are removed, all tissues, whether of animals, plant or microbial origin, should be extracted without delay to minimize any possibility of degradation of their lipid constituents. Membranes are rapidly disrupted in dead tissues, for example, and their lipid components are exposed to enzymes from which they are normally protected, or with which they exist in a state of dynamic equilibrium in life. The most important of these enzymes are lipases, especially the phospholipases A, C and D, which hydrolyze diacylphospholipids to lysophospholipids (and free fatty acids), diacylglycerols and phosphatidic acid respectively. These hydrolysis products are usually minor components of tissues, and if they are found in appreciable amounts in a sample, it can often be a sign of careless or faulty storage and extraction. For example, lysophosphatidylcholine was long thought to be a major constituent of the lipids of the chromaffin granules in adrenals, but it is now known to be an artefact, resulting from a unsuitable tissue handling procedures [35]. If it is not feasible to extract a tissue at once, it should be frozen rapidly and stored under chloroform in a glass container at $-20°C$ [327]. Some workers have recommended plunging tissues into boiling water for brief periods [239, 298] or boiling them with dilute acetic acid [651, 652] to deactivate the enzymes and to kill any microbial contaminants; unfortunately, such methods do not appear to have been tested independently with a variety of different samples. Tissues and their lipid constituents continue to deteriorate, even at low temperatures, especially in microorganisms [302], and as a further example, a 33% increase in the lysophosphatidylcholine content of heart tissue was found after 2 weeks of storage at $-20°C$ [542].

Blood lipid compositions can be affected by the nature of the anticoagulant used during collection. Ethylenediaminetetraacetic acid (EDTA) at a concentration of 1 mg per ml of blood appeared to have least effect on the lipid components and has been recommended, but the erythrocytes and plasma (or serum) should be separated as quickly as possible and the lipids extracted [566].

Frozen tissues should not be allowed to thaw before being homogenized with the extracting solvent.

Lipid extracts should not be left in the dry state, but should be dissolved in a small volume of a relatively non-polar (aprotic) solvent, such as chloroform, and stored at $-20°C$ in a glass (never plastic) container, from which air is excluded by flushing out with a stream of nitrogen. For long-term storage, the containers should ideally be sealed under vacuum.

2. *Minimizing autoxidation*

If they are not protected, polyunsaturated fatty acids will autoxidize very rapidly in air, and it may not be possible to obtain an accurate analysis by chromatographic means. The mechanism of autoxidation involves attack by free radicals and is exacerbated by strong light and by metal ions. Once initiated, the reaction proceeds autocatalytically. Linoleic acid is autoxidized twenty times as rapidly as oleic acid, and each additional double bond in a fatty acid can increase the rate of destruction by two- to three-fold. As the reaction causes double bonds to migrate, forming conjugated systems which absorb strongly at wavelengths in the UV region of the spectrum, autoxidized lipids can interfere when lipids are analysed by means of HPLC with UV detection. Conversely, peroxidized lipids *per se* can often be analysed very easily and specifically by such a technique.

Natural tissue antioxidants, such as the tocopherols, afford some protection to lipid extracts, but it is usually advisable to add further synthetic antioxidants, such as BHT ("butylated hydroxy toluene" or 2, 6-di-*tert*-butyl-*p*-cresol) to solvents as a level of 50 to 100 mg/l [891]. This compound need not interfere with chromatographic detection, although it does absorb strongly in the UV region, as it is relatively volatile and can be removed, sometimes inadvertently, together with solvents when they are evaporated in a stream of nitrogen; it is also rather non-polar and tends to elute at the solvent front, ahead of most lipids, in many liquid chromatography system. In contrast, excessive amounts of added antioxidants can sometimes act as prooxidants!

Wherever possible, lipids should be handled in an atmosphere of nitrogen. On the other hand, it is rarely necessary to go to the length of constructing a special nitrogen box to contain all the equipment used in the handling of lipids. Usually it is sufficient to ensure that nitrogen lines are freely available so that the air can be flushed out of glass containers or reaction vessels.

When it is necessary to concentrate lipid extracts, large volumes of solvents are best removed by means of a rotary film evaporator at a temperature, which in general should not exceed about 40°C. The flask containing the sample should not be too large, otherwise the lipid can spread out over a large area of glass and so be more accessible to oxygen. At the start of evaporation, it may be advisable to flush out the equipment with nitrogen, but the solvent vapours eventually will displace any air. Small volumes of solvent can be evaporated by carefully directing a stream of nitrogen onto the surface of the solvent. This should not be done too vigorously or at too high a temperature, since the more volatile fatty acid derivatives, including methyl esters of fatty acids, may also be lost by evaporation or by physical transport as an aerosol.

PLEASE NOTE!

As constant repetition is tedious, it will be assumed in all the subsequent discussion of methodology in this book that precautions will be taken at all times to minimize autoxidation. For example, if lipid classes are isolated by preparative HPLC, all antioxidants originally present will probably have eluted with the void volume, and it will be necessary to add more.

3. Contaminants and artefacts in extraction procedures

All solvents, including from time to time those grades that are nominally of high purity, can contain contaminants, and as large volumes of solvent may be used to obtain small amounts of lipids, any such impurities can be troublesome. The higher quality grades of solvent may have to be checked periodically to ensure that they meet the required standards, while those of poorer quality should be redistilled before use. It should not be forgotten that water is an important solvent with many uses in the chromatographic analysis of lipids, and that it too should meet a high standard. Microbial growth in water of insufficient purity can introduce contaminants, which may block filters and frits in HPLC equipment. Similarly, buffers prepared for use in mobile phases and stored for lengthy periods in refrigerators will gradually accumulate a substantial microbial population. Some extraneous substances, for example antioxidants, are added deliberately by manufacturers to minimize peroxide formation in ethers; these need not cause problems, provided their presence is recognized.

Other extraneous lipid-like materials can be introduced accidentally into lipid samples from a variety of sources. Plastic ware of all kinds (other than that made from Teflon™ or PTFE) can be especially troublesome and is best avoided, since plasticizers (diesters of phthalic acid usually) are very easily leached out. They tend to co-chromatograph with lipids, so they may spread confusion and obscure compounds of interest in chromatograms, since they absorb very strongly in the UV range. Wet animal tissues in contact with plastic can leach out small but significant amounts of plasticizers, while organic solvents can extract large amounts. Conversely, it has been shown that lipids can themselves dissolve in some plastics, leading to selective losses of a proportion of the less polar constituents [483].

Manufacturers of fine chemicals, like all human kind, are fallible, and all laboratory reagents can on occasion contain impurities, which may cause problems in analytical procedures. It is necessary to exercise vigilance to detect and eliminate these at an early stage. Further contaminants can arise from fingerprints and from a host of materials in everyday use in laboratories, including cosmetics, hair preparations, hand creams, soaps, polishes, the exhausts from vacuum pumps, lubricants and greases, if they are used carelessly.

Under optimum conditions, lipids should not change in composition and structure during extraction or storage. However, there are some combinations of conditions that can give rise to unwanted alterations. For example, if any methanolic extracts or solutions of lipids are left in contact for lengthy periods with small amounts of sodium carbonate or bicarbonate derived from tissues, transesterification of lipids can occur and methyl esters of the fatty acids will then accumulate. This was first noticed in bile, and was overcome by adjusting the pH of the aqueous medium to about 4 to 5 [492]. It is possible that the same problem may arise if the pH is too low, and the reaction may even be catalyzed enzymatically by transacylases. The phenomenon has recently been studied systematically [421, 836], and is troublesome not only because losses of lipids occur but because simple esters of this type may be present as minor natural constituents of tissues and could be obscured. If it is intended to study such compounds, it is necessary to extract them from the tissues in a non-alcoholic solvent, such as acetone. On the other hand, there are circumstances when acetone should be avoided. For example, it was demonstrated that it brought about some dephosphorylation of polyphosphoinositides [174, 871], and that it could react with phosphatidylethanolamine to form an imine derivative [28, 311].

Artefactual enzymatic hydrolysis of lipids, catalyzed by tissue enzymes, can be promoted by the solvents used for extraction. This is especially troublesome with plant tissue, in which phospholipase D activity (both hydrolytic and transphosphatidylase) is stimulated by some solvents, but especially by butanol saturated with water [154]. The problem is usually circumvented by heating or by a pre-extraction with isopropanol, which deactivate the enzyme (see below). It is also possible to obtain an artefactual enzyme-catalyzed acylation of some lipids, such as glycosyldiacylglycerols, in certain circumstances [308].

Some rearrangement of plasmalogens was found to occur when they were stored in methanol for long periods [854], especially if acidic conditions were employed during extraction, when lysophospholipids accumulated as artefacts [543]. Other unwanted lipid by-products can be produced during derivatization procedures, and these are discussed at the appropriate points in later chapters.

4. Hazards

Some of the problems of toxicity and flammability of solvents have been discussed briefly already (Chapter 2). All solvent should be used with care in well-ventilated areas, or in fume cupboards if at all possible. Solvents should never be evaporated or distilled on an open bench. No solvent can be considered to be entirely safe if inhaled, and benzene especially should be excluded entirely from the laboratory. Some operations generate more vapour than others, and filtration is probably the procedure which produces most. When not in use, all solvents should be stored in well-stoppered bottles, made of dark glass, and in flameproof cabinets. No more solvent than is required for immediate needs should be stored in the laboratory. Ethers develop peroxides on storage, especially in bright light, and many explosions have resulted as these were concentrated when large volumes of ethers were distilled. It may seem self evident, but there should be no naked flames in laboratories in which solvents are handled.

The concentrations of chloroform vapour permitted in the laboratory atmosphere have steadily been reduced in recent years as regulations have been revised. Unfortunately, there appears to be no adequate substitute in many analytical problems that face lipid analysts. As supplied, chloroform usually contains 0.25 to 2% of added ethanol, which acts as a stabilizer, but also has a considerable effect on the chromatographic properties of the solvent. It can be removed, if there is a need, by a simple washing procedure [133], but photochemical formation of the highly toxic substance, phosgene, can then occur and the destabilized solvent should not be stored for any length of time. Chloroform–methanol mixtures are powerful irritants when they come in contact with the skin.

Many other reagents to be found in laboratories are known to have toxic properties, some of which may take some time to manifest themselves, and the catalogues of suppliers are often informative on the subject as should be the labels on containers. The toxicity of numerous other reagents has yet to be investigated, and it is best to err on the safe side and assume that there is some unknown hazard associated with all chemicals. They should then be handled accordingly. Similarly, the hazards associated with strong mineral acids should be well known to analysts. They should not be stored in the same cupboard as solvents.

C. Extraction of Lipids from Tissues

1. The principles of solvent extraction procedures

Lipids occur in tissues in a variety of physical forms. The simple lipids are often part of large aggregates in storage tissues, from which they are relatively easily extractable. On the other hand, the complex lipids are usually constituents of membranes, where they occur in a close association with such compounds as proteins and polysaccharides, with which they interact, and they are not extracted so readily. Factors affecting the solubilities of lipids in organic solvents have been reviewed [133, 911, 915]. It appears certain that the complex lipids do not in general form covalent bonds with the other constituents of the tissue matrix, although some proteins containing covalently-bound fatty acids have been found in nature, and the lipopoly-saccharides of cell walls are similar in that the covalently-bound lipid portion is but a small proportion of the whole. These exceptions are not considered further here. Generally, lipids are linked to other cellular components by weak hydrophobic or Van der Waals' forces, by hydrogen bonds and by ionic bonds. For example, the hydrophobic aliphatic moieties of lipids interact with the non-polar regions of the amino acids constituents, such as valine, leucine and isoleucine, of proteins to form weak associations. Hydroxyl, carboxyl and amino groups in lipid molecules, on the other hand, can interact more strongly with biopolymers via hydrogen bonds. Finally, the strongest bonds of all are the ionic linkages between the acidic phosphate groups on lipids such as the polyphosphoinositides and metal ions, which may in turn be bound similarly to the cellular proteins or polysaccharides.

Pure lipids will dissolve in a variety of solvents, depending on the relative strengths of the interactions between the solvent and either the hydrophobic or the hydrophilic regions of the molecules. Lipids with functional groups of low polarity only, such as triacylglycerols or cholesterol esters, are very soluble in hydrocarbon solvents like hexane, cyclohexane or toluene, and also in solvents of somewhat higher polarity, such as chloroform or ethers. They tend to be rather insoluble in polar solvents such as alcohols, and methanol especially; solubility increases as the chain-lengths of the fatty acid moeties in these lipids decrease or as the chain-length of the solvent alcohol increases. In contrast, the polar complex lipids tend to be only sparingly soluble in hydrocarbon solvents, though dissolution can be aided by the presence of other lipids, but they do dissolve readily in more polar solvents such as chloroform, methanol and ethanol. Acetone is a good solvent for glycolipids, but not for phospholipids, and it can even be used to precipitate the latter from solution.

In order to extract lipids from tissues, it is necessary to find solvents which will not only dissolve the lipids readily but will overcome the interactions between the lipids and the tissue matrix. The energy required to overcome the weak forces is always less than 2 kcal/mole, while that for hydrogen bonds can be as much as 12 kcal/mole; ionic bonds can only be overcome by lowering the pH of the medium or by increasing its ionic strength. In addition, some lipids can be physically trapped within a tissue matrix; lysophosphatidylcholine, for example, is contained within starch macromolecules in wheat grains in the form of an inclusion complex. Also, the cell walls in some organisms are less permeable than others to solvents; water then assists the extraction by causing swelling of the biopolymers and it is an essential component of any extractant. In some circumstances, it may be necessary to effect an appreciable denaturation of the other constituents of the cell walls by some means before a thorough extraction of the lipids is possible.

Various solvents or solvent combinations have been suggested as extractants for lipids, and currently some interest is being shown in isopropanol–hexane (2:3 by volume), because its toxicity is relatively low [283, 684], but it does not yet appear to have been tested with a sufficiently wide range of tissues. It does not extract gangliosides quantitatively. Most lipid analysts use chloroform–methanol (2:1 by volume), with the endogeneous water in the tissue as a ternary component of the system, to extract lipids from animal, plant and bacterial tissues. Usually, the tissue is homogenized in the presence of both solvents, but better results may be obtained if the tissue is first extracted with methanol alone before the chloroform is added to the mixture. With difficult samples, more than one extraction may be needed, and with lyophilized tissues, it may be necessary to rehydrate prior to carrying out the extraction. The homogenization and extraction should be peformed in a Waring blender, or better in equipment in which the drive to the blades is from above, so that the solvent does not come into contact with any lubricated bearings. Generally, there is no need to heat the solvent to facilitate the extraction, although there may be times when this is necessary.

The extractability of tissues and of particular lipids is variable, and there are many instances when alternative or modified procedures must be used. Butanol saturated with water appeared to be by far the most useful solvent mixture to disrupt the inclusion complexes of lipids in starch and gave the best recoveries of lipids from cereals [154, 522, 554]. This solvent combination has also been recommended for the quantitative recovery of lysophospholipids, which are more soluble in water than are many other common phospholipids of tissue homogenates [78]. Bacterial cells can be especially difficult to extract and various procedures that involve vigorous extraction, or a pretreatment with chemical or enzymatic reagents, have been described [534, 535, 695, 857]. Similarly, if quantitative recovery of the highly-complex glycosphingolipids of intestinal cells is required, it has been recommended that that the tissue be partially digested by alkali, RNAase, DNAase and a protease prior to extraction with chloroform–methanol [772]. Acidic extraction conditions are occasionally suggested for quantitative recovery of particular lipid classes, but can bring about some hydrolysis of plasmalogens as mentioned above.

Difficulties have been experienced by many workers in the quantitative extraction of the biologically-active polyphosphoinositides from tissues. Many solutions to the problem have been suggested, and some workers favoured lowering the pH of the medium [250], while others prescribed extraction in the presence of calcium chloride, followed by acidification [297]. More recently it was demonstrated that if the ion-pairing reagent, tetrabutylammonium sulphate, was added to the extraction medium, the recovery of phosphatidylinositol 4, 5-bisphosphate was much better and was independent of the concentration of calcium ions [256]. Similar problems can be encountered in the extraction of other minor acidic lipids, and some traditional and newer methods have recently been compared [435].

As discussed above, plant tissues should be pre-extracted with isopropanol to minimize artefactual degradation of lipids by tissue enzymes.

Lipid extracts from tissues, obtained in the above manner, also tend to contain appreciable amounts of non-lipid contaminants, such as sugars, amino acids, urea and salts. These must be removed before the lipids are analysed. Many procedures have been suggested for the purpose, including dialysis and adsorption or cellulose ion-exchange chromatography. A recent survey of the available methods concluded that none was entirely satisfactory, but that elution from deactivated alumina held particular promise [915].

Most workers use a simple washing procedure, devised originally by Folch, Lees and Stanley [218] (one of the most frequently cited of all research papers), in which a chloroform–methanol (2:1 by volume) extract is shaken and equilibrated with one fourth its volume of a saline solution (i.e. 0.88%

potassium chloride in water). The mixture partitions into two layers, of which the lower phase is composed of chloroform–methanol–water in the proportion 86:14:1 (by volume) and contains virtually all of the lipids, while the upper phase consists of the same solvents in the proportions of 3:48:47 (by volume) respectively, and contains much of the non-lipid contaminants. It is important that the proportions of chloroform, methanol and water in the combined phases should be as close as possible to 8:4:3 (by volume), otherwise selective losses of lipids may occur. If a second wash of the lower phase is needed to remove any remaining contaminants, a mixture of similar composition to that of the upper phase should be used, i.e. methanol–saline solution (1:1 by volume).

Any gangliosides present in the sample partition into the upper layer, together with varying amounts of oligoglycosphingolipids. It was shown, however, that they could be recovered from this layer by dialyzing out most of the impurities of low molecular weight, and then lyophilizing the residue [396]. Alternatively, gel-filtration has been used for this last step. Although there has been much debate among the experts in this field as to the best procedure for quantitative recovery of gangliosides, most appeared to have settled upon some variant of the above method [99, 103, 111, 268, 470, 809, 812, 876].

It has also proved possible to carry out the washing procedure by means of liquid–liquid partition chromatography. In this technique, the aqueous washing phase was immobilized on a column of a dextran gel such as Sephadex G-25™, while the organic phase was passed through the column [765, 870, 894]. Although such methods are more elegant, and possibly more complete, they are also more time-consuming and have not been widely adopted. Modified partition procedures may be of value for the isolation of gangliosides [103, 765], and on a small scale of phospholipids [216]. In the latter instance, small prepacked reversed-phase columns (Bond-Elut™, Analytichem International, Harbor City, California) were used for the quantitative recovery of isotopically-labelled phospholipids, free of their water-soluble precursors. The lipid incubation mixtures, were passed through the column in a solvent equivalent to the "Folch" upper phase, and the column was washed with water to remove contaminants, and then with chloroform–methanol (2:1 by volume) to recover the required lipids.

A further novel procedure for eliminating non-lipid contaminants, which may repay investigation, consisted in pre-extracting tissue homogenates with dilute acetic acid [651, 652]. Potential contaminants were apparently solubilized, while the lipids remained in the tissue for extraction in a subsequent step.

2. *Recommended procedures*

Many modifications of the basic extraction procedure have been devised for use in particular circumstances, and the analyst must decide what he requires of a method. One which extracts all of the more minor lipid classes exhaustively is obviously desirable for many applications, but may be too tedious and time-consuming for routine use. On the other hand, a method which is suited to the quantitative extraction of the main lipid classes in large numbers of samples in a routine manner by relatively inexperienced staff, may not give complete recoveries of certain trace components of biological importance. The modified "Folch" procedure [866], which follows, probably falls somewhere between these extremes.

> "The tissue (1 g) is homogenized with methanol (10 ml) for 1 minute in a blender, then chloroform (20 ml) is added and the homogenization continued for 2 minutes more. The mixture is filtered, when the solid remaining is resuspended in chloroform–methanol (2:1 by volume, 30 ml) and homogenized for 3 minutes. The mixture is filtered again and re-washed with fresh solvent. The combined filtrates are transferred to a measuring cylinder, one fourth of the total volume of 0.88% potassium chloride in water is added, and the mixture is shaken thoroughly before being allowed to settle. The aqueous (upper) layer is drawn off by aspiration, one fourth the volume of lower layer of methanol–saline solution (1:1 by volume) is added and the washing procedure repeated. The bottom layer, containing the purified lipid, is filtered before the solvent is removed on a rotary film evaporator. The lipid is stored in a small volume of chloroform at −20°C, until it can be analysed."

A more exhaustive procedure has been described [711], and references to procedures suited to specific samples or lipid classes are cited above, while further methods for particular tissues or membranes have been reviewed [911]. In contrast, a method devised by Bligh and Dyer [83] is recommended for large samples, which contain a high proportion of endogenous water.

With plant tissues, it is necessary to extract first with isopropanol, in order to deactivate the enzymes, as follows [585, 586].

> "The plant tissues are homogenized with a hundred fold excess (by weight) of isopropanol. The mixture is filtered, the residue is re-extracted with fresh isopropanol, and finally is shaken overnight with isopropanol–chloroform (1:1 by volume). The filtrates are then combined, most of the solvent is removed on a rotary evaporator, and the lipid residue is taken up in chloroform–methanol (2:1 by volume) and given a "Folch" wash as above."

A more exhaustive procedure for the extraction of difficult plant tissues has been described elsewhere [915].

In any extraction procedure, it is important that the weight of fresh tissue extracted is recorded, together with the weight of lipid obtained from it. For some purposes, it may be desirable to determine the amount of dry matter in the tissue, so that the weight of lipid relative to that of dry matter can be calculated.

D. Preliminary Fractionation of Lipid Extracts

1. Objectives

Great strides have been made in the development of methods for the separation of most of the individual simple and complex lipids of a sample in a single analytical step (see Chapter 5). However the complexity of natural lipid extracts in such that it is not possible to claim that all the lipid classes of a sample can be separated in one operation. In addition, few analysts as yet have available to them the HPLC equipment necessary to achieve some of the more comprehensive analyses that have been described, or they may wish for a variety of reasons to pursue other approaches to analysis or lipid separation. It is then of value to be able to isolate distinct simple lipid, phospholipid or glycolipid fractions for analysis. For example, it is frequently easier technically to isolate small amounts of pure lipid classes preparatively by means of HPLC (or other methods), after a preliminary fractionation has been carried out. Unfortunately, no procedure appears yet to have been described that is satisfactory in all respects. Nonetheless, some useful methods are available. Most workers use a simple column separation with silica gel as adsorbent, but a mild partition procedure is also available for semiquantitative work. A chemical method is of special value for the isolation of glycosphingolipids. Ion-exchange chromatography has advantages as a preparative-scale method for isolating particular groups of complex lipids especially.

2. Adsorption column chromatography

The simplest small-scale procedure for isolating groups of lipids consists in making a short column of silica gel (about 1 g), in a glass disposable Pasteur pipette say, and about 30 mg of lipid can be applied to this. Elution with chloroform or diethyl ether (10 ml) yields the simple lipids, acetone (10 ml) gives a glycolipid fraction, and methanol (10 ml) yields the phospholipids. Different brands or batches of silica gel tend to vary somewhat in their properties and some cross-contamination of fractions may be found. For example, the acetone fraction may contain some of the acidic phospholipids, especially phosphatidic acid and diphosphatidylglycerol but occasionally phosphatidylethanolamine even; if this is observed to occur, it can be minimized by adding some chloroform to the acetone prior to elution. Indeed for many purposes, there may be no need to include an elution step with acetone, as some tissues contain negligible amounts of glycolipids. On the other hand, it is possible to insert an additional elution step with methyl formate before the acetone wash, to obtain a fraction that contains most of the prostaglandins in the extract (together with some of the glycolipids) [725].

It may now be more convenient to use small proprietary prepacked cartridges of silica gel for these small-scale group separations [76, 77, 375, 430, 708, 907]. In one application, milk fat samples (100 mg), high proportions of which consist of triacylglycerols, were applied to Sep-Pak™ cartridges of silica gel (Waters Associates, Milford, USA); non-polar lipids were recovered by elution with hexane-diethyl ether (1:1 by volume, 40 ml), while the complex lipids were recovered by elution first with methanol (20 ml), and then with chloroform–methanol–water (3:5:2 by volume, 20 ml) [77]. In other work with these cartridges, chloroform (40 ml) was used to elute the simple lipids, acetone–methanol (9:1 by volume; 160 ml) gave the neutral glycosphingolipids, and chloroform–methanol (1:1 by volume; 80 ml) eluted the phospholipids [907]. Aminopropyl-bonded phase cartridges (Bond-Elut™) have also been used for the isolation of simple and complex lipid fractions [395], but others [149] obtained very poor recoveries of acidic phospholipids, such as phosphatidylinositol, with these.

Somewhat larger-scale procedures but with similar objectives have been described for the isolation of animal [248, 710], plant [857] and bacterial [756] lipid fractions. By careful selection of an adsorbent of the correct grade and activity, alumina can also be used for the purpose [915].

The individual simple lipid classes can then be further resolved, if need be, on columns of silica gel or Florisil™ (a preparation of magnesium oxide and silica gel produced by the Floridin Company (Pittsburg, USA)), eluted with hexane containing increasing amounts of diethyl ether in a stepwise manner [108, 133]. A summary of the optimum elution conditions for each lipid class is given in Table 4.1. Again, these can only be taken as a guide because of variations in the properties of the adsorbent or in local conditions of temperature and humidity. After elution of the simple lipids, the complex lipids are recovered from silica gel by elution with methanol, but they tend

TABLE 4.1

Suggested conditions for the separation of simple lipid classes on columns of silica gel or Florisil™ by elution with mixtures of diethyl ether in hexane [133]

	Silica gel		Florisil™	
Lipid class	Solvent[a]	Column volumes	Solvent[a]	Column volumes
cholesterol esters	2	15	5	10
triacylglycerols	5	10	15	10
diacylglycerols + cholesterol	15	10	30	15
monoacylglycerols	100	10	100	10

[a]percent by volume of diethyl ether in hexane.

to adhere very strongly to Florisil™, from which quantitative recovery is rarely possible. Alternatively, the complex lipids can also be partially factionated on silica gel [133]. With plant lipids, monogalactosyldiacylglycerols are recovered by elution with chloroform-acetone (1:1 by volume, 10 column volumes), and digalactosyldiacylglycerols by elution with acetone (10 column volumes) [857]. Then as a rough guide, phosphatidic acid and diphosphatidylglycerol can be eluted with chloroform–methanol (19:1 by volume, 10 column volumes), and the same solvents but with increasing proportions of methanol are used to elute the remaining phospholipids, i.e. phosphatidylethanolamine and phosphatidylserine (4:1, 20 column volumes), phosphatidylinositol and phosphatidylcholine (1:1, 20 column volumes), and sphingomyelin and lysophosphatidylcholine (methanol alone, 20 column volumes). Single lipid classes can be isolated subsequently from these fractions with relative ease by means of HPLC or other chromatographic methods.

Alumina is rarely used as an adsorbent for the separation of phospholipids because of its basic nature, which can lead to some hydrolysis. However, neutral alumina has been put to good use for the rapid isolation of fractions enriched in particular components, such as phosphatidylcholine [768] or phosphatidylinositol [495].

3. *Partition procedures*

Lipid extracts can be separated into broad simple and complex lipid fractions with relative ease and with only a little cross-contamination by a solvent partition procedure, as follows [228].

> "87% aqueous ethanol is equilibrated with hexane, and the hexane layer is put into two separating funnels (45 ml each). The lipid sample (up to 10 g) in the ethanolic phase (15 ml) is added to the first funnel, and after a thorough shaking the bottom layer is run into the second funnel and the layers are again shaken together. The bottom layer is then set aside. A further aliquot (15 ml) of fresh ethanolic phase is added to the first funnel and the procedure is repeated. After six further extractions of this kind, the combined ethanol layers contain all the complex lipids, while most of the simple lipids are in the combined hexane layers."

The method appears to be suited best to lipid samples that contain a high proportion of simple lipids relative to phospholipids, such as in milk or adipose tissue lipids. A little cross-contamination of the fractions probably occurs, but an appreciable enrichment is obtained and the lipid classes present in each fraction are representative of those in the unfractionated lipid extracts in their fatty acid compositions, for example. A more comprehensive partition procedure that makes use of a counter-current distribution apparatus has been described, but is impracticable for routine use [24].

4. *Ion-exchange chromatography*

Column chromatography on DEAE-cellulose is a valuable method for the isolation of particular groups of complex lipids in comparatively large amounts [133, 711]. The principle of the separation process is complex, involving partly ionic interactions between the packing material and the ionic regions of complex lipids, and partly adsorption effects with the polar regions of the molecules. In practice as a rough guide, about 300 mg of complex lipids can be applied to a 30 × 2.5 cm column to yield fractions with distinctive compositions and little cross-contamination. Although it has been used mainly on this scale, it is also possible to use the technique with much smaller columns and proportionately less lipid.

It is first necessary to convert the packing material to the acetate form by washing sequentially with 1 M aqueous hydrochloric acid, distilled water, 0.1 M aqueous potassium hydroxide and water once more; the ion-exchange medium is left overnight in glacial acetic acid, then packed into a column in a slurry of this solvent. Finally, the acetic acid is washed out by elution with methanol, chloroform–methanol (1:1 by volume), and chloroform alone. An elution scheme suitable for use in the separation of lipid extracts from animal tissues is shown in Table 4.2. Chloroform elutes the simple lipids. All the choline-containing phospholipids are eluted with a chloroform–methanol mixture of relatively low polarity, while a much higher proportion of methanol

TABLE 4.2
Elution scheme for the fractionation of lipid extracts from animal tissues by means of DEAE-cellulose column chromatography [133, 711].

Fraction	Lipids eluted	Solvents	Column volumes
1	simple lipids	chloroform	10
2	phosphatidylcholine lysophosphatidylcholine sphingomyelin ceramide monohexoside	chloroform–methanol (9:1, v/v)	10
3	phosphatidylethanolamine ceramide oligohexosides lysophosphatidylethanolamine	chloroform–methanol (1:1, v/v)	10
4	—	methanol	10
5	phosphatidylserine	glacial acetic acid	10
6	—	methanol	4
7	phosphatidic acid diphosphatidylglycerol phosphatidylglycerol phosphatidylinositol cerebroside sulphate	ammonium acetate solution[a]	10
8	—	methanol	10

[a]Chloroform–methanol (4:1 by volume) made 0.5 M with respect to ammonium acetate and with 20 ml of 28% ammonium hydroxide added per litre.

is required to recover the ethanolamine-containing phospholipids; phosphatidylserine is eluted with glacial acetic acid, and a solvent of high ionic strength is required to recover the more acidic phospholipids (the salts can be removed later by a "Folch" washing step). Further separation of the individual components of particular fractions can later be achieved by means of HPLC or TLC, more easily than with the unfractionated extract. With plant lipid extracts, monogalactosyldiacylglycerols tend to elute with the chloroform fraction, but digalactosyldiacylglycerols can be recovered on their own if care is taken [587]. At the end of the analysis , it is an easy matter to regenerate the column for re-use, though it is important not to move too abruptly from solvents of high to low polarity.

Quaternary triethylammonium (QAE) groups covalently-bound to controlled-pore glass (Glycophase™) appear to afford similar separations to those obtained with DEAE-cellulose, but the former have better packing and compressibility properties in low-pressure column chromatography applications, and will probably be more widely used in the future [49]. Triethylaminoethyl (TEAE)-cellulose has been used in an analogous way and gives somewhat different fraction, complementing those obtained on DEAE-cellulose [133, 711]. Carboxymethyl (CM)-cellulose has been recommended for the isolation of acidic lipids, such as phosphatidylserine, as they can be eluted under comparatively mild conditions [156]. It has not, however, given reproducible results in the isolation of less polar phospholipids in the author's hands (unpublished results).

5. *Methods for the isolation of a sphingolipid or glycolipid fraction*

Procedures for the extraction and isolation of a glycolipid fraction from tissues have been reviewed by Hakomori [268]. A crude glycosphingolipid fraction can be obtained simply by means of adsorption chromatography on silica gel and elution with acetone as described above (Section D.2). A similar fraction can be isolated with Florisil™ as adsorbent and elution with very dry solvents (obtained by the incorporation of some 2, 2-dimethoxypropane) [709]. Ceramide is eluted with chloroform–methanol–dimethoxypropane (95:5:5 by volume, 10 column volumes), while ceramide monohexosides and cerebroside sulphate are eluted with the same solvents in the relative proportions of 70:30:5 respectively by volume (20 column volumes). Glycosyldiacylglycerols, free of phospholipids, were isolated from plant lipid extracts by a related procedure [596]. In addition, distinct fractions containing either acidic or neutral glycosphingolipids have been obtained by column chromatography on DEAE-Sephadex™ [482, 909] or DEAE-silica gel [462, 463].

A more comprehensive separation of glycosphingolipids from phospholipids was obtained when the former were acetylated with acetic anhydride–

pyridine (5:1 by volume), prior to chromatography on Florisil™ [718]; 1,2-dichloroethane eluted the simple lipids and 1,2-dichloroethane-acetone (1:1 by volume) gave a clean glycosphingolipid fraction, after de-acetylation with sodium methoxide in methanol as described below.

It also proved possible to eliminate any non-sphingolipid component from a concentrate of these compounds by mild alkaline transesterification, by means of which any lipids containing O-acyl fatty acids were converted to methyl esters and water-soluble products, while the glycosphingolipids and sphingomyelin (which contain amide-bound fatty acids) were not affected [872]. A suitable method is as follows.

> "The glycolipid concentrate (up to 100 mg) in 0.5 M sodium methoxide in methanol (3 ml) is heated for 10 min at 50°C. Acetic acid (0.15 ml) is added, together with chloroform (6 ml) and water (2.25 ml). The mixture is shaken thoroughly, the upper layer is removed by means of a Pasteur pipette, and the lower layer is washed twice more with methanol–water (1:1 by volume, 2 ml portions). Following evaporation of the solvent, the glycolipids in the lower layer are purified by silicic acid chromatography; chloroform elutes the methyl esters, chloroform–methanol (1:1 by volume) elutes the glycolipids and methanol elutes any sphingomyelin present."

A disadvantage of the method is that any glycosyldiacylglycerols in the crude glycolipid starting material are removed, and any of the rare sphingolipids containing O-acyl bound fatty acids are partially hydrolyzed.

CHAPTER 5

Methods for the Separation of Simple Lipids, and of Both Simple and Complex Lipid Classes in a Single Step

A. HPLC Procedures for the Separation of Simple Lipid in General

In this section, methods applicable to the separation of individual simple lipid classes from a simple lipid fraction, prepared as described in Chapter 4 and devoid of complex lipids, are considered. Simultaneous fractionation of specific lipids into molecular species is seldom necessary or desirable at this stage, and this aspect of lipid analysis is described later (Chapter 8). Many different types of column packing materials, eluents and detectors have been used for lipid class separations by means of HPLC. Of these, the nature of the detectors available to the analyst has generally been a dominant factor, governing the approach to the problem. For example, refractive index, ultraviolet (200–210 nm) and to a lesser extent infrared detectors are well suited to the isolation of particular lipid classes on a small scale for analysis by other procedures, since they are non-destructive (although the refractive index detector can only be used with isocratic elution). They tend to be less useful for quantitative analysis of lipid classes, although some workers have employed them in this way after careful calibration. Some of the more successful quantitative analyses of simple lipid by means of HPLC have made use of detectors which operate on the transport-flame ionization principle, but these and other "universal" or "mass-selective" detectors are still to be found in a few laboratories only.

In most of the published separations of simple lipid classes, adsorption chromatography with columns of silica gel has been used with a variety of different elution/detector systems. Several applications of the Pye Unicam (Cambridge, UK) LCM2 transport detector, which is no longer manufactured, were among the first reports of separations of this kind. A column (2.1 mm i.d. × 500 mm) of microPak™ SH-10 silica gel and a gradient of ethanol into hexane–chloroform (9:1 by volume) were employed to effect good separations of triacylglycerols, diacylglycerols, sterols, free fatty acids and monoacyl-

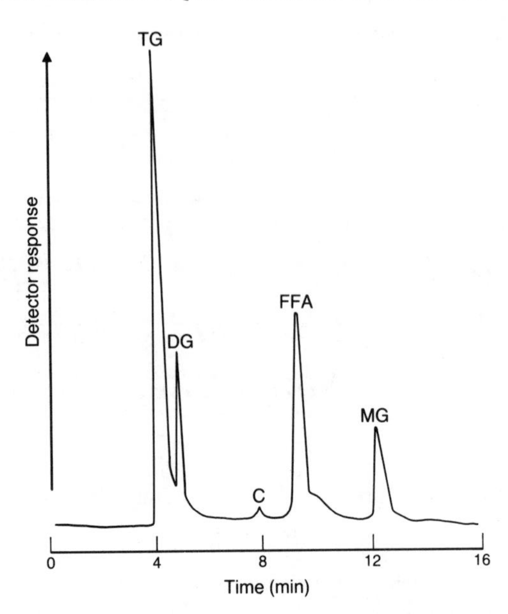

FIG. 5.1. HPLC separation of simple lipids on a column (2.1 × 500 mm) of MicroPak™ SH-10 silica gel, eluted with a gradient of ethanol into hexane–chloroform (9:1 by volume) with a Pye Unicam LCM2 transport-flame ionization detector [426]. (Reproduced by kind permission of the authors, and of the *Journal of Chromatographic Science* and Preston Publications Inc., and redrawn from the original publication.) Chapter 1 contains a list of abbreviations.

glycerols in about 15 min, although a lengthy re-equilibration period was necessary before the next sample could be analysed [426]. The nature of this separation is illustrated by Fig. 5.1. The method was applied to the simple lipids of soybean preparations, and the detector response to authentic standards was found to be rectilinear, permitting excellent reproducibility in quantitative analyses. In a further study from this group, a slightly modified elution system was used [428]. The same detection system, with a column of Lichrosorb™ SI-60 silica gel, was utilized in the analysis of the waxes of human sebum and similar lipids [19]. In this instance, a complex ternary solvent system, producing two gradients, was used, i.e. carbon tetrachloride–isooctane (34:66 by volume) (A), chloroform–dioxane–hexane (40:11:49 by volume) (B) and chloroform–methanol–diisopropylether (34:36:30 by volume) (C), with B into A initially then C into B. Squalene was eluted first, followed by wax esters (together with cholesterol esters), methyl esters (formed prior to chromatography by reaction of the free fatty acids with diazomethane), triacylglycerols, 1,2- and 1,3-diacylglycerols, cholesterol and monoacylglycerols. Monoolein diacetate was added as an internal standard, and excellent quantification was obtained both with standards and with natural lipid extracts. Later, this HPLC system was applied in the analysis of the partial hydrolysis products of triacylglycerols [14], while concise details of a comparable analytical system were reported elsewhere [275]. A similar separation, making use of a mass detector, has been described in brief [644].

A modified Pye transport detector was used in conjunction with gel-permeation chromatography on a column (4 × 1000 mm) packed with a polystyrene gel (Poragel™ 200A to 60A), maintained at 45°C, and eluted with aqueous acetone (4 to 7.5% water) in order to separate simple lipid classes [478]. In this instance, separation was achieved essentially on the basis of molecular weight, with monoacylglycerols and free fatty acids eluting ahead of diacylglycerols and cholesterol, and these were in turn eluted before triacylglycerols. Although reproducible results in quantitative analyses were obtained, the system appears only to have been employed for standard mixtures of relatively-simple composition, and it might not operate successfully with lipid classes containing many different fatty acids constituents. A similar system, but with RI detection, is described below. Gel-permeation chromatography is perhaps better suited to the analysis of lipid polymers (see Section B.3 below).

Comparable analyses to the above could no doubt be obtained now with the new commercial transport-flame ionization detector (Tracor Instruments), with the Privett detector, or with a mass spectrometric detector (see Chapter 2).

A refractive index detector was utilized with a column (4.6 × 250 mm) of Ultrasil™ Si (5 μm silica gel) and isocratic elution with isooctane–tetrahydrofuran–formic acid (90:10:0.5 by volume) to separate most of the common simple lipid classes encountered in animal tissue extracts, such as those of liver [249]. A typical separation is shown in Fig. 5.2. and it was noteworthy that cholesterol esters, triacylglycerols and cholesterol all gave symmetrical peaks. The system could be of some value for the isolation of specific lipid classes on a small scale (1 to 2 mg) for further analysis. Although an attempt was made to use the technique quantitatively, the results were not convincing, as negative solvent peaks tended to interfere, as the linearity of the response for each lipid class was not determined rigorously, and as some variation in response with fatty acid composition was observed. However, other showed that this elution system could give acceptable accuracy with relatively simple mixtures, such as those obtained in commerce by glycerolysis of seed oils, if an internal standard (ricinoleic acid) was used and a careful calibration was performed [693]. The detector response was found to be rectilinear for up to 1 mg of glyceride.

RI detection has been used with gel-permeation chromatography to separate triacylglycerols and their lipolysis products [151]. Columns (7 × 250 mm each) of LiChrogel™ PS_4 and LiChrogel™ PS_1 were used in series, with toluene at 0.5 ml/min as the mobile phase. In this instance also, the RI detector gave excellent quantitative results, both with real samples and with standard mixtures, provided that substantial correction factors were employed to compensate for differing degrees of unsaturation of the fractions.

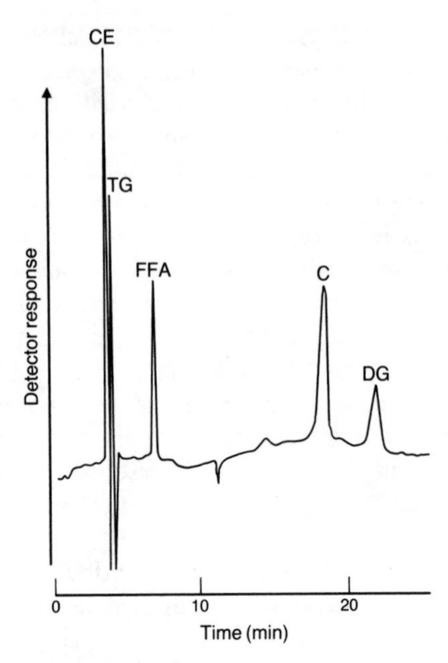

FIG. 5.2. HPLC separation of simple lipids on a column (4.6 × 250 mm) of Ultrasil™ Si silica gel, eluted isocratically with isooctane–tetrahydrofuran–formic acid (90:10:0.5 by volume), and with refractive index detection [249]. (Reproduced by kind permission of the authors and of *Analytical Biochemistry*, and redrawn from the original publication.) Chapter 1 contains a list of abbreviations.

When refractive index detection is used, it is strongly advised that the sample be dissolved in a solvent identical to the mobile phase for application to the column, in order to minimize any disturbance to the recorder base-line at the start of the chromatographic run.

Relatively few applications of UV spectrophotometry at 200–210 nm in the separation of simple lipids have been described, possibly because this form of detection is of limited value for quantification purposes (see Chapter 2 and 6 for a fuller discussion). However, detection at 206 nm and a column (3.9 × 300 mm) of Porasil™ silica gel were used in the separation of simple lipids prepared from serum and liver extracts [271]. Isocratic elution with either hexane–isopropanol–acetic acid (100:0.5:0.01 by volume) or hexane–*n*–butyl chloride–acetonitrile–acetic acid (90:10:1.5:0.01 by volume) at a flow-rate of 2 ml/min gave good separations of cholesterol esters, triacylglycerols, free fatty acids and cholesterol, and some unidentified compounds (possibly carotenoids). Some partial fractionation according to the nature of the fatty acid constituents of each lipid class was observed with the second solvent system, possibly as a consequence of partition effects with a layer of solvent molecules adhering to the silica gel adsorbent (see Chapter 2), and an even more marked effect of this kind was apparently observed by others

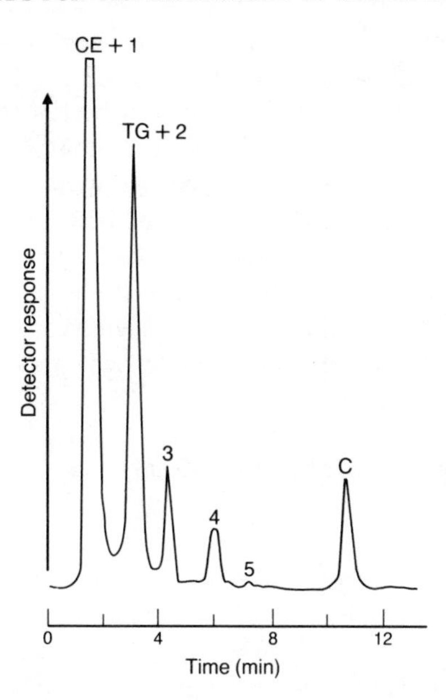

FIG. 5.3. HPLC separation of the simple lipids of sheep liver on a column (5 × 100 mm) of Radial-PAK™ CN (nitrile-bonded phase), eluted isocratically with 0.05% isopropanol in hexane at 1 ml/min, and with spectrophotometric detection at 210 nm [617]. (Reproduced by kind permission of the authors and of *Analytical Biochemistry*, and redrawn from the original publication.) Chapter 1 contains a list of abbreviations. In addition, 1 = BHT, retinol esters, dolichol esters; 2 = vitamin E; 3 = ubiquinone; 4 = dolichol; 5 = phytol.

[728]. Whether such partial separations have any analytical value is doubtful. Recoveries of lipid classes from the column eluent were shown to be essentially complete, so fractions could be collected for quantification or for the analysis of the fatty acid constituents. Ultraviolet detection has also been used in the fractionation of some leaf waxes [910].

Isocratic elution of a cyanopropyl-bonded phase (Waters Radial-PAK™ CN) column (5 × 100 mm) with 0.05% isopropanol in hexane at 1 ml/min gave excellent resolution of the simple lipids of sheep liver, as shown in Fig. 5.3 [617]. As spectrophotometric detection at 210 nm was used, unsaturated terpenoid "lipids", such as retinol, vitamin E, dolichol, ubiquinone and their esterified forms showed up especially prominently. Much better reproducibility of retention times was observed in this work, than with the more usual silica gel adsorbent, and this also been the author's experience (unpublished results). Unfortunately, the free fatty acid fraction and other acidic lipids appeared to be adsorbed rather strongly to such column packing materials, and after 50 injections it was necessary to wash the column with polar solvents in the reverse direction to remove contaminants. Others have described the elution of simple lipids from a column of Spherisorb™ CN during the separation of algal and bacterial carotenoids [243].

Infrared detection (at 5.72 to 5.75 μm) has been little used in the analysis of lipid classes, although some applications have been described. For example, commercial shortenings and the products of partial hydrolysis of seed oils (free fatty acids were methylated with diazomethane prior to analysis) were separated on a column (4.6 × 250 mm), packed with an amino-cyano-bonded phase (Partisil™ PXS 10/25 PAC), and eluted with a gradient of acetonitrile–hexane–chloroform (35:25:65 by volume) into hexane–chloroform (60:65 by volume) [636]. Methyl esters were eluted well ahead of triacylglycerols, and these were in turn clearly separated from diacylglycerols and then monoacylglycerols, which each gave a sharp symmetrical peak. On the other hand, appreciable base-line drift was evident. Methyl dihydroxystearate was employed as internal standard, and the IR detector gave a good rectilinear response to each lipid class, although the slopes of the calibration lines were markedly different.

IR detection was also utilized in the separation of wax constituents by adsorption chromatography on a column (4 × 400 mm) of Spherisorb™ S5W silica gel, maintained at 40°C [38]. For relatively simple mixtures (as in Fig. 5.4), a gradient of tetrahydrofuran into hexane was used, but a complex ternary gradient in which the three solvent reservoirs contained 0.05% isopropanol in hexane, tetrahydrofuran and acetonitrile was required when

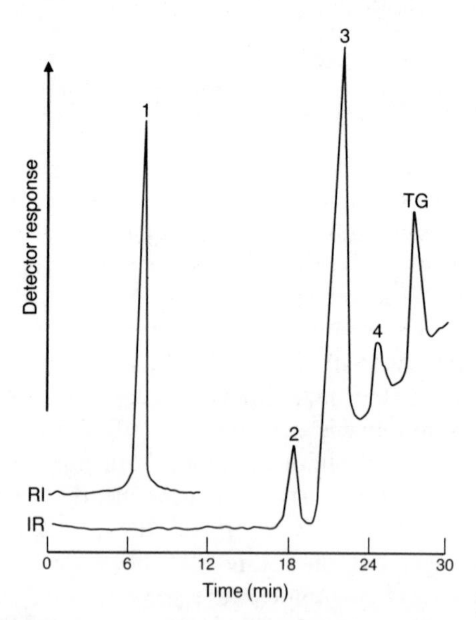

FIG. 5.4. HPLC separation of wax components on a column (4 × 400 mm) of Spherisorb™ S5W silica gel at a temperature of 40°C, and eluted with a gradient of tetrahydrofuran into hexane [38]. (Reproduced by kind permission of the authors and of *Chromatographia*, and redrawn from the original publication.) Detection was partly by means of a differential refractometer (RI), but mainly by infrared (IR) spectrophotometry at 5.75 μm. 1 = eicosane; 2 = octadecyl octanoate; 3 = methyl caprate; 4 = tridecan-2-one; TG = triacylglycerols.

more lipid components were present. Eicosane (detected by means of a differential refractometer, as it has no carbonyl function), octadecyl decanoate, methyl stearate, and methylundecyl ketone were eluted separately ahead of triacylglycerols and other simple lipids. Retention times and peak areas were reproducible, although a high degree of base-line drift was again observed. Although the technique had little value for analysis *per se*, in a modified form it did give representative triacylglycerols for fatty acid analysis at least [71].

Stewart and Downing [790] reported difficulties in the separation of wax esters and cholesterol esters from skin lipids by various conventional chromatographic means, but were successful with HPLC and a column packed with magnesium hydroxide as adsorbent (and with refractive index detection). Hexane containing 2% ethyl acetate was the mobile phase.

A great deal of work has been done to establish conditions for particular simple lipid separations using HPLC, but inevitably many compromises have had to be made because of detector limitations. It is to be hoped that a more widespread availability of mass-selective detectors will lead to further developments, especially if these can be used in tandem with spectrophotometric detection. Silica gel chromatography has been much used, but some advantages of nitrile-bonded phases in specific applications have become apparent. In addition, many other types of polar bonded phase are now manufactured, and their value to the lipid analyst ought to be explored.

B. HPLC Analyses of Specific Simple Lipid Classes

1. Cholesterol

Methods for the analysis of cholesterol in body fluids such as plasma have assumed particular importance in recent years with the realization that the concentration of this compound may have a bearing on the diagnosis of heart disease. Chromatographic methods, other than HPLC, and chemical and enzymatic methods have been reviewed elsewhere [133]. HPLC methods for the analysis of steroids in general have also been reviewed comprehensively [307, 753], and they are not discussed further here. The free cholesterol in tissues can often be determined directly by HPLC of a simple lipid extract, while the total amount of cholesterol (free plus esterified) is determined similarly after hydrolysis of the sample. Reversed-phase separations on columns packed with ODS phases have generally been preferred.

For example, cholesterol in a simple lipid extract of plasma has been determined by HPLC on a column of *micro*Bondapak™ C_{18} (4 × 300 mm), with spectrophotometric detection at 200 nm [197]. Isopropanol–acetonitrile (1:1 by volume) was used as eluent to separate and measure the total amount of cholesterol (after a hydrolysis step), and to obtain a profile of cholesterol and the molecular species of triacylglycerols and cholesterol esters. In the analysis of the free cholesterol, isopropanol–acetonitrile–water (6:3:1 by volume) was the solvent used. For this purpose, only 50 μl of plasma were required and microgram amounts of cholesterol could be accurately quantified. Others [109] employed a column containing Zorbax™ ODS, UV detection at 213 nm, and a linear gradient of water (3 to 0%) in acetonitrile–tetrahydrofuran (65:35 by volume) to separate and quantify both cholesterol and the molecular species of cholesterol esters in plasma samples (separations of molecular species of cholesterol esters are discussed in detail in Chapter 10). Similar separations to the last were carried out isocratically with refractive index detection in another laboratory [640]. The total cholesterol concentration in milk fat and other foods was determined by chromatography of a partially-purified non-saponfiable lipid fraction on a column of Alltech™ ODS, with hexane–isopropanol (99:9:0.1 by volume) as mobile phase and spectrophotometric detection at 205 nm; a 2% coefficient of variation with 20 μg of cholesterol was obtained [344, 345].

While satisfactory results can be obtained with untreated samples, as above, more sensitive and specific estimations can be made by converting the cholesterol to an appropriate derivatized form. For example, a method for determining cholesterol in food stuffs was described, in which the benzoate ester derivative of cholesterol was prepared by reaction of the non-sponifiable fraction of a lipid extract with benzoyl chloride in pyridine [582]. An aliquot of the reaction products was chromatographed on a column of *micro*-Bondapak™ ODS, with methanol as the mobile phase at 2 ml/min, and with UV detection at 230 nm. Amounts of cholesterol benzoate of as little as 10 ng could be determined. Following derivatization with 3-chloroformyl-7-methoxycoumarin, both cholesterol and cholestanol in human serum were estimated with great sensitivity by spectrofluorimetry, after chromatographic separation on a column of an ODS phase with methanol–tetrahydrofuran–acetic acid (13:1:0.28 by volume) as eluent [518].

Other workers have determined cholesterol in an HPLC eluent as a means of confirming the separation of lipoprotein fractions from a gel-permeation column (see Chapter 10 for further discussion); the detection and quantification involved an enzymatic assay in a post-column reaction chamber [285, 603, 608, 609].

Cholesterol and its oxidation products have been separated by HPLC, and detected by enzymatic assay [597] and by differential refractometry [757]. An HPLC procedure for the rapid isolation of sterols as a class from lipid extracts, for further analysis, involved elution from a column of silica gel with hexane-ethyl acetate (9:1 by volume) as the mobile phase [335].

2. *Free (unesterified) fatty acids*

The free fatty acid fraction of plasma has considerable biological importance as it turns over very rapidly, providing components to peripheral tissues for a variety of purposes. Also, the free fatty acid concentration can be an indicator of spoilage in foods, such as milk. Methods for the determination of free fatty acids, other than by HPLC, have been reviewed elsewhere [133]. Generally, the preferred HPLC approach to the problem has been to prepare fatty acid derivatives, which contain a chromophore that permits sensitive and specific spectrophotometric detection, followed by separation of the individual components on a reversed-phase column; the total amount of the fraction and its composition are thereby determined in a single analysis. As Chapter 7 deals with the HPLC separation of fatty acid derivatives at some length, brief details only of the chromatographic aspects are dealt with here.

In one of the first applications of HPLC to the problem, the phenacyl derivatives of the free fatty acid fraction of human fingerprints were prepared, and components up to C_{30} in chain-length were separated and quantified by reversed-phase chromatography with spectrophotometric detection at 254 nm [101]. At almost the same time, the analysis of the free fatty acid fraction of plasma by a procedure of the same kind was described [171]; the method was applicable over a range of 0 to 300 ng of fatty acids. The free fatty acids in butter were determined, essentially by the same approach, by conversion to the *p*-bromophenacylesters via a crown ether-catalyzed reaction, with no requirement for a preliminary separation from the other constituents of butter fat, and they were subjected to HPLC on a column containing an ODS phase; the limits of detection varied from 7 ng for butyric acid to 45 ng for linoleic acid [687]. Other workers have used analogous methods in the analysis of the free fatty acid fraction in various biological samples [63, 342, 348, 625].

Similar procedures, but involving the preparation of fluorescent derivatives, which can be analysed with great sensitivity, have been used by several groups. For example, the free fatty acids in as little as 10 μl of human serum were determined at the pmole down to fmole level by conversion either to the 9-aminophenanthrene derivatives [349], to the 4-bromomethyl-7-acetoxycoumarin and related compounds [835, 855], or to other derivatives [538, 899, 900] for reversed-phase separation of the individual constituents on columns containing ODS or similar bonded phases. In what was probably the simplest derivatization procedure of this kind, 9-anthryldiazomethane was used to prepare derivatives of the free fatty acids, but especially of arachidonic acid, in animal tissue extracts for fluorimetric detection and quantification after separation in the reversed-phase mode [240, 296, 572]. In addition, an extraction and derivatization (dansyl-semipiperazide) procedure has been developed that may be used with a wide range of fatty acids, but which is claimed to be especially suited to the analysis of components of shorter chain-length [904].

A procedure for the isolation of oxygenated fatty acids from plant tissues and large volumes of natural waters for HPLC analysis has been described [703].

3. Miscellaneous simple lipid classes

Triacylglycerols, either as the pure compounds or as components of lipoprotein fractions, in eluents from HPLC columns have been determined continuously in post-column reaction chambers by means of enzymatic (see Chapter 10) [286, 446] and chemical (see Chapter 8) [438] procedures.

Polymerized triacylglycerols, i.e. dimers and trimers formed by free radical-catalyzed cross-linking of the fatty acids in different molecules, are formed in a variety of circumstances, but especially in heated cooking oils. They have been fractionated according to molecular weight by gel-permeation chromatography in comparatively low-pressure systems, which require lengthy analysis times [12, 738]. In contrast in comparable HPLC methods with a macroporous styrene–divinylbenzene copolymer as the gel (P-EPL Gel™ or Ultrastyragel™), tetrahydrofuran as the eluent, and refractive index detection, polymers and dimers were clearly separated from monomeric triacylglycerols, oxidized material and partial glycerides in only 20 min [373, 467, 645, 746].

Diacylglycerols, which can have great metabolic importance, were isolated from lipid extracts of hepatocytes by HPLC on a column (3.9 × 300 mm) of *micro*Porasil™, eluted isocratically with hexane–isopropanol–acetic acid (100:1:0.1 by volume) and with refractive index detection [86]; they eluted as a discrete but broad band (presumably because some acyl migration has occurred to give a mixture of 1,2- and 1,3-isomers) after free cholesterol. A similar procedure was used in the estimation of the same compounds from the forebrain of the gerbil [1].

During structural analysis of triacylglycerols by means of hydrolysis with pancreatic lipase, 2-monoacyl-*sn*-glycerols are produced and must be isolated so that their fatty acid compositions can be determined. This has been achieved, on the 1 to 2 mg scale, by HPLC on a column (5 × 250 mm) of Hypersil™ silica gel, by means of isocratic elution with isooctane–tetrahydrofuran (3:2 by volume) and refractive index detection [138]. 2- and 3-Monopalmitoyl-*sn*-glycerols were separated from each other, the 2-isomer eluting first, on a column (4.6 × 250 mm) containing Altex™ C_{18} phase, with a gradient of tetrahydrofuran–water–isopropanol–acetonitrile (from 7.6:23:5:64.4 to 3.2:0:70:26.8 by volume over 30 min) at 1.21 ml/min [434]. The diacetate derivatives of isomeric 1(3)- and 2-oleoylglycerols, and the corresponding palmitoyl compounds, were eluted in this order and were clearly separated from each other preparatively (5 mg scale) on a column (5 × 250 mm) containing a diol-bonded phase, i.e. Nucleosil™ 7OH [146]. Isocratic elution with isooctane–tetrahydrofuran (95:5 by volume) at 1 ml/min and refractive index detection were used. Resolution of enantiomeric mono- and di-acylglycerols is discussed in Chapter 8 (Section C.2).

Di- and tetra-ether glycerols, obtained by hydrolysis of the lipids of Archaebacteria, were separated on a column of Spherisorb™ amino phase, eluted with hexane-*n*-propanol (99:1 by volume); they were detected by differential refractometry [507].

The hydrocarbon squalene is an important intermediate in the biosynthesis of cholesterol. A fraction enriched in this compound was isolated from a non-saponifiable lipid fraction of various rate tissues by preparative HPLC on a column (12 × 250 mm) of Merck SI 60–7 silica gel, with hexane–isopropanol–water (98:1.986:0.014 by volume) as the mobile phase and with spectrophotometric detection at 215 nm [803]. Squalene was then quantified on the 0.1 to 1.5 μg scale by rechromatography on an analytical column (4.6 × 250 mm) of Merck SI 60–5 silica gel, eluted with hexane.

Many plants contain unusual phenolic lipids, consisting of phenols with complex alkyl chains, which are apparently derived biosynthetically from fatty acids. The best known are those of the shell of the cashew nut (*Anacardium occidentale*). HPLC methods, involving both adsorption chromatography on silica gel and reversed-phase chromatography with an ODS phase, for the separation and quantitative analysis of such compounds have been described; the aromatic chromophore permitted specific spectrophotometric detection at 275 nm [839].

Methods for the analysis of tocopherols and related lipophilic substances have been reviewed elsewhere [827].

C. HPLC Methods for the Analysis of Both Simple and Complex Lipids in One Step

When lipid analysts first became aware of the potential of HPLC, arguably the most important application that was envisaged was the separation and quantification of the wide range of lipid classes, varying in polarity from cholesterol esters at one end of the spectrum to lysophosphatidylcholine at the other, that are found in animal and plant tissues. Remarkable success in this endeavour was obtained by Privett and his colleagues at the Hormel Institute from 1973 onwards, mainly because of the unique transport-flame ionization detectors, which they designed and built (see Chapter 2) [680, 797]. Unfortunately, few other analysts have had access to comparable equipment, and it has taken until now for the rest of the world to show signs of beginning to catch up.

In the first separations demonstrated with this detector [206, 679, 796], Corisil™ II (37–55 μm silica gel particles) was the adsorbent, and it was treated overnight with concentrated ammonium hydroxide, prior to activation and packing into a column (2.8 × 1000 mm). The pumping system had four pressurized solvent reservoirs, which permitted two quite distinct solvent gradients to be created. Pentane was the initial eluent and a gradient with diethyl ether was produced to separate first cholesterol esters and then triacylglycerols; the mobile phase was then changed to chloroform, and a gradient with 8% ammonium hydroxide in methanol was used to elute each

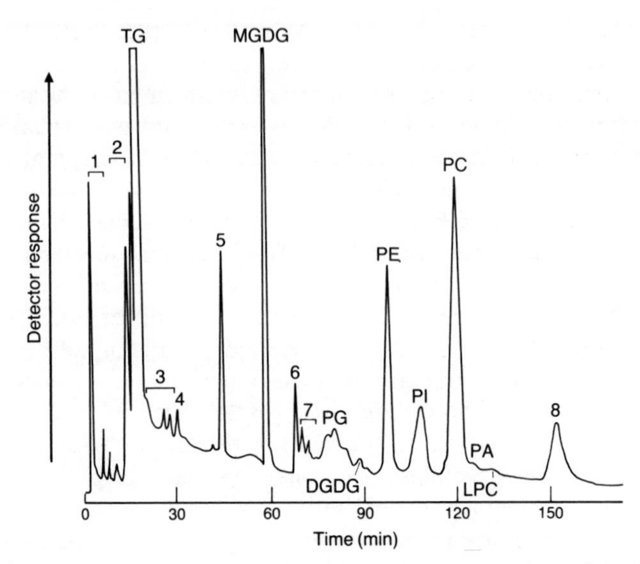

FIG. 5.5. HPLC separation of the lipids of mature soybeans on a column (2.8 × 1000 mm) of Bio Rad™ minus 325 mesh silica gel, treated with ammonium hydroxide. Elution was with a double binary-gradient system (see text for details), and a transport-flame ionization detection system was used [679]. (Reproduced by kind permission of the authors and of the *Journal of the American Oil Chemists' Society*, and redrawn from the original publication.) Chapter 1 contains a list of abbreviations. In addition, 1 = waxes, hydrocarbons and pigments; 2 = sterol esters; 3 = terpenes, tocopherols, alcohols; 4 = sterols; 5 = esterified sterol glucoside; 6 = sterol glucoside; 7 = glycosphingolipids; 8 = highly polar lipids. Many components were not identified.

of the polar lipids in turn at a flow-rate of only 0.1 ml/min. Figure 5.5 [679] illustrates the nature of the separations that were achieved with a lipid extract from mature soybeans. Many more components were separated than could be identified, but hydrocarbons and waxes eluted first, followed by triacylglycerols, sterols, the glycolipid constituents and each of the individual phospholipids, ending with lysophosphatidylcholine. The separation took 150 min, and a lengthy re-equilibration period with solvents of decreasing polarity was necessary before the next sample could be analysed. With lipid extracts from animal tissues, equally impressive separations were obtained [796], and indeed these are still the most comprehensive ever to have been achieved, although the work was published as long ago as 1973.

Subsequently [653], the adsorbent was changed to Spherisorb™ S-GP (8 μm silica gel particles), and it was again pre-treated with concentrated ammonium hydroxide before being packed into a column (2 × 450 mm). Hexane–dichloromethane (1: by volume) was used to elute the non-polar lipids, then chloroform–dichloromethane (1:3 by volume) was employed to elute cholesterol, before a gradient of 6% concentrated ammonium hydroxide in methanol into the latter solvent mixture was used to elute each of the complex lipids. The separation was completed in an hour. In addition, a novel feature of this particular work on lipid extracts from animal tissues was that the eluent emerging from the column was passed through a spectrofluorometric detector immediately ahead of the transport-flame ionization detector. The two signals were displayed together on a recorder so that fluorescent lipophilic or "lipofuscin" substances could easily be identified.

In a more recent paper on this topic from Privett's laboratory [649], the same type of column and a slightly modified solvent system was used to speed up the separation somewhat. The response of the latest modification of the detector was determined for each lipid class, and in general was found to deviate only a little from the rectilinear. Accordingly, response factors were calculated of the form

$$Y = A \cdot X^{B}$$

where $Y =$ the detector response, $A =$ a constant, $X =$ the weight of sample (microgram), and $B =$ the response index, which takes into account the slight deviation from direct proportionality [221, 748]. When such factors were used in analyses of animal tissue lipids, duplicate analyses generally agreed within 1%, and were similar to results obtained by longer-established chromatographic methods. Although these analyses were carried out with microgram amounts of lipid, the detector has the potential to be used at the nanogram level.

In a further exciting development, Erdahl and Privett [205] have linked the transport system of their detector to a mass spectrometer, apparently enabling simultaneous mass spectral and flame-ionization detection (although only the former was illustrated in their paper). A total ion-current chromatogram was obtained, in order to demonstrate the nature of the separation, while the chemical-ionization mass spectra gave information on the nature and quantitative composition of the fatty acid constituents of each of the lipid classes.

A few other groups have attempted separations of lipids of markedly different polarity. For example, the simple lipids and some unidentified complex lipids of soybean "lecithin" were separated by HPLC on a silica gel column, eluted with a ternary gradient system; detection was by means of the Pye Unicam LCM2 transport-flame ionization detector [426]. An alternative approach was to use isocratic elution but with flow-programming [224]. In this instance, triacylglycerols, sterols, free fatty acids, phosphatidylethanolamine and phosphatidylcholine were somewhat imperfectly separated on the column (4.6 × 125 mm) of Hypersil™ silica gel, eluted with hexane–isopropanol–0.1 M (pH 4.66) acetate buffer (8:8:1 by volume) with the flow-rate changed in a stepwise manner at programmed intervals from 0.4 to 3 ml/min; spectrophotometric detection at 206 nm was employed.

One of the more recent developments in the separation of lipids differing widely in polarity has come in the author's own laboratory [136]. Here, the objective was to separate and quantify the more abundant lipid classes in animal tissues, on the 0.2 to 0.4 mg scale, in as short a time as could conveniently be managed. The ACS mass detector (see Chapter 2) was utilized, together with a ternary solvent delivery system and a short (5 × 100 mm) column, packed with Spherisorb™ silica gel (3 μm particles). In selecting a mobile phase, the choice of solvents was constrained by the need for sufficient volatility for evaporation in the detector under conditions that did not cause evaporation of the solute, and by the necessity to avoid inorganic ions, which would not evaporate. Similar restrictions apply to detectors operating on the transport-flame ionization principle. It was necessary to use a complicated ternary-gradient elution system with eight programmed steps, starting with isooctane to separate the lipids of low polarity and ending with a solvent containing water to elute the phospholipids; a solvent of medium polarity was then needed to mediate the transfer from one extreme to the other, and mixtures based on isopropanol gave satisfactory results. The three solvent mixtures selected by trial and error were isooctane–tetrahydrofuran (99:1 by volume) (A), isopropanol–chloroform (4:1 by volume) (B) and isopropanol–water (1:1 by volume) (C). In essence, a gradient of B into A was created to separate each of the simple lipids, then a gradient of C into A plus B was produced to separate each of the complex lipids; finally, a gradient in the reverse direction was generated to remove most of the bound water and to re-equilibrate the column prior to the next analysis. A relatively-high flow-rate (2 ml/min) appeared to assist the separation greatly, perhaps compensating for the absence of strong acid or inorganic ions, which others have found necessary for the separation of phospholipids (see Chapter 6). However, in the longer term, it was found to be advantageous to add small amounts of organic ionic species (see below) [139].

The nature of the separation achieved with a lipid extract from rat liver is shown in Fig. 5.6. In spite of the abrupt changes in solvent composition at various points, no base-line disturbance was apparent, and each of the main simple lipid and phospholipid classes was clearly resolved in only 20 min. Only the highly acidic lipids, phosphatidic acid and to a lesser extent phosphatidylserine, did not give satisfactory peaks. There was no "solvent peak" at the start of the analysis, as is often seen with other detectors, and BHT added as an antioxidant evaporated with the solvent so did not interfere. After a further 10 min of elution to regenerate the column, the next sample could be analysed.

One additional point worthy of comment was the observation that to obtain reproducible resolution of the cholesterol esters and triacylglycerols, it was necessary to inject the sample in as small a volume as possible (5 μl or less) of a solvent of comparatively-low polarity; chloroform–isooctane (1:1 by volume) was found to be satisfactory.

When using the mass detector, it is necessary to work out the optimum conditions for the desired separations first and then to carry out a calibration

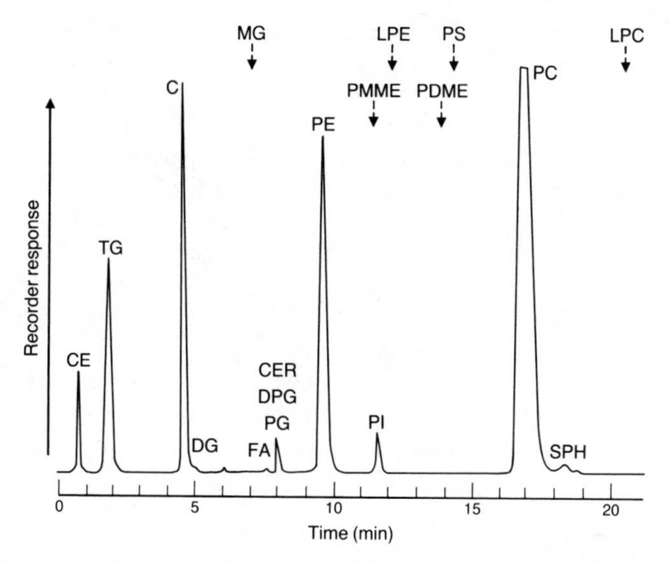

FIG. 5.6. Separation of rat liver lipids (0.35 mg) by HPLC on a column (5 × 100 mm) of Spherisorb™ silica gel (3 μm particles), with mass detection; the elution conditions are described in the text [136]. (Reproduced by kind permission of the *Journal of Lipid Research.*) The elution times of some lipids not present at detectable levels in the sample are indicated. Chapter 1 contains a list of abbreviations.

with authentic lipid standards. The detector operating parameters, i.e gas pressure, evaporator temperature, and attentuation, must also be rigorously standardized. If the separation conditions or detector settings have to be changed subsequently for any reason, a tedious recalibration would probably then be necessary. Calibration curves for some of the main lipid classes in the analysis described above are shown in Fig. 5.7. For most lipids, the detector response was approximately rectilinear in the range 50 to 200 μg, but it tended to fall off rapidly below 10 μg.

In later work [139], it was observed that much better resolution of the minor acidic components was obtained by adding small amounts of ionic species to the aqueous component of the eluent. The life time of the column was also greatly extended by this simple step. In practice, the optimum results were obtained with 0.5 to 1 mM serine buffered to pH 7.5 with triethylamine. Ionic species at this concentration had no effect on the base-line of the detector, although the response changed a little and recalibration was necessary. In addition, hexane replaced isooctane in the mobile phase, in order to reduce the maximum operating pressure required.

Much remains to be done to improve the conditions for the simultaneous separation of simple and complex lipids by means of HPLC. Many more solvent combinations must be tried, and many more column packing materials require to be tested for the purpose. However, as mass-selective detectors

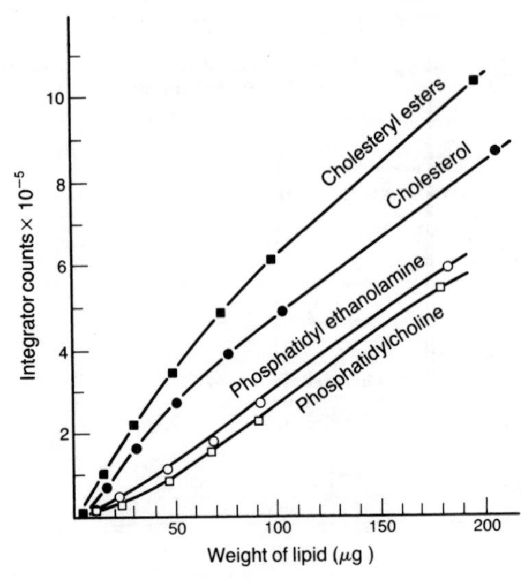

FIG. 5.7. Calibration curves of detector response against amount of sample injected for some lipid classes separated by HPLC with mass detection [136]. (Reproduced by kind permission of the *Journal of Lipid Research*.)

appear in more laboratories, there is now a solid foundations upon which to build.

D. Alternative Methods to HPLC for the Analysis of Simple Lipids

Methods other than HPLC for the separation and analysis of simple lipids, or of simple and complex lipids together, have been reviewed comprehensively by the author in another book [133]. However, it may be of value to consider them briefly here, so that HPLC methods can be viewed in a wider perspective. Preparative-scale column chromatography procedures are discussed in Chapter 4.

TLC procedures with silica gel G (containing calcium sulphate as binder) layers have been employed most frequently for lipid class separations. Commonly, the solvent elution system used is hexane–diethyl ether–formic acid (80:20:2 by volume), and this gives the separations shown in Fig. 5.8. Cholesterol esters migrate to the solvent front, and they are followed by triacylglycerols, free fatty acids, cholesterol, diacylglycerols, monoacylglycerols and phospholipids (with other polar lipids). For small-scale preparative purposes (2 to 20 mg), the author prefers glass plates (20 × 20 cm) coated with a layer 0.5 mm thick of silica gel G. Bands are then most conveniently detected by spraying with an 0.1% (w/v) solution of 2′,7′-dichlorofluorescein in 95% methanol and viewing the dried plate under

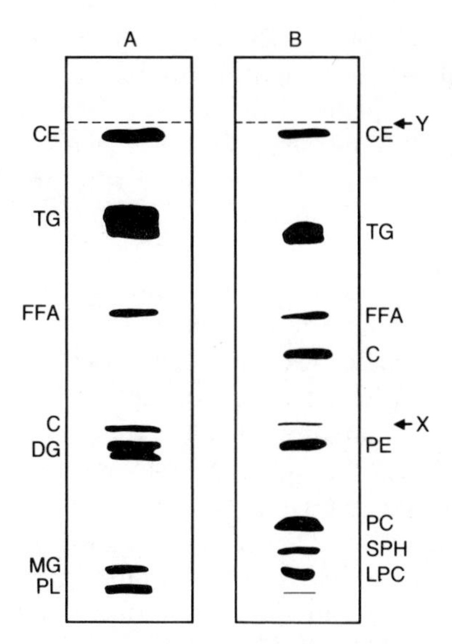

FIG. 5.8. Schematic TLC separations. A. Simple lipids on a silica gel G layer. Hexane–diethyl ether–formic acid (80:20:2 by volume) was the developing solvent. B. Simple lipids and phospholipids of plasma on high-performance silica gel; first and second developments as far as "X" with chloroform–methanol–water (60:30:5 by volume), and a third development to "Y" in hexane–diethyl ether–acetic acid (80:20:1.5 by volume) [466]. Chapter 1 contains a list of abbreviations.

ultraviolet light; the lipids appear as yellow spots against a dark background, and they can be recovered from the adsorbent by elution with solvents for further analysis. For example, it is possible to add an internal standard, such as the methyl ester of an odd-chain fatty acid not present naturally in the sample, to transesterify, and to determine the fatty acid composition and the amount of lipid (relative to the standard) simultaneously by GLC analysis [150]. If the lipid class composition only is required, layers of silica gel 0.25 mm thick are preferred, less lipid (< 1 mg) is needed, and the separated lipids can be quantified by fluorimetry, or by charring/densitometry. For the latter purpose, scanning photodensitometers are available, or the author can recommend a published procedure in which the charred bands were suspended in an emulsifier-scintillant cocktail and determined by measuring the amount of quenching, relative to standards, on counting in the external standard mode in a liquid-scintillation counter [754]. The degree of quenching was proportional to the amount of lipid present.

Both the simple and complex lipid classes in a plasma sample have been separated on a single plate by high-performance TLC, as is shown also in Fig. 5.8 [466]. Commercially-available precoated plates were used for the purpose, and they were first given a double development for part of the way in a polar solvent to separate the phospholipids, and then for the full length of the plate in a less polar solvent to separate the simple lipids. In comparable work, more than 20 different classes of lipids from animal tissues, including simple lipids, glycolipids and phospholipids, were separated in a single dimension on HP-TLC plates with three or four consecutive developing solvents, and by carefully controlling variables such as atmospheric humidity [907]. Others have used a similar procedure with brain lipids [497], while so-called "three-dimensional" TLC systems have been developed for the same purpose [194, 445].

In the Iatroscan TLC analysis system (Iatron Laboratories, Japan), the TLC medium is a thin quartz rod to which a layer of silica gel is fused by means of a sintering process (Chromarods™). Lipids are applied to these rods, and they are separated in a conventional manner by means of development in a solvent mixture. Then the rods are dried and placed in an apparatus that feeds them through a flame-ionization detector, producing a signal, which is amplified and passed to a recorder and integrator. The rods are regenerated in the process and can be re-used up to a hundred times. Ackman [2] has reviewed the use of this equipment for lipid separations in general. Many applications to the separation of simple lipids have been described [143, 212, 444, 514, 622, 623, 824].

The first models of the Iatroscan analyser were rather unsophisticated and the Chromarods were of uneven quality, but both have been improved somewhat. Initially, it appeared that the apparatus did not give acceptable reproducibility in quantitative analyses, but now it has a number of devotees who have shown that if rigorous calibrations are performed, if each analysis is replicated a sufficient number of times, and if an internal standard is employed, then satisfactory results can be achieved. For example, Kramer and colleagues compared the results of analyses performed with the Iatroscan analyser with those obtained by means of longer-established chromatographic methods, and showed examples where the former technique was apparently preferable [444]. It is obvious, however, that many problems in using the technique for quantification purposes remain and require to be resolved [163, 623].

As an alternative to TLC procedure, Kuksis and colleagues have developed a method, which has proved to be of great value for the analysis of plasma lipids especially, and consisted in subjecting them, after suitable derivatization, to separation by means of high-temperature GLC. In this procedure, the plasma lipid extract was first reacted with the enzyme phospholipase C to convert phosphoglycerides to diacylglycerols and sphingomyelin to ceramides; the reaction mixture was then reacted with a trimethylsilylating reagent to convert the free hydroxyl groups of diacylglycerols, ceramides, cholesterol and monoacylglycerols to the trimethylsilyl ether derivatives, and free fatty acids to the trimethylsilyl esters. Cholesterol esters and triacylglycerols were unchanged, and tridecanoin was added as an internal standard. Initially, short packed columns, containing a low-polarity stationary phase such as SE-30, were utilized in the final GLC separation step, and they were temperature-programmed from 200°C to about 340°C

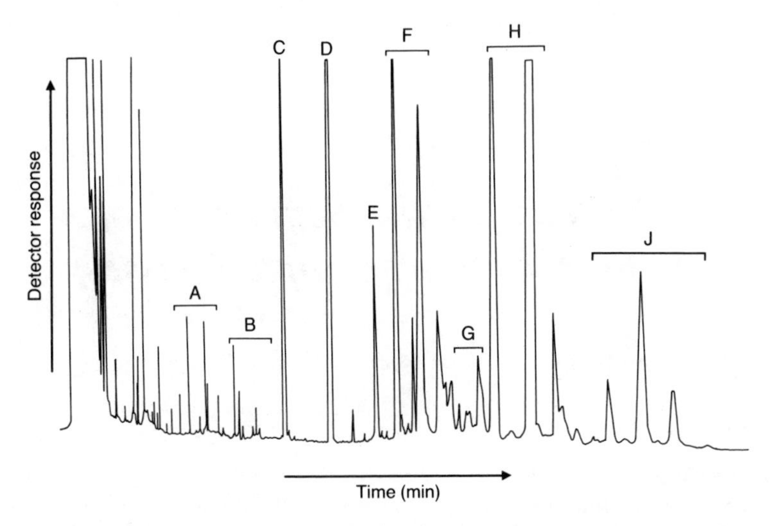

FIG. 5.9. High-temperature GLC of derivatized plasma lipids on a fused-silica capillary column (8 m × 0.30 mm) coated with SE-54, and temperature-programmed to a maximum of 340°C [564]. Abbreviations: A, trimethylsilyl (TMS) ester derivatives of free acids; B, TMS ethers of monoacylglycerols; C, TMS ether of cholesterol; D, tridecanoin (internal standard); E, TMS ether of 16:0 amide of sphingosine; F, TMS ethers of diacylglycerols (derived from phosphoglyceride); G, TMS ethers of ceramides (derived from sphingomyelin); H, cholesterol esters; J, triacylglycerols. (Reproduced by kind permission of the authors and of the *Journal of Biochemical and Biophysical Methods*, and redrawn from the original publication.)

[459, 461]. Lipids were separated essentially according to the total number of carbon atoms in the molecules, so information was obtained not only about the total amount of each lipid class in the sample, but also on the chain-length distributions of the fatty acid constituents in each.

More recently, it was established that much more information could be obtained if a fused-silica open-tubular (capillary) column, with SE-54 as stationary phase and with "on-column" injection, was employed in the separation of the lipid species, as shown in Fig. 5.9 [564]. As a consequence of the increased resolution possible with such columns, some separation was achieved both according to the chain-lengths and the degree of unsaturation of the constituent fatty acids of the lipid classes. The derivatives of the free fatty acids eluted first, followed by those of monoacylglycerols, cholesterol and phospholipids, and finally by the cholesterol esters and triacylglycerols. Very little overlap of peaks from different lipid classes was observed. With this method, which appears to be well suited to routine analyses of large numbers of samples, a remarkable amount of information could be gathered from as little as 1 μl of plasma. The main disadvantage is that no information on the relative compositions of the different classes of phosphoglyceride can be obtained. Although the procedure seems only to have been used with plasma lipids, there appears to be no reason *a priori* why it should not be applied to other tissues.

CHAPTER 6

Methods for the Separation of Individual Phospholipid Classes

A. HPLC – General Isocratic Elution Procedures

1. Introductory comments

In this chapter, HPLC procedures are described for the separation and analysis of individual phospholipid classes either from a concentrate of these compounds, prepared as outlined in Chapter 4, or in such a way that any simple lipids present migrate as a single distinct band at the solvent front. Separations of simpler molecular species of phospholipids are discussed in Chapter 8 rather than here. Methods for the analysis of phospholipids by HPLC or by other means have been reviewed elsewhere [16, 133, 137, 377, 501, 502, 503]. In comparison to TLC procedures, which until very recently have been preferred for phospholipid analysis in most laboratories, HPLC can give better and more consistent separations of minor components in skilled hands, while larger sample loads can be applied to columns before an appreciable loss of resolution is seen. No oxidation of the unsaturated fatty acid constituents need occur during fractionation on an HPLC column. HPLC is of course a cleaner technique, and quantification can often be accomplished more conveniently. Whatever the separation method employed, accurate quantification is essential and often the only suitable method is phosphorus assay, which can also be invaluable as an objective check on other procedures (see Section E below).

Many lipid extracts contain so many different complex lipids that gradient elution procedures may be obligatory, if effective separations are to be achieved. On the other hand, isocratic elution techniques can frequently be applied to less complicated mixtures, or to facilitate the isolation or determination of particular components. The equipment necessary is much simpler and thus less costly than that required for gradient elution, and a wider range of solvents and detection systems can then be used. As discussed in relation to simple lipids in the last chapter, the approach of the analyst to phospholipid separations has been governed to a larger extent by the nature of the detection systems available to him. As mass-selective detectors are obtained by more laboratories, many of the present HPLC procedures will be superceded, although the information and experience acquired on elution conditions will continue to be of great value. Those elution schemes that successfully avoid the use of buffers of high ionic strength or of strong acids are worthy of special note.

The author has not attempted to recommend particular HPLC procedures, because of the wide discrepancies in the availability of specific items of equipments to analysts, but intends to present sufficient information so that informed choices can be made. Also, it appears that some elution systems work well with a newly-packed column in phospholipid separations, but that the resolution deteriorates rapidly as the column ages in continued use. Various reasons for this have been advanced, but a definitive explanation is still awaited. Suggested "cures", discussed below, include repacking of the top 2 mm or so of the column at intervals, the use of acids, bases or buffers as ion suppressants, high column temperatures (35° to 90°C), high flow-rates (about 2 ml/min) and column washing procedures as cautioned earlier (Chapter 2), it is essential that polar solvents should be flushed out of columns at the end of each session of work to minimize the rate of deterioration. It is not always easy for an independent observer to evaluate each new "improvement" to published methods, since it can take some time for problems of column deterioration to become apparent.

So many different combinations of solvents have been employed in mobile phases that the reader may quickly become somewhat confused. To summarize briefly, if solvents such as hexane, isopropanol, methanol, acetonitrile and water are the only components of the eluent, then UV detection at 200–210 nm is possible. With most other solvents, an alternative detection system will probably be necessary. There have been two main lines of approach to the design of mobile phases, and these confer distinctive selectivities to the separations, i.e. to use mixtures based on acetonitrile–methanol–water or hexane–isopropanol–water. With the former and columns of silica gel, the acidic phospholipids are eluted rather easily, and phosphatidylcholine and sphingomyelin are particularly well resolved. With the latter type of mobile phase, the acidic phospholipids usually migrate between phosphatidylethanolamine and phosphatidylcholine, and simple lipids and glycolipids appear to be more readily distinguished. The addition of other solvents to these primary mixtures may assist with the resolution of particular components. Good separations can be obtained with many combination of pure solvents, when a new column of silica gel is available. On the other hand in prolonged use, it is probably desirable to utilize ionic species in the mobile phase in order to lengthen column life and to obtain reproducible elution volumes for individual lipid classes. While strong acids or bases have been used by some analysts for this purpose, equally good results have been obtained by others with the addition of as little as 0.5 mM serine to the aqueous component of the mobile phase, for example.

2. *Adsorption chromatography on silica gel*

Adsorption chromatography on silica gel has been employed most frequently for the separation of individual phospholipids by means of HPLC, although polar bonded phases also have value. The most popular elution systems have been based on acetonitrile–methanol–water mixtures, since these solvents are essentially transparent in the 200 to 210 nm region of the spectrum, where some of the functional groups, but especially isolated double bonds, of lipids exhibit a weak absorbance (see Chapter 2). In the first application of such as system to lipids, a column (2.1 × 500 mm) of Micro-Pak™ SI-10 silica gel, elution with acetonitrile–methanol–water (61:21:14 by volume initially, but modified somewhat for particular samples) at 1 ml/min, and detection at 203 nm were employed [380]. Phosphatidylethanolamine and phosphatidylserine emerged together near the solvent front, but phosphatidylcholine and sphingomyelin were cleanly separated. Nowadays, such separations do not look impressive, but at the time they were undoubtedly a great step forward. It was shown that the response of the detector both to phosphatidylcholine and sphingomyelin was rectilinear upto at least 60 μg, and that direct quantification was possible, provided that the apparent extinction coefficient for each component was determined accurately. As the adsorption spectra of individual lipid classes can be very different (as shown in Fig. 6.1), mainly because of the differing degree of unsaturation of each, it is essential that the standards used for calibration should resemble the compounds to be analysed in the composition of their acyl moieties as closely as possible.

Elution procedures of this kind appear to give much better separations of phosphatidylcholine and sphingomyelin than many others, which have been

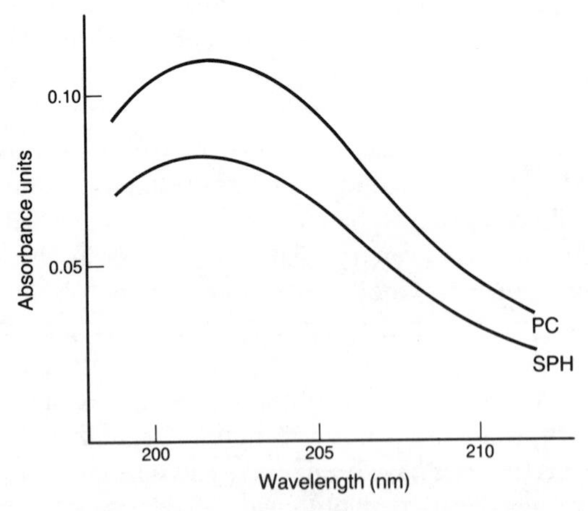

FIG. 6.1. Apparent absorption spectra of phosphatidylcholine (PC) and sphingomyelin (SPH) [380]. (Reproduced by kind permission of the authors and of the *Biochemical Journal*, and redrawn from the original publication.)

described. The ratio of these compounds in amniotic fluid is of special importance as an indicator of fetal lung maturity. A method, which made use of the above solvent system at a flow-rate of 1.9 ml/min, a column (3.9 × 300 mm) containing *micro*Porasil™ silica gel and detection at 205 nm, gave an excellent separation in only 10 minutes, and satisfactory quantitative results for the required ratio [175]. Similar procedures were employed for heart lipids [376], and for lipid extracts from cells of patients with Niemann–Pick disease [385]. In addition, a virtually identical system was used for the determination of phosphatidylcholine in food products, such as chocolate [346, 677].

An isocratic mobile phase, i.e. acetonitrile–methanol–water (50:45:6.5 by volume), at a flow-rate of 0.4 ml/min, and a column (3.9 × 300 mm) of *micro*Porasil™ silica gel (with guard column) were utilized to effect separations of a wider range of individual phospholipid classes [393]. In this instance, a rather novel feature was that detection and quantification was by means of a continuous phosphorus analyser, which removed aliquots at 30-sec intervals, oxidized them to inorganic phosphorus and reacted them with molybdenum blue reagent for colour measurement and comparison with standards. The nature of the separation achieved with a reference phospholipid mixture is seen in Fig. 6.2. Phosphatidylinositol was eluted

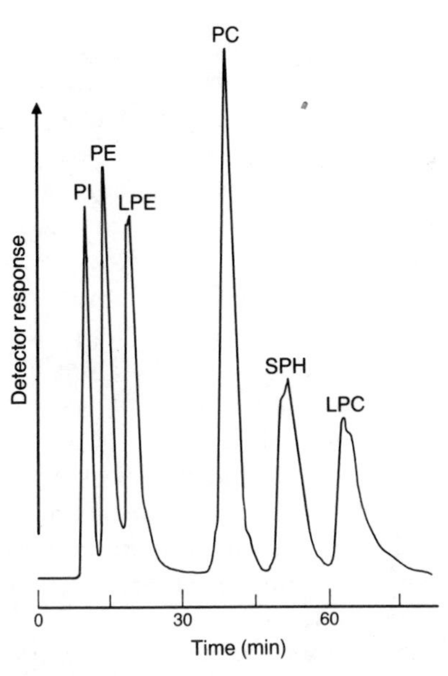

FIG. 6.2. Separation of a reference phospholipid mixture on silica gel column, and by isocratic elution with acetonitrile–methanol–water (50:45:6.5 by volume) at a flow-rate of 0.4 ml/min. Detection was by phosphate analysis [393]. (Reproduced by kind permission of the authors and of *Analytical Chemistry*, and redrawn from the original publication.) Chapter 1 contains a list of abbreviations.

ahead of phosphatidylethanolamine, and all the choline-containing phospholipids were in turn cleanly separated from each other.

The same elution system, but with refractive index detection, was used for the isolation of the products of phospholipase. A hydrolysis of phosphatidylcholine, i.e. lysophosphatidylcholine and free fatty acids, so that their fatty acid compositions could be determined [144]. (This has also been accomplished by reversed-phase HPLC with synthetic compounds containing one specific fatty acid [166]). In addition, a rather similar elution system (with the solvents in the proportions 50:45:2 by volume respectively) and refractive index detection were employed to isolate the individual phospholipids of bovine milk for analysis of their fatty acid constituents [104]. Here, it was clearly established that both phosphatidylserine and phosphatidylinositol were eluted separately (in this order) ahead of phosphatidylethanolamine, a finding that could be of value in many circumstances.

Although these separations appear perfectly satisfactory for the major phospholipid components of animal tissues, others have claimed improved results by incorporating strong acids into the mobile phase as ion suppressants. This may be especially important when columns are utilized repeatedly for the same purpose. For example, acetonitrile–methanol–85% phosphoric acid (130:5:1.5 by volume) at 1 ml/min, a column (4 × 300 mm) of Micro-PAK™ SI10 silica gel, and detection at 203 nm were used with this mind [120]. The order of elution was again the acidic phospholipids first, followed in turn by the ethanolamine- and choline-containing phospholipids. In this instance, a rather surprising finding, and one worthy of note, was that sphingomyelin eluted after rather than before lysophosphatidylcholine. The method had a number of disadvantages. For example, it was necessary to equilibrate the mobile phase with the stationary phase for a lengthy period, degradation of plasmalogens almost certainly occurred, and lipids collected from the eluent could not readily be determined by a phosphorus assay, because of the inorganic phosphate in the fractions. Others have used this elution system with some modifications for the analysis of the phospholipids of platelets (the use of fluorescein as an internal standard was a novel feature of this work) [410], and of lung surfactant [185]. The excellent resolution of the choline-containing phospholipids has led to the use of this procedure for the isolation of 1-O-alkyl-2-acetyl-*sn*-glycero-3-phosphorylcholine (platelet-activating factor) from tissues; this lipid class tended to migrate just ahead of lysophosphatidylcholine and sphingomyelin [364, 861]. In this instance, the drawbacks of the procedure could be ignored, as they had no effect on the elution of the compound of interest.

The same order of elution of individual phospholipids was obtained with acetonitrile–methanol–sulphuric acid (100:3:05 by volume) as the isocratic mobile phase [392]. Again, plasmalogens were degraded, but phosphorus assay could be used for quantification purposes. From some of the work described below, it now appears that better results could be obtained with much less corrosive additives as ion suppressants.

Mobile phases based on hexane–isopropanol–water mixtures are also transparent in the 200 to 210 nm region of the spectrum. In the first separations of phospholipids with a solvent mixture of this kind, described by research workers from the Netherlands [237, 300], gradient elution was used as discussed below (Section B). Others have since built on this work and have developed isocratic elution conditions which can be applied in many circumstances.

For example, in a rapid method for the quantitative estimation of phosphatidylcholine in soyabean "lecithin", a column (3 × 250 mm) of Lichrosorb™ SI 60 silica gel, maintained at 30°C, elution with hexane–isopropanol–water (1:4:1 by volume) at 2 ml/min, and detection at 206 nm were utilized [575]. The compound of interest was eluted in only 3 minutes, and was quantified from the detector response after careful calibration with authentic standards. With the same column but elution with hexane–isopropanol–acetate buffer (pH 5.8) (8:8:1 by volume) as the mobile phase, with the flow-rate programmed from 0.3 to 2.3 ml/min, phosphatidylethanolamine, phosphatidylcholine, phosphatidylinositol and phosphatidic acid were eluted in sequence and were partially resolved [576]. With hexane–isopropanol–0.2 M acetic acid (8:8:1 by volume) as the mobile phase, the acidic lipids eluted with phosphatidylethanolamine, and well ahead of phosphatidylcholine. Although an elution system of this kind has been used to determine the ratio of phosphatidylcholine to sphingomyelin in amniotic fluid [29], the system based on an acetonitrile–methanol–water mixture, described above, appears to be preferable.

In one of the most comprehensive and successful isocratic elution procedures to have been described, hexane–isopropanol–25 mM buffer-ethanol–acetic acid (367:490:62:100:0.6 by volume) (see the original paper for the method of mixing) was the mobile phase, at a flow-rate of 0.5 ml/min for the first 60 min then increased to 1 ml/min [630]. The column (4.6 × 250 mm) contained LiChrospher™ Si-100 silica gel, and detection was at 205 nm. Figure 6.3 illustrates the nature of the separation achieved with a rat liver extract. Phosphatidylethanolamine eluted just after the neutral lipids, and was followed by each of the acidic lipids, i.e. phosphatidic acid, phosphatidylinositol and phosphatidylserine, then by diphosphatidylglycerol and by the individual choline-containing phospholipids. Only the phosphatidylcholine and sphingomyelin overlapped slightly. As each component was eluted, it was collected, washed to remove the buffer, and determined by phosphorus analysis. Excellent agreement with the results of a TLC procedure were obtained. In addition, the fatty acid composition of each lipid class could be obtained with relative ease, by GLC analysis after transmethylation of the fractions.

Elution systems based on chloroform–methanol–water mixtures have been used with refractive index detection for the separation of phospholipid classes. For example, with these solvents in the ratio 178.5:64:5 (by volume) respectively at 2 ml/min as the mobile phase, and with a column (4 × 300 mm)

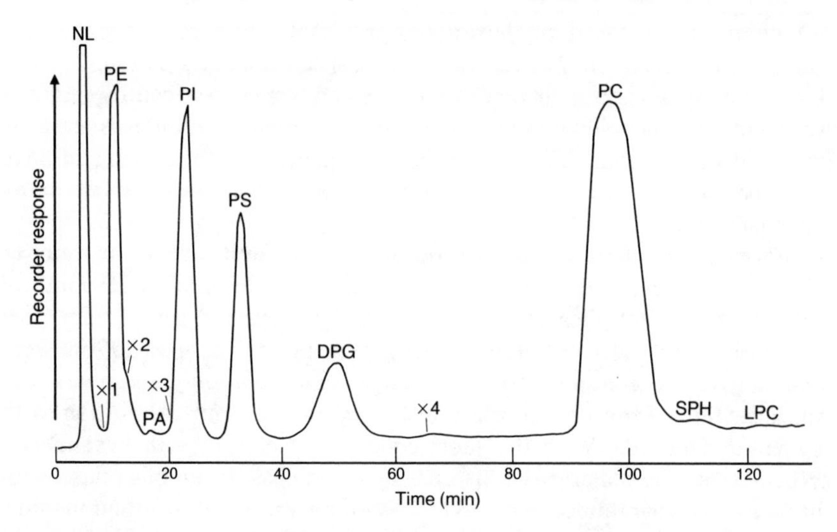

FIG. 6.3. Isocratic elution of rat liver phospholipids from a column of silica gel with hexane–isopropanol–25 mM phosphate buffer–ethanol–acetic acid (367:490:62:100:0.6 by volume) as mobile phase at a flow-rate of 0.5 ml/min for the first 60 min then of 1 ml/min, and with spectrophotometric detection at 205 nm [630]. (Reproduced by kind permission of the authors and of the *Journal of Lipid Research*, and redrawn from the original publication.) Chapter 1 contains a list of abbreviations; in addition, NL, neutral lipids; X1, X2, X3 and X4, unidentified lipids.

of *micro*Porasil™ silica gel, the phospholipids in fetal amniotic fluid eluted in the order phosphatidylethanolamine, phosphatidylglycerol, phosphatidylinositol, phosphatidylcholine and sphingomyelin [627]. The sensitivity appeared to be comparable to UV detection systems. On the other hand, even with careful calibration, the results were judged to be not sufficiently reliable to supplant TLC methods in quantitative analyses. Difficulties were experienced with variable sensitivity, noisy base-lines and temperature fluctuations, all apparently inherent faults of refractive index detection. Comparable mobile phases and adsorbents have been used in preparative-scale HPLC equipment with wide-bore columns for the purification of gram quantities of digalactosyldiacylglycerols, phosphatidylcholine and phosphatidylethanolamine [238, 626].

In a novel HPLC approach to phospholipid analysis, but one which may be of more academic than practical interest, deuterated solvents, i.e. chloroform–acetonitrile–deuterated methanol–deuterium oxide (136:25:34:5.9 by volume), were utilized as the mobile phase for infrared detection at a wavelength of 5.75 μm (for glycerophospholipids) and at 6.15 μm (for sphingomyelin) [122].

In other work of this kind, acid or base has been employed as an ion suppressant with chloroform-containing solvent mixtures. Elution with chloroform–methanol–acetic acid–water (14:14:1:1 by volume) at 2 ml/min, a column of *micro*Porasil™ silica gel, and refractive index detection were used in the determination of phosphatidylcholine in crude soyabean "lecithin" [691]. On the other hand, a wider range of phospholipids were separated on a column (4.8 × 1800 mm) of Corasil™ II silica gel, eluted isocratically with chloroform–methanol–ammonia (50:35.9:7 by volume) at a flowrate of 2.78 ml/min, conditions optimized by means of a two-dimensional simplex approach [685]. In this instance, a Pye transport-flame ionization detector was used. Lipids eluted in the order phosphatidylethanolamine, phosphatidylserine, phosphatidylinositol, phosphatidylcholine, phosphatidic acid, sphingomyelin and lysophosphatidylcholine.

Relatively simple mixtures of phosphatidylcholine and lysophosphatidylcholine were separated on various columns containing LiChrosorb™ SI 60 silica gel, eluted with ethanol–water–14 M ammonia mixtures (e.g. 78:21:1 by volume), and they were quantified by refractive index detection [18]. When the method was in continuous use, it was observed that the silica gel in the column dissolved slowly in the alkaline mobile phase. It proved possible to rectify the problem by repacking the top 2 to 4 mm of silica gel at intervals of 1 to 2 weeks. The author has also observed that this simple procedure can be highly efficacious in prolonging column life and resolution, even with 3 μm silica gels (but not necessarily at such frequent intervals) [139].

3. *Chromatography on bonded phases*

By employing stationary phases onto which various functional groups have been bonded by means of chemical reactions, it has proved possible to impart a degree of selectivity to certain phospholipid separations. In some circumstances, particular phospholipid classes can be eluted under comparatively mild conditions.

The bonded material closest to silica gel in its properties is probably a "diol" phase, i.e. 1,2-dihydroxypropyl moieties linked covalently to silica gel, giving a surface with a relatively-uniform layer of hydroxyl groups. Although gradient elution was employed in some of the first reports of the use of this packing for phospholipid separations (see Section B.2), some isocratic applications have also been described. For example, a column (4.6 × 250 mm) of LiChrosorb™ DIOL was eluted with *tert*-butylmethyl ether–methanol–water–ammonia (200:100:10:0.02 by volume), with UV detection at 210 nm, to separate phosphatidylcholine, sphingomyelin and lysophosphatidylcholine, and to isolate fractions enriched in platelet-activating factor and its lyso derivative [506]. In systematic study of the properties of a phase of this type, it was shown that isocratic elution with hexane–isopropanol–aqueous piperazine acetate (5×10^{-3} M, pH 6.9) (6:8:0.15 by volume) gave good separations of phosphatidylethanolamine, phosphatidylglycerol, lysophosphatidylethanolamine, phosphatidylcholine and sphingomyelin [112]. In a systematic study of the properties of this stationary phase, it was observed that it was necessary to change the proportions of the solvents to 6:8:0.6 respectively before lysophosphatidylcholine, phosphatidylinositol (and phosphatidylserine) and phosphatidic acid were eluted in a reasonable time. This was still a less polar mobile phase than might be required with columns of silica gel for the same purpose.

In contrast, a column containing non-polar bonded phase (LiChrosorb™ RP-2) was utilized for the separation of the lipids of green leaves, by means of elution with acetonitrile–0.01 M phosphate buffer (pH 7.7; 9:1 by volume; the acidic lipids were seemingly separated from the glycolipids and from phosphatidylcholine and phosphatidylethanolamine [692]. Unfortunately, the paper lacks detail, but the use of such stationary phases may be worthy of further investigation.

Rather different selectivity is obtainable by making use of stationary phases which have a surface layer of amine groups covalently bound to a silica gel support. In one of the first reported applications, a column (6.4 × 300 mm) of *micro*Bondapak™-NH_2 was eluted with various chloroform–methanol–water mixtures, in order to separate phospholipid standards similar in nature to those found in soybean "lecithin" [427]. The separated components were detected and quantified by means of a Pye LCM2 transport-flame ionization detector. Retention times and the order of elution were highly dependent on the amount of water in the mixture. In particular, it was noted that phosphatidylcholine eluted before phosphatidylethanolamine, reversing the order found with adsorption chromatography on silica gel. Others used a column (4.6 × 150 mm), containing an aminopropyl-bonded phase of their own preparation, isocratic elution with acetonitrile–methanol–0.05 M acetic acid (10:5:0.4 by volume) at a flow-rate of 2 ml/min and spectrophotometric detection at 206 nm, in order to separate a standard mixture of phospholipids [112]. Figure 6.4 illustrates the nature of the separation. Phosphatidylcholine was separated from sphingomyelin and lysophosphatidylcholine, and they were followed by the ethanolamine-containing phospholipids. With a column containing a diamine-bonded silica gel, the retention time of phosphatidylethanolamine increased greatly, but the resolution of the choline-containing phospholipids did not improve. Marked changes in the selectivity of elution were observed for some phospholipids with an acetamido-bonded phase, when the nature and pH of the eluent were changed. A wider range of phospholipids have been separated on amine-bonded phases in gradient applications (see Sections B2 and C3 below).

One further type of bonded phase to have been used in phospholipid separations with isocratic elution had a benzene sulphonate residue as the functional group. A column (4.6 × 250 mm) of Partisil™-SCX and elution with acetonitrile–methanol–water (400:100:34 by volume) at a flow-rate of 2.5 ml/min were used to effect separation of the main ethanolamine- and choline-containing phospholipids of animal tissues as shown in Fig. 6.5 [168, 253]. Of those components not illustrated, phosphatidylinositol eluted

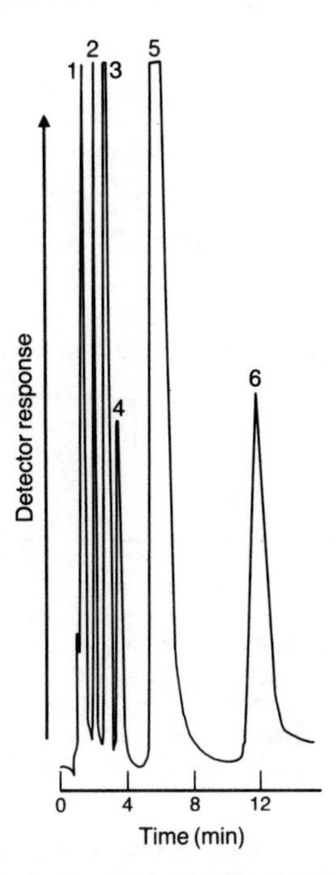

FIG. 6.4. Separation of phospholipid standards by HPLC on a column containing an amine-bonded phase, eluted isocratically with acetonitrile–methanol–0.05 M acetic acid (10:5:0.4 by volume) at 2 ml/min, and spectophotometric detection at 206 nm [112]. (Reproduced by kind permission of the authors and of *Annali di Chimica*, and redrawn from the original publication.) Abbreviations: 1, solvent front; 2, phosphatidylcholine; 3, sphingomyelin; 4, lysophosphatidylcholine; 5, phosphatidylethanolamine; 6, lysophosphatidylethanolamine.

with the solvent front, phosphatidylserine emerged as a broad "hump" underneath the phosphatidylethanolamine, and disphosphatidylglycerol was apparently not sufficiently soluble in the mobile phase to be eluted. While spectrophotometry at 203 nm was used to detect the components, phosphorus assay was preferred for quantification purposes. Recoveries were 91 to 95% of the material applied to the column, but it would not be surprising if some hydrolysis or transesterification of lipids occurred, as this has been observed by others with a stationary phase containing a similar functional group, but under circumstances that were admittedly somewhat different [6]. On the other hand, it did appear that a plasmalogen standard was not adversely affected, and good results were obtained in analyses of the lipids of rabbit myocardium.

An additional feature of Fig. 6.5, which should be noted, is that sphingomyelin eluted as a double peak, i.e. with two distinct maxima. This was because more separation according to the chain-lengths of the fatty acid constituents had occurred. Thus, molecular species containing predominantly C_{22} to C_{24} fatty acids were in the first peak, and those containing mainly

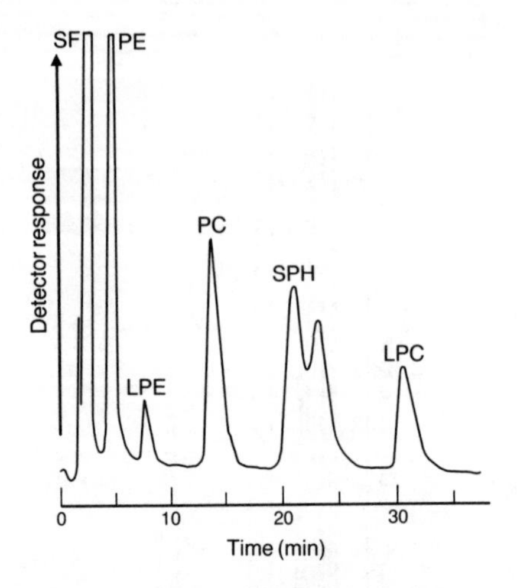

FIG. 6.5. HPLC separation of phospholipid standards on a benzene sulphonate-bonded phase, eluted isocratically with acetonitrile–methanol–water (400:100:34 by volume) at 2.5 ml/min, and with spectrophotometric detection at 203 nm [253]. (Reproduced by kind permission of the authors and of the *Journal of Chromatography*, and redrawn from the original publication.) Chapter 1 contains a list of abbreviations; SF, solvent front.

C_{16} fatty acids were in the second. The phenomenon occurs with many different elution systems (cf. Figs 6.6 and 6.7, for example).

B. HPLC – General Gradient Elution Procedures

1. Adsorption chromatography on silica gel

Gradient elution procedures offer some potential for the separation of a wider range of phospholipids in less time than has proved possible under isocratic conditions, although it is doubtful whether that potential will be fully realized until mass-selective detection systems become more widely available.

Some gradient elution procedures in which the individual simple lipid and phospholipid constituents of tissues were separated in a single analysis are described in Chapter 5. Although further detailed discussion would be be redundant here, it should be noted that certain of these methods contain some of the most comprehensive separations of phospholipids published to date. Indeed such methods can be useful, even when the analyst has no interest in the simple lipid constituents. For example, the phospholipids of milk comprise only about 0.5% of the total lipids, and in order to analyse them effectively, it was necessary to first obtain a concentrated fraction, by one of the methods described in Chapter 4, and then to subject this to HPLC as shown in Fig. 6.6 [149]. The chromatographic system employed the ACS mass detector and a ternary-gradient elution scheme as described elsewhere

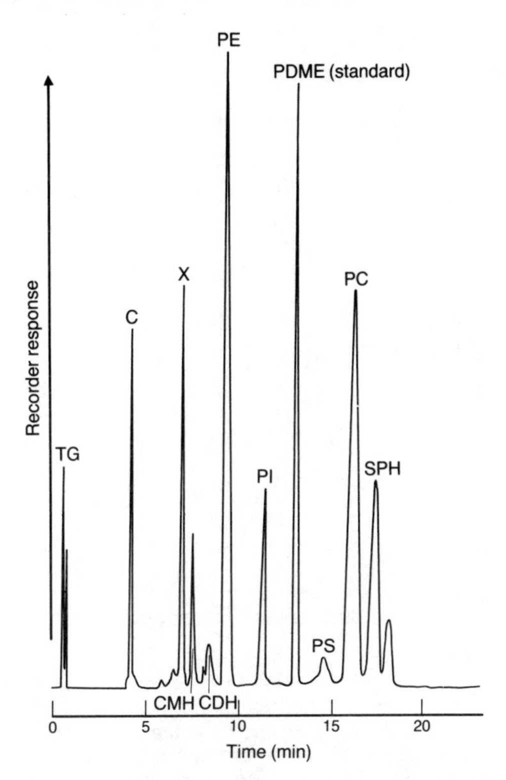

FIG. 6.6. HPLC separation of a phospholipid fraction from cows' milk on a column of silica gel, eluted with a ternary-gradient system (see Chapter 5 [136]) and with mass detection [149]. (Reproduced by kind permission of the authors and of the *Journal of the Society of Dairy Technology*.) Chapter 1 contains a list of abbreviations; X = an unidentified lipid.

[136, 139] (and also discussed in Chapter 5). A further novel feature of this work was the use of a synthetic phospholipid, phosphatidyldimethylethanolamine (dipalmitoyl), as an internal standard, in order that the absolute amount of phospholipid in the extract could be determined in addition to the relative compositions. Although this lipid is found naturally in some tissues, it is generally present at very low levels indeed so endogenous material does not interfere with the standard.

The acetonitrile–methanol–water elution system, described in Section A above in an isocratic application, gave somewhat better separations when adapted to gradient use [393]. A simple gradient elution system, consisting of water programmed in a linear manner from 2.5 to 15% into acetonitrile at a flow-rate of 1 ml/min, and a column (4 × 250 mm) of Bio-Sil™ HP-10 silica gel were used to remarkably good effect to separate the phospholipids of semen and spermatozoa, as shown in Fig. 6.7 [591, 592]. Although the base-line obtained with spectrophotometric detection at 203 nm was untidy, most of the phospholipids of interest were clearly separated, and with

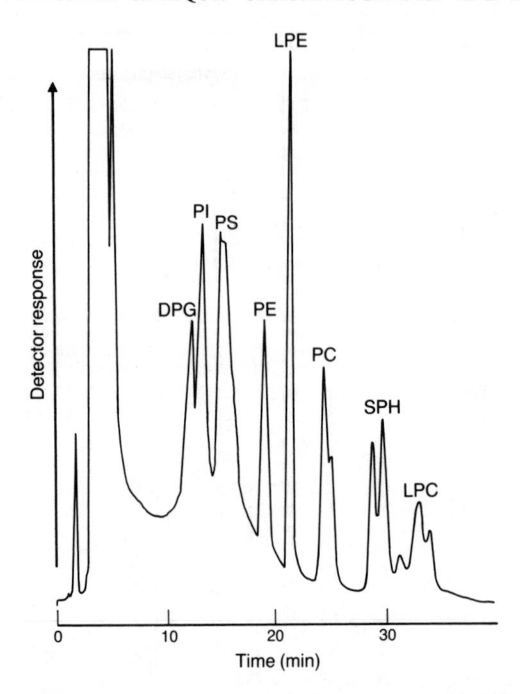

FIG. 6.7. HPLC separation of phospholipid standards on a column of silica gel, eluted with a gradient of water into acetonitrile, and with spectrophotometric detection at 203 nm [591, 592]. (Reproduced by kind permission of the authors and of *Fette Seifen Anstrichmittel*, and redrawn from the original publication.) Chapter 1 contains a list of abbreviations.

careful calibration they could be determined directly or they could be collected for subsequent analysis of their fatty acid components. Others successfully adapted a system of this type for the quantitative analysis of the phospholipids in human lung lavage [657], and with the addition of a little ammonia to the separation of various standard phospholipid mixtures [386].

More workers have tended to employ the alternative UV-transparent hexane–isopropanol–water mixtures in gradient elution. In the first such application [237, 300], these solvents were present initially in the mobile phase in the ratio 6:8:0.75 respectively by volume, and they were changed by a linear programme to 6:8:1.4. A column (6 × 250 mm) of LiChrosorb™ Si-60 was used, and the acidic phospholipids, which tend in general to cause most difficulties, eluted between phosphatidylcholine and phosphatidylethanolamine. Other workers have used this or closely related systems to separate, isolate and quantify (by a variety of methods) the main phospholipid components from tissue extracts of both animal and plant origin [37, 82, 119, 181, 196, 368, 376, 694, 728, 786]. In a systematic study [694] of how many different phospholipids (and their monoacyl forms) eluted with differing proportions of hexane, isopropanol and water (but no ionic species) in the mobile phase, it was concluded that one limiting factor would always be the inherent heterogeneity of the acyl group compositions in natural lipids, since this tended to cause broadening of peaks. In addition, it was observed that the column performance changed with time, as the resolution deteriorated and the retention times of phospholipids were reduced (a problem not confined to this elution system). Soyabean "lecithin" was the natural sample studied in this work. Some workers were apparently able to restore the activity of the silica gel after every 15 to 30 runs by eluting with methanol [196]; as the temperature of the column here was maintained at above ambient, this may also have assisted.

Other analysts have found it to their advantage to add sulphuric acid (0.02 to 0.1%) to the aqueous component of the solvent system, in order to suppress the multiple ionic forms of certain of the minor acidic phospholipids and thus to sharpen the separations of these components especially, or to alter their elution times [905, 906]. Unfortunately, plasmalogens were hydrolysed under these conditions. Although the quality of the resolution of phosphatidylinositol and phosphatidylserine tended to deteriorate with time, the column was restored by an acidic washing procedure, a finding confirmed elsewhere [694]; on the other hand, the author (unpublished observation) found that some residual acid on the silica gel caused breakdown of plasmalogen, when the column was used later with neutral mobile phases.

Equally good separations, but with much less corrosive ionic species in the mobile phase, have been reported by a number of other workers. For example, with a column (4.6 × 300 mm) of Partisil™ silica gel and a similar solvent gradient, but with a 1 mM acetate buffer (pH 6.0) as an ion-suppressant in the aqueous component, excellent resolution of isotopically-labelled phospholipids (containing ^{3}H-arachidonic acid) from incubations of animal tissue preparations was achieved, but presumably with no degradation of plasmalogen [259]; in this instance, aliquots of the eluent were removed for liquid-scintillation counting, which was also the detection system, as illustrated in Fig. 6.8. Eight different phospholipid classes were clearly resolved. As an alternative, a small amount of acetic acid (0.005%) has been incorporated into the aqueous component of the mobile phase, to assist in the isolation of platelet-activating factor especially, although good separations of other phospholipids were obtained at the same time [82].

Phospholipids labelled isotopically *in vitro* with ^{32}P tended to incorporate the label rapidly into some of the minor components, which are not easy to detect under normal circumstances. These were then separated rather well by a complex ternary gradient system, consisting of chloroform–propanol–acetic acid–water in various proportions, on a column (4.6 × 250 mm) of LiChrosorb™ Si 60–10, and they were monitored continuously by liquid-scintillation counting [84]. Only phosphatidylglycerol was poorly resolved, but an alternative gradient elution scheme, in which toluene, propanol, water and acetic acid were utilized, was devised to solve this problem. Because of

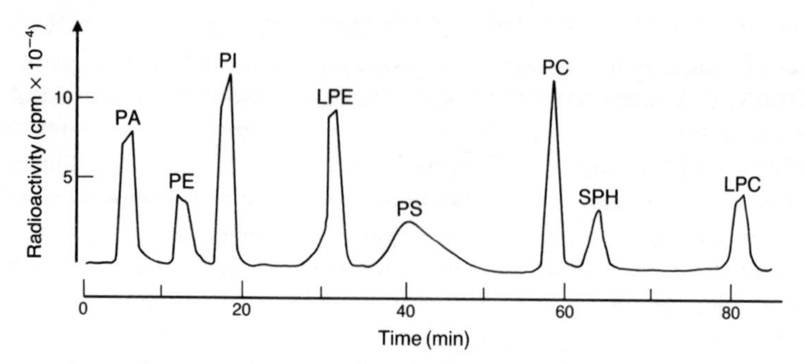

FIG. 6.8. HPLC separation of phospholipids containing ^{3}H-labelled arachidonic acid on a column of silica gel, eluted with gradients of mixtures of chloroform, isopropanol and 1 mM acetate buffer (pH 6.0); detection was by means of liquid-scintillation counting [259]. (Reproduced by kind permission of the authors and of the *Journal of Chromatography*, and redrawn from the original publication.) Chapter 1 contains a list of abbreviations.

the relatively large volume of the solvent mixing-chamber, a high flow-rate (2 ml/min) was required to ensure uniform formation of the gradient. A similar approach to detection (but with both continuous and discontinuous counting), but with a stepwise change in the composition of the mobile phase, has been employed by other workers [21, 84]. In essence, propanol–ethyl acetate–benzene–water (13:8:3:2 by volume) was utilized as the mobile phase to elute in order prostaglandins, phosphatidylethanolamine and phosphatidylinositol; then propanol–toluene–acetic acid–water (93:110:15:15 by volume) was used to bring off phosphatidylserine, phosphatidylcholine, sphingomyelin, platelet-activating factor and lysophosphatidylcholine (the flow-rate was also increased at various points in the elution sequence to speed up the analysis time).

HPLC with mass spectrometric detection of lipids, including phospholipids, was mentioned briefly in Chapter 5. In addition, this technique has recently been utilized with great success by Jungalwala, Evans and McCluer [381] to separate phospholipids and simultaneously to identify various molecular species. In the HPLC separation, a Brownlee cartridge (2 × 60 mm) of silica gel was used, with gradient elution with dichloromethane–methanol–water (93:6.5:0.5 by volume) (A) and dichloromethane–methanol–water–15 M ammonia (65:31:4:0.2 by volume) (B) as the components of the mobile phase, programmed from 12% B to 45% in 10 min then to 100% B in a further 2 min. The sample was equilibrated with a solvent containing ammonia prior to injection onto the column. The order of elution, indicated by the total ion current, differed somewhat from most others to have been described, i.e. phosphatidylethanolamine, phosphatidylcholine, phosphatidylinositol, sphingomyelin and phosphatidylserine (together with lysophosphatidylinositol). With repeated injections, reproducible retention times were obtained, although this could not apparently be achieved with solvents containing acetic acid. The HPLC eluent was applied as a fine stream to the moving belt of a Finnigan HPLC-MS interface, and was thence introduced into the mass spectrometer for chemical ionization (both positive and negative ion) with ammonia and methane as the reagent gases. Ions were obtained which were characteristic of the base component and perhaps more importantly of the diacylglycerol moieties of the phosphoglycerides, and of the ceramide moieties of sphingomyelin. Indeed sufficient structural information was obtained to permit identification and quantification of many distinct molecular species. For example, nineteen different molecular species of the phosphatidylcholine from lung were identified from its mass spectrum. Related work is discussed further in Chapter 8 (Section D.4), and it has been reviewed elsewhere [377, 503].

Although it will be many years before such equipment is available routinely to analysts, this work can perhaps be considered as pointer to the future.

2. *Chromatography on bonded phases*

Useful separations of phospholipids have been achieved on a column (4.6 × 250 mm) containing a diol-bonded phase, i.e. LiChrosorb™ DIOL, maintained at 35°C, and eluted with a gradient of water into acetonitrile (2.5 to 15%) at a flow-rate of 2 ml/min; UV detection at 203 nm was employed [92]. In less than 20 min, phosphatidylglycerol, phosphatidylinositol, phosphatidylethanolamine, phosphatidylcholine, sphingomyelin and lysophosphatidylcholine were clearly separated from each other, although phosphatidylserine tended to elute as a broad hump, partly under phosphatidylethanolamine (this order was very different from that obtained with hexane–isopropanol–water mixtures as the mobile phase (see Section A.3) [112]. The method could be used to determine phosphatidylglycerol and the ratio of phosphatidylcholine to sphingomyelin in amniotic fluid. Subsequently it was shown that much cleaner separations could be obtained with such columns by using 0.005 M sodium dihydrogen phosphate buffer (pH 5) as the aqueous component; in this way, sharper peaks were obtained with all the acidic phospholipids, including phosphatidylserine [449]. A similar order of elution, but with poorer resolution, was obtained in a study in which both a diol- and a nitrile-bonded phase were utilized [29]; the latter was preferred because it was apparently more versatile and was re-equilibrated with the initial mobile phase more readily, and because a lower proportion of water was required for elution of the most polar constituents. The use of nitrile-bonded phases, in particular, in phospholipid separations might well repay further investigation.

Amine-bonded stationary phase have also been used for phospholipid separations with gradient elution. For example, simple mixtures of choline- and ethanolamine-containing phospholipids were separated on a column containing Ultrasil™-NH_2, eluted with hexane–isopropanol–methanol–

water mixtures, in which the proportions of water and methanol were increased [282]. The order of elution was as described above (Section A.3), i.e. choline-containing phospholipids first, and the same was true when acetonitrile–methanol–water mixtures were the mobile phase [121]. Neither research group were able to obtain elution of the acidic phospholipids, which were probably held back by strong ionic bonds, from such columns; this was achieved by others [165] with 10 mM aqueous ammonium acetate as the mobile phase. Lipids tentatively identified as cerebrosides eluted ahead of phosphatidylcholine [282]. In continued use, an amine-bonded stationary phase gradually lost its activity and phospholipids were eluted with solvents of lower polarity, possibly because peroxides, ketones or aldehydes reacted slowly with the amine groups, rendering them inactive, or because of a steady accumulation of material which was strongly adsorbed [121]. It has been recommended that a silica precolumn be used to confer some protection to the amine phase [18].

C. Some Specific HPLC Separations

1. Phosphatidylcholine and related lipids

Procedures for the analysis or isolation of phosphatidylcholine [18, 575, 691], lysophosphatidylcholine [18, 144], and platelet-activating factor [21, 82, 84, 364, 506, 861], and for the determination of the ratio of phosphatidylcholine to sphingomyelin in amniotic fluid and other tissues [29, 92, 175, 376, 385, 627, 657] have been described in various sections above. In addition, methods for the purification of phosphatidylcholine, phosphatidylethanolamine and lysophosphatidylcholine on a large-scale by means of HPLC have been published [209, 238, 626].

A selective method for the determination of choline-containing phospholipids in lipoprotein fractions, separated by means of HPLC (see Chapter 10), involving a continuous reaction with an enzyme preparation (a commercial kit) and colour development in a post-column reaction-chamber, has been described [267, 600, 601].

2. Phosphatidylethanolamine and phosphatidylserine

By preparing suitable derivatives of the amine groups of phosphatidylethanolamine and phosphatidylserine, it is possible to separate them by HPLC for quantification by sensitive, specific spectrophotometric means. For example, the biphenylcarbonyl derivatives of these lipids, having molar extinction coefficients of about 23,000 at 268 nm, were prepared and separated on a column of silica gel [387]; isocratic elution with dichloromethane–methanol–15 M ammonium hydroxide (92:8:1 by volume) as the mobile phase was used for the phosphatidylethanolamine derivatives, and the same solvents in the ratio 80:15:3 were used for the derivatives of phosphatidylserine and lysophosphatidylethanolamine. Lipids were determined down to the 10 nmole level, but this could probably have been reduced futher by employing microderivatization techniques. In comparable work, trityl (and less usefully benzoyl) derivatives of phosphatidylethanolamine and phosphatidylserine were prepared for analysis [519].

More sensitive analysis still is possible if derivatives which can be detected by fluorimetry are prepared. For example, the dansyl derivatives of amino lipids were made and separated by means of an elution scheme similar to that just described, for detection at 342 nm (excitation) and 500 nm (emission) [123]. In this instance, the lower limit of detection was about 20 pmole with lipid extracts from rat brain. The succinimidyl 2-naphthoxyacetate derivatives were also prepared, and they were analysed in the same manner, with fluorescence detection at 228 nm (excitation) and 342 nm (emission), but faster elution conditions were possible [124].

3. Phosphatidylmono- and phosphatidyldimethylethanolamine

The partially methylated lipids, formed biosynthetically from phosphatidylethanolamine, have important biological functions. They can be separated with relative ease from each other and from the ultimate product, phosphatidylcholine, by means of HPLC on columns of silica gel eluted with chloroform–methanol–water–ethanolamine (77.8:20:2:0.2 by volume) [722, 723] or with hexane–isopropanol–water mixtures [136, 546]. With the first of these systems, the lipids were eluted in the order phosphatidylethanolamine, phosphatidyldimethylethanolamine, phosphatidylmonomethylethanolamine and phosphatidylcholine. With the latter mobile phase, the order of elution of the partially methylated compounds was reversed. These lipids have also been separated on a column containing an amine-bonded phase, eluted with acetonitrile–methanol–water mixtures [121]. In this instance, phosphatidylmonomethylethanolamine eluted ahead of phosphatidylethanolamine; acidic lipids would be strongly retained and would not interfere with the analysis. As the methylated intermediates tend to be present in tissues at very low levels indeed, detection has presented a problem and most workers have utilized radioactive tracers, with collection of discrete fractions of the HPLC eluent for liquid-scintillation counting, to identify fractions.

4. Determination of the plasmalogen content of phospholipids

A base-line separation of the ether and diacyl forms of phosphatidylethanolamine has been achieved in Horrocks's laboratory [196]. A column (4.6 × 250 mm) of Zorbax™ SIL, maintained at 34°C, was eluted with hexane–isopropanol–water (83:15.9:1.1 by volume), with UV detection at 205 nm. Indeed, the alkyl and alkenyl forms were partially resolved. Unfortunately, this approach is not likely to be successful for many other lipid classes.

The plasmalogen or vinyl ether bond in a phospholipid is hydrolyzed with relative ease by brief exposure to strong acid, with formation of an aldehyde and a lysophospholipid (see Chapter 3); with care, the diacyl and alkylacyl forms of phospholipids are not affected. By briefly exposing lipid extracts to fumes of hydrochloric acid, the plasmalogen bonds were hydrolyzed and the unchanged lipids, together with the lysophospholipids, were converted to UV-absorbing derivatives for separation and quantification by HPLC as described above (Section C.2) [387]. The relative proportions of the plasmalogen to diacyl (plus alkylacyl) forms in the phosphatidylethanolamine and phosphatidylserine of rat tissues were determined in this way. Subsequently, fluorescent derivatives of these lipids were analysed by the same technique [123, 124], as were native phosphatidylcholine and phosphatidylethanolamine [591, 592]. In an analogous approach, the proportions of phosphatidylcholine and phosphatidylethanolamine of rat heart were determined before and after acid treatment relative to the endogenous cholesterol in the extract, which was treated as an internal standard; the amount of the plasmalogen and non-plasmalogen form of each could then be calculated [136].

An HPLC procedure for the isolation and quantification of alkenylacyl, alkylacyl and diacyl forms of phospholipids as the diradylglycerol acetate derivatives is described in Chapter 8 below (Section E).

5. *Phosphatidylglycerol and diphosphatidylglycerol*

The content of phosphatidylglycerol in amniotic fluid, like the ratio of phosphatidylcholine to sphingomyelin, is an important indicator of fetal lung maturity. The elution scheme of Patton *et al.* [630] for isocratic separation of phospholipid classes has been adapted to the analysis of phosphatidylglycerol specifically [354]. In this work, an important feature was the use of a microbore column and the appropriate HPLC technology, i.e. a column (1 × 500 mm) of silica gel (10 μm particles) was employed, with the flow-rate of the mobile phase set at only 90 μl/min. Thus small amounts only of sample were needed, and the total consumption of solvent was not more than 500 ml per week. UV detection at 210 nm was utilized, no loss of sensitivity over conventional columns was observed, and good quantification was possible with careful calibration. The column lasted for 3 months and was used for 1000 samples before replacement.

Diphosphatidylglycerol from bovine heart tissue was purified on a preparative scale in the form of its calcium salt by partition column chromatography on silica gel, followed by preparative HPLC (300 mg scale) on a column (22 × 250 mm) of LiChrosorb™ Si 60–5 silica gel, eluted with isopropanol–cyclohexane–water (45:50:5 by volume) at a flow-rate of 17 ml/min, and with refractive index detection [773].

6. *Miscellaneous phospholipids*

Synthetic thio analogues of phosphatidylethanolamine and lysophosphatidylethanolamine were separated on a column of Ultrasil™ NH_2, eluted with acetonitrile–methanol–water (66.5:28.5:5 by volume) and monitored by UV spectrophotometry at 206 nm [613].

Phosphatidylmannose (and other phospholipids) was isolated by anion-exchange chromatography on a column of Mono Q HR 5/5™, with a gradient of 1 mM to 10 mM ammonium acetate in water as the mobile phase [165].

Synthetic phosphatidic acid (isotopically labelled with ^{32}P) was purified by HPLC on a column of silica gel with hexane–isopropanol–1.4 M phosphoric acid (90:10:0.5) at a flow-rate of 1 ml/min as the mobile phase [412]. The compound was detected by its absorbance at 210 nm, and by on-line liquid-scintillation counting.

Brief details have been published of a procedure for the separation of the acidic phospholipid fraction from a DEAE-cellulose column, i.e. phosphatidylinositol, phosphatidylserine and phosphatidylglycerol, as the benzoylated derivatives, by HPLC on a column of silica gel [377, 386].

Water-soluble intermediates in the biosynthesis of phospholipids (and the products of their catabolism) have been separated by ion-exchange chromatography, in the reversed-phase mode, or by adsorption chromatography on silica gel, and by elution with mobile phases in which buffers of increasing polarity and ionic strength were incorporated [394, 471, 489, 521]. Inositol cyclic phosphates, produced by the reaction of phospholipase C with polyphosphoinositides, were isolated by ion-exchange HPLC on a column of, for example, Partisil™ 10 SAX [158, 523, 880]. These compounds together with various glycerophosphoinositols have also been successfully separated by HPLC on a bonded-amine phase [75].

Sphingomyelin was separated from ceramide phosphorylethanolamine and from the phosphonolipids, ceramide 2-aminoethylphosphonate and ceramide *N*-methylaminoethylphosphonate, by means of HPLC on a column of Zorbax™ SIL silica gel, with acetonitrile–methanol–water (72:40:10 by volume) as the mobile phase and with spectrophotometric detection at 207 nm [356]. Using this method, the four components could routinely and reproducibly be identified and quantified in lipid extracts prepared from freshwater shellfish.

In a procedure reminiscent in many ways of the high-temperature GLC method described in Chapter 5 (Section D), the phospholipids of plasma were hydrolysed with phospholipase C and derivatized; the resulting mono-acylglycerol, diacylglycerol and ceramide derivatives, derived from monoacyl-glycerophospholipids, diacylglycerophospholipids and sphingomyelin respectively, together with cholesterol esters, were then separated by HPLC in the reverse-phase mode on a column containing an ODS-bonded phase, with detection by UV spectrophotometry at 230 nm [68] or by chemical-ionization mass spectrometry [453]. The nature of such separations is discussed in some detail in Chapter 8, but in brief information is obtained simultaneously both on the amounts and on the compositions of the molecular species of each lipid group.

7. *Plant glycosyldiacylglycerols and phospholipids*

The complex lipids in plants, especially those in leaf tissue, tend to contain (in addition to the phospholipids) appreciable amounts of glycolipids, including mono- and digalactosyldiacylglycerols and sulphoquinovosyldiacylglycerol. These present an additional complication in analysis. A number of procedures for the separation of components of the crude complex lipid fraction sold commercially as soyabean "lecithin" are described above, and glycosyldiacylglycerols would probably also be present. Again, some of the most comprehensive separations of plant complex lipids have been described by Privett and colleagues, and these are discussed in Chapter 5.

Mono- and digalactosyldiacylglycerols were found to elute in this order ahead of the individual phospholipids from a column of silica gel, with a gradient of isopropanol–hexane (4:3 by volume) to isopropanol–hexane–water (8:6:1.5 by volume), and with detection at 205 nm [181]. A similar elution scheme was employed for the analysis of soyabean complex lipids [786]. On isocratic elution with acetonitrile–methanol–sulphuric acid (135:5:0.2 by volume) as the mobile phase from a column of silica gel, phosphatidylinositol eluted first followed by monogalactosyldiacylglycerol, *N*-acylphosphatidylethanolamine, *N*-acyllysophosphatidylethanolamine, phosphatidylethanolamine, and digalactosyldiacylglycerols [515]. In addition, a preparative-scale HPLC method for the separation of digalactosyldiacylglycerols has been described [238]. Other workers used a column (4 × 250 mm) of LiChrosorb™ NH_2, eluted with a complex gradient of hexane, isopropanol, methanol and water [306]. In this instance, the monogalactosyldiacylglycerols eluted ahead of phosphatidylcholine, which in turn eluted some distance ahead of the digalactosyldiacylglycerols. The acidic lipids were of course retained on the column. UV detection at 200 to 205 nm was used in the above separations, and components were collected for analysis, or peak area measurements were utilized for quantification after careful calibration with real samples. HPLC in the reversed-phase mode has also been used for the purpose, but the elution pattern was complicated as molecular species were separated simultaneously (see Chapter 8) [838].

D. Alternatives to HPLC for the Analysis of Phospholipid Classes

1. TLC separation procedures

Methods for the separation, identification and determination of phospholipids, other than by HPLC, have been described comprehensively elsewhere [133]. On the other hand as in the previous chapter, it may be of interest to consider them briefly here, so that the HPLC methods can be viewed in a broader context. Preparative-scale column chromatography on silica gel and ion-exchange celluloses, procedures of value for the isolation of fractions enriched in particular components, are described in Chapter 4. Methods for the simultaneous separation of simple lipids and phospholipids are described in Chapter 5. Although the pattern is changing, TLC is still the most widely used technique for the separation of phospholipids on an analytical scale. One-dimensional TLC procedures can be recommended for the analysis of natural mixtures with relatively simple compositions for the most abundant components, for rapid group separations, and for small-scale preparative purposes (when it is intended that the fatty acid compositions of individual phospholipids be determined, for example).

When acidic phospholipids are present in a sample at low levels only, the common phospholipids may be separated on layers of silica gel G (with calcium sulphate as binder) by using chloroform–methanol–water (25:10:1 by volume) as the solvent mixture for development. The nature of the separation is shown in Fig. 6.9 (plate A). The minor acidic lipids, such as

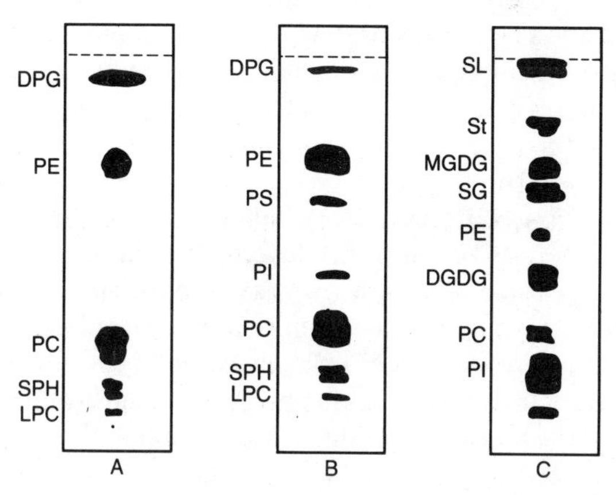

FIG. 6.9. Schematic TLC separations of phospholipids. Plate A, silica gel G layer and development with chloroform–methanol–water (25:10:1 by volume); Plate B, silica gel H layer and development with chloroform–methanol–acetic acid–water (25:15:4:2 by volume) [771]; Plate C, plant complex lipids on silica gel G and development with diisobutyl ketone–acetic acid–water (40:25:3.7 by volume) [585]. Chapter 1 contains a list of abbreviations; St, sterols; SG, sterol glycosides.

diphosphatidylglycerol, migrated ahead of phosphatidylethanolamine and the choline-containing phospholipids; phosphatidylserine tended to elute with phosphatidylethanolamine while phosphatidylinositol co-chromatographed with phosphatidylcholine.

In most circumstances, however, it is preferable to employ layers of silica gel H (without a binder) as sharper separations of most of the individual phospholipid classes, but especially of the minor acidic components, are obtained in the absence of metal ions. Unfortunately, silica gel H layers tend to be more fragile than those prepared from silica gel G, and different brands can vary greatly in their elution characteristics. The most popular separation system of this kind (especially for the lipids from animal tissues) consisted of a layer of silica gel H, made in a slurry with 1 mM sodium carbonate solution to render it slightly basic, and developed in chloroform–methanol–water–acetic acid (25:15:4:2 by volume); the separation is shown in Fig. 6.9 (plate B) [771]. Diphosphatidylglycerol and phosphatidic acid moved to the solvent front, and phosphatidylserine and phosphatidylinositol migrated between the most abundant phospholipid constituents. If simple lipids were present in the sample, they could be run to the top of the plate by elution with acetone-hexane (1:3 by volume), prior to separation of the phospholipids [769]. Many modifications of this system have been described, often to compensate for local conditions of temperature and humidity, or for changes in the properties of particular makes of silica gel. Problems are most often manifested in the separation of phosphatidylserine and phosphatidylinositol from the other lipids, and some commercial brands of silica gel H appear to be better than others for the purpose; the author has recently obtained good results with Kieselgel D (Riedel-De Haen, Hanover). Others have reported improved separations by incorporating either ammonium sulphate [408] or EDTA [23] into the silica gel H layer. The main disadvantage of the method is that phosphatidic acid, phosphatidylglycerol and diphosphatidylglycerol are not separated from each other, but a modified elution system has been described for this purpose [770].

One recently described procedure, which did appear to afford distinctive separations, made use of silica gel layers containing boric acid, which presumably forms complexes with hydroxyl groups such as those in phosphatidylinositol, retarding the rate of migration of this lipid class; the order of elution was thus very different from those obtained with other systems, and both phosphatidylserine and phosphatidylinositol especially were separated with relative ease from the other phospholipids [217].

One-dimensional TLC systems have been used less often with plant lipid extracts, as glycolipids tend to co-chromatograph with phospholipids when many of the common elution systems are used. Nonetheless, some valuable separations have been described [554, 585, 665], and one is illustrated in Fig. 6.9 (plate C). Mono- and digalactosyldiacylglycerols each eluted as distinct bands.

The applications and limitations of the Iatroscan TLC Analyser are described in Chapter 5, specifically with respect to simple lipid classes. The equipment has also been utilized for the separation and quantification of individual phospholipid classes with apparent success in some hands [51, 219, 319, 352, 443, 845].

The principles and practice of two-dimensional TLC in general have recently been reviewed [912]. In most instances, many more distinct phospholipid classes can be separated by two- than by one-dimensional TLC procedures, and this is of special value with some of the biologically-important phospholipids, which are often present in tissues at rather low levels. In the former technique, the sample is applied as a spot to the bottom left-hand corner of the plate, which is developed in one direction as normal before the plate is dried, turned anticlockwise through 90° and developed again with a different mobile phase. The most useful separations are achieved when contrasting solvent mixtures are employed for development in each direction. For example, a neutral or basic solvent in the first direction may be followed by a development with an acidic solvent in the second direction. Rouser and colleagues [711] have described a number of solvent combinations of particular applicability to animal lipids, that appear to have stood the test of time. They favoured four solvents mixtures:

chloroform–methanol–water	(65:25:4 by volume)
butanol–acetic acid–water	(3:1:1)
chloroform–methanol–28% aqueous ammonia	(13:7:1)
chloroform–acetone–methanol–acetic acid–water	(10:4:2:2:1)

In preparing the TLC chromatogram shown in Fig. 6.10, a layer of silica gel H containing 10% magnesium silicate as binder was employed, and the first of these solvents was used in the first direction while the second solvent was used in the second direction.

Very many such systems have been described, too many indeed to be evaluated critically, although methods with specific applications in the separation of phosphatidic acid [702] and phosphatidylglycerol [670] may be worthy of further examination; layers impregnated with boric acid have also been used in this way [217]. In addition, smaller than usual (10 × 10 cm) high-performance TLC plates have been utilized to obtain excellent two-dimensional separations of brain lipids [669]. Distinctive two-dimensional TLC procedures have been developed to separate the complex phospholipids and glycolipids of plant tissues [586, 766], and such elution systems have tended to be applied to lipid extracts of bacterial origin also.

With adsorbent layers of 0.5 mm in thickness, up to 3 mg of phospholipid can be separated on a single 20 × 20 cm TLC plate, either in one- or two-dimensional systems for the isolation of specific components. On the other hand, adsorbent layers of 0.25 mm in thickness are preferable in analytical separations, when microgram amounts of lipids can be quantified.

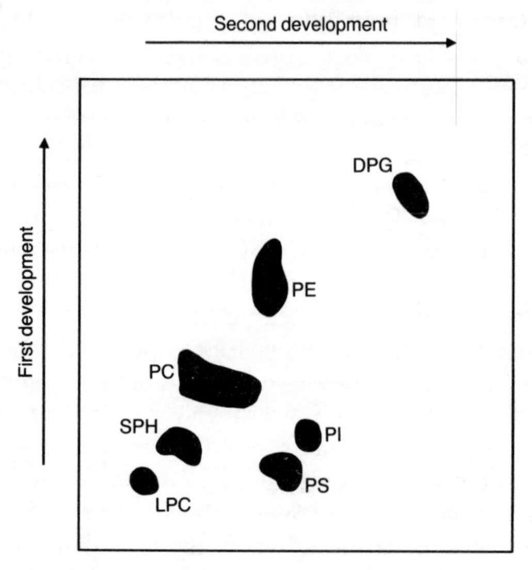

FIG. 6.10. Schematic two-dimensional TLC separation of phospholipids on a silica gel H plate containing 10% magnesium silicate (w/w) as binder. The mobile phase in the first direction was chloroform–methanol–water (65:25:4 by volume), and that in the second was *n*-butanol–acetic acid–water (3:1:1 by volume) [711]. Chapter 1 contains a list of abbreviations.

The principal disadvantages of two-dimensional TLC techniques is that the precise orientation of spots of lipid on the plate can vary greatly with small changes in the properties of the adsorbent, or in the temperature or humidity in the laboratory. In addition, the double development is time-consuming so that the technique is used by many research groups only for preliminary investigations of samples or for checks on purity of lipids isolated by other methods, while one-dimensional techniques are preferred for routine analyses.

2. *Methods for the identification and quantification of phospholipids separated by TLC*

Procedure for locating and quantifying lipids separated by TLC are discussed in Chapter 5 and elsewhere [133], and many of these can be applied to phospholipids. In addition, phosphorus assay is an important and specific method for the determination of phospholipids (see Section E below). As highly polar non-volatile solvents are used in phospholipid separations, great care must be exercised to remove all of these by evaporation in a stream of nitrogen or by placing the plate in a vacuum oven before attempting to locate the lipids. Excess of acid can be removed by neutralization with an ammonia spray. When non-destructive spray reagents such as 2′7′-dichloro-fluorescein solutions are used, the areas of silica gel containing each lipid can be scraped from the plate and the components recovered by elution with chloroform–methanol–water (5:5:1 by volume) for determination or for further analysis (e.g. of their fatty acid constituents). One virtue of TLC procedures is that a variety of different spray reagents can be employed to detect and identify particular structural moieties in phospholipids. For example, a reagent containing molybdate is commonly used to confirm the presence of the phosphate group [717], and to distinguish between phospholipids and phosphonolipids [791]. Phospholipids containing a choline moeity can be identified by means of the Dragendorff reagent [860], while a ninhydrin spray is specific for the free amine group in phospholipids such as phosphatidylethanolamine and phosphatidylserine. A periodate-Schiff reagent can be used to detect vicinal diol groups, as in phosphatidylinositol and phosphatidylglycerol [755], while a number of reagents have been employed to detect the carbohydrate moieties in the glycophospholipids of bacteria, or to distinguish between the more common glycolipids and phospholipids [371, 765, 808].

3. *Some special problems*

Methods for the quantitative extraction of the biologically-important polyphosphoinositides, i.e. phosphatidylinositol 4-phosphate and phosphatidylinositol 4, 5-bisphosphate, from tissues are discussed in Chapter 4. As these compounds are present in tissues at very low levels, separation and detection can be difficult. Often, it is necessary to use lipids labelled isotopically with ^{32}P and to locate the compounds on TLC plates by radiography. Many different TLC elution systems have been described [133], but one which appeared particularly promising used a one-dimensional approach, but with a double development; the plate was placed first in chloroform–methanol–ammonium hydroxide–water (60:35:2.5:2.5 by volume) to move the monophosphorylated lipids into the upper part of the plate, and then it was developed in *n*-propanol–4.3 M ammonium hydroxide (13:7 by volume) to separate each of the polyphosphoinositides [256]. A rather more elegant method utilized small columns of neomycin reductively-coupled to glass beads; most phospholipids were eluted from such columns with solvents of low ionic strength, then each of the polyphosphoinositides was eluted in turn with solvents of increasing ionic strength [727].

Methods for the separation of phosphonolipids from phospholipids have been reviewed by Moschidis [555].

The separation of the alkyl and alkenyl forms of phospholipids from the generally more abundant diacyl form is rather difficult task, and often it can only be accomplished following chemical modification of the original lipid. Such methods have been reviewed in some detail elsewhere [133, 170], and they can only be considered briefly here. An HPLC procedure is described above, and this is also discussed further below (Chapter 8, Section E). Hydrolysis procedures have been used to isolate the various types of alkyl moeity for quantification, or the lipids have been converted to non-polar derivatives and each form has then been isolated by TLC with multiple developments. In a promising newer approach, phospholipids were hydrolyzed to the diradylglycerols by means of phospholipase C, before the alkylacyl, alkenylacyl and diacyl forms were separated from each other by gel-permeation chromatography on a column of Lipidex-5000™, eluted with heptane–toluene (19:1 by volume) [169].

E. Phosphorus Determination

Phospholipids separated chromatographically can be estimated by determining their phosphorus content. The traditional methods have involved digestion of the phospholipids by refluxing in perchloric acid in order to release inorganic phosphate; this was reacted to form phosphomolybdic acid, which was reduced to a blue compound for spectrophotometric determination [55, 133, 189]. Because of the real hazards to the analyst associated with digestion with such a strong acid, many workers have attempted to develop modified procedures which avoid this step. The following method of this kind has so far been tested with a relatively limited range of samples, but it should have wider applicability [343, 720].

> "To prepare the chromogenic solution, *solution 1* was made by dissolving ammonium molybdate (16 g) in water (120 ml), and *solution 2* was made by shaking *solution 1* (80 ml) with concentrated hydrochloric acid (40 ml) and mercury (10 ml), and decanting the supernatant; the remaining *solution 1* was diluted with concentrated sulphuric acid (200 ml), it was combined with *solution 2* to give *solution 3*, and a portion of the last (25 ml) was added to methanol (45 ml), chloroform (5 ml) and water (20 ml). This final chromogenic solution could be stored at 4°C for long periods. Phospholipid extracts, blanks and standards were dried down, then they were dissolved in chloroform (0.4 ml), the chromogenic solution (0.1 ml) was added, the solutions were thoroughly mixed and they they were heated at 100°C for 1.25 min. The samples were allowed to cool to room temperature, remixed and left for 5 minutes, before nonane (2 ml) was added and the whole left for a further 15 min. After centrifuging the tubes at 2500 rpm for 3 min, the absorbance of the supernatant fluid at 730 nm was compared with the blank. The results were related to the amounts of lipid originally present by reference to suitable calibration curves, prepared from known phospholipid standards."

The coloured complex formed during the last stage of the procedure was stable for at least 12 hr, so readings did not have to be taken immediately. The method was not found to be affected by cholesterol, glycerol or inorganic phosphorus.

CHAPTER 7

The Analysis of Fatty Acids

A. Introduction

Fatty acids are the distinctive structural components of lipids, and as their compositions and biological functions can vary widely in tissues, methods for their isolation and analysis are of great significance. It was the invention of gas–liquid chromatography that revolutionized the analysis of fatty acids, and it is the author's opinion that this will remain the technique of choice, for routine analytical purposes with the more common fatty acids especially, for some time to come. The equipment required can seem inexpensive, when compared to HPLC, and the running costs are relatively low; modern open-tubular GLC columns of fused silica offer very high resolution, together with high sensitivity and excellent reproducibility in quantitative analyses. On the other hand, it should be recognized that individual fatty acids can only be identified tentatively by GLC alone, and that it is often advisable to confirm the identity of particular components by unequivocal degradative and spectroscopic procedures. To this end, it is first necessary to isolate the individual pure fatty acids, and until recently this has been the province of preparative GLC, or of TLC with adsorbent layers of silica gel or of silica gel impregnated with silver nitrate. The first of these techniques has many disadvantages, and now need only be considered as a last resort, while the second still has many applications but can be somewhat messy. Nowadays, HPLC methods should almost certainly be preferred for the isolation of specific fatty acids on a small scale, whenever the appropriate equipment is available.

In addition, because the separation are carried out at ambient temperature, HPLC methods do have a number of valuable applications in analysis *per se*. For example, fatty acids with labile functional moieties, such as hydroperoxide groups or cyclopropene rings, can often be safely analysed by this means, whereas unwanted rearrangements occur at the high temperatures required for GLC analysis. High specificity and sensitivity is possible if fluorescent derivatives of fatty acids are prepared for HPLC analysis (down to femtomole levels), and this can sometimes be the best method available for very small samples.

Before the fatty acids from an extract or a purified lipid can be separated and analysed, they must first either be hydrolyzed to the free (unesterified) form or be converted directly or indirectly to suitable derivatives of low molecular weight and polarity. The methyl ester derivatives are the simplest in structural terms, and they have proved of great value with GLC, where their properties are very well documented and understood. However, ester derivatives with UV-absorbing or fluorescent chromophores in the alcohol moiety may be of greater utility in HPLC analysis. The ideal derivative should have a high molar extinction coefficient (above 10^4) at an absorption maximum of about 254 nm, and the alcohol moiety should be as non-polar and as unobtrusive chromatographically as possible. Generally, this means an aromatic chromophore with a suitable functional group attached. HPLC in one of two modes then tends to be used for the separation of the fatty acid derivatives, i.e. with silver ion complexation to separate according to degree of unsaturation, or by the reversed-phase technique (almost always on an ODS phase) to separate both by chain-length and degree of unsaturation. On their own, each of these techniques has certain disadvantages; together, they make a rather powerful combination.

Once individual fatty acids or their derivatives have been isolated, by whatever means, a variety of methods are available for structure elucidation and some of the more valuable of these are described below. At all stages in the analysis of fatty acids, it is important to take steps to minimize autoxidation by handling samples under a blanket of nitrogen and by the use of antioxodants, and this is especially true of purified lipids or fatty acids where the natural tissue antioxidants may have been removed during chromatography. These precautions are discussed at greater length in Chapter 4. Methods for the separation, identification and quantification of fatty acids have been reviewed elsewhere [133, 452, 485].

B. Hydrolysis of Lipids and Preparation of Fatty Acid Derivatives

1. Hydrolysis

Fatty acids can be released from glycerolipids with relative ease by hydrolysis (or saponification) of the ester bonds under basic conditions. Any lipid-soluble non-saponifiable materials (e.g. cholesterol, glycerol ethers, hydrocarbons) can be removed if necessary, when the reaction has been completed, by extraction of the basic hydrolysis mixture with diethyl ether, prior to acidification to obtain the free fatty acids. The following method is suitable.

> "The lipid sample (up to 100 mg) is dissolved in a solution of 1 M potasium hydroxide in 95% ethanol (2 ml), and the whole is refluxed for 1 hour (or is left overnight at room temperature). After cooling, water (5 ml) is added, and the solution is extracted with diethyl ether (10 ml) to remove any non-saponifiable material, with centrifugation if necessary to break any emulsions which might form (this step can be

omitted if the unwanted by-products are soluble in the aqueous layer). The aqueous layer is acidified with dilute hydrochloric acid before it is extracted with diethyl ether (3 × 5 ml). The combined ether layers, which then contain the free fatty acids, are washed with water (5 ml), and are dried over anhydrous sodium sulphate before the solvent is removed."

A somewhat longer reaction time might be necessary to effect complete hydrolysis of cholesterol esters. The volumes of the reagents can be scaled to suit the size of the sample. In an alternative procedure, free fatty acids and non-saponfiables were separated on an ion-exchange column [916]. The amide bonds in glycosphingolipids are virtually unaffected by the above method, but procedures suitable for these lipids are available when required (see Chapter 9).

2. *Preparation of methyl and other alkyl esters*

There is no need to hydrolyze glycerolipids prior to preparation of the methyl ester derivatives, as most can be transesterified by simple direct methods, which have been reviewed in some detail elsewhere [130, 133, 173]. No single reagent can be used in all possible circumstances, however. Both acidic and basic transesterification procedures are available, with the former being preferred when unesterified fatty acids are present at appreciable concentrations in samples. On the other hand, basic transesterification is particularly rapid and convenient, and it is especially suited to lipids, which have already been purified by chromatographic means. Care should be taken in removing solvents by evaporation in a stream of nitrogen, as appreciable amounts of esters of fatty acids up to C_{14} in chain-length can be lost if this is done too vigorously. The author developed the following alkaline transesterification procedure, and has been using it successfully for some years [134].

"The glycerolipid (1 to 10 mg) is dissolved in sodium-dried diethyl ether (1 ml) in a stoppered test-tube, and methyl acetate (25 μl) is added to suppress the competing hydrolysis reaction. 1 M sodium methoxide in dry methanol (25 μl) is added, and the solution is swirled briefly at room temperature to mix the reactants. The solution generally becomes cloudy as sodium-glycerol salts are precipitated. After 5 min, the reaction is stopped by adding acetic acid (6 μl), and the mixture is centrifuged at about 1500 g for about 2 min to precipitate the solid by-products. The solvent is removed in a gentle stream of nitrogen, taking care not to blow the precipitate from the tube, then hexane is added and an aliquot can be taken directly for injection onto a chromatographic column, or the solution containing the methyl esters can be decanted and stored until it is required."

A longer reaction time (1 hr) is necessary for cholesterol esters. If esters with short chain-lengths ($< C_{12}$) are present, the step involving evaporation of solvent should be omitted. The above method has also been used to prepare butyl and benzyl esters, by substituting the appropriate alcoholic reagents [142].

Acidic transesterification reagents are required to esterify free fatty acids, and to transesterify lipid samples containing such constituents. The following method is suitable.

> "The lipid is dissolved in toluene (1 ml) in a stoppered test-tube, and methanol (2 ml) containing 1% of sulphuric acid is added. The mixture is left overnight at 50°C, or alternatively it is refluxed for 2 hr. When the reaction is completed, the mixture is cooled and diethyl ether (10 ml) is added, followed by water (5 ml). After thorough shaking, the ether layer is removed, and it is washed with dilute potassium bicarbonate solution (5 ml), then with water (5 ml). After it is dried over anhydrous sodium sulphate, the solvent is evaporated, and the required methyl esters are taken up in hexane for storage or analysis."

Again, other alkyl esters can be prepared by substituting the appropriate alcohol into the transesterification reagent.

3. *Phenacyl and substituted-phenacyl esters*

A large number of different types of fatty acid ester, that contain UV-absorbing or fluorescent chromophores in the alcohol moiety, have been prepared to simplify the problem of detection in separation and analysis by means of HPLC. There is no true consensus as to which of these is the most suitable, but increasing numbers of analysts appear to favour phenacyl or related derivatives, such as *p*-bromo- or *p*-methoxyphenacyl esters, because of the ease of preparation of such compounds and because of their relatively-high molar extinction coefficients in the UV region of the spectrum. The preparation of these derivatives only is discussed in detailed here, and it has also been reviewed by Lam and Grushka [474]; methods for UV-absorbing derivatives in general have been reviewed elsewhere [388] and with many, similar reaction conditions are employed to those required for phenacyl derivatives. Two methods are in general use and in common with most others, it is necessary for the fatty acids to be in the unesterified state initially. The first method makes use of a tertiary amine to catalyze the reaction between the acid and phenacyl bromide and to scavenge hydrogen bromide from the medium, as shown in Fig. 7.1. The following method is suitable for the purpose.

> "Stock solutions of phenacyl bromide, purified by recrystallization from pentane if necessary, in acetone (10 mg per ml) and of triethylamine in acetone (10 mg per ml) are prepared. These reagents (25 μl of each) are added to the fatty acids (0.1 mg) in a test-tube fitted with a PTFE

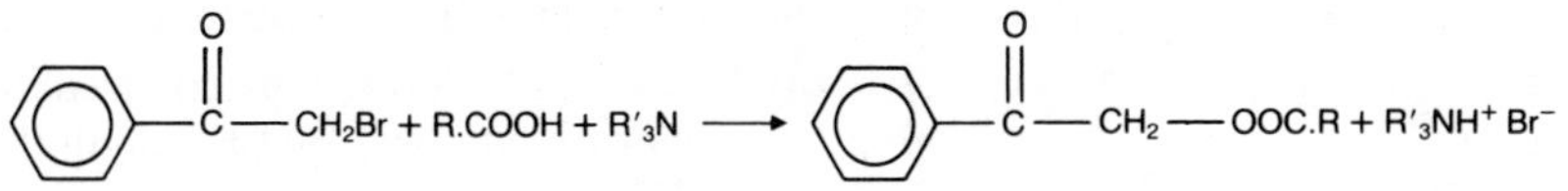

$$C_6H_5\text{—}\overset{O}{\overset{\|}{C}}\text{—}CH_2Br + R.COOH + R'_3N \longrightarrow C_6H_5\text{—}\overset{O}{\overset{\|}{C}}\text{—}CH_2\text{—}OOC.R + R'_3NH^+ Br^-$$

FIG. 7.1. The preparation of phenacyl derivatives of fatty acids.

screw cap. The mixture is heated in a boiling water bath at 100°C for 15 min, then acetic acid (3.5 μl) is added to react with the excess reagent, and the whole is heated for a further 5 min. The solvents are evaporated in a stream of nitrogen, and the required phenacyl esters are redissolved in an appropriate solvent, e.g. acetonitrile, for analysis."

In the alternative method, a crown ether is used as a phase-transfer catalyst to solubilize potassium salts in an organic solvent; the anions formed are highly reactive and so are readily alkylated [198]. The reaction is carried out in the following manner.

"The alkylating agent consists of phenacyl bromide and 1, 4, 7, 10, 13, 16-hexaoxacyclooctadecane ("18-crown-6") in a molar ratio of 10:1 in acetonitrile solution (1 mg/ml). This reagent (1 ml) and potassium carbonate (20 mg) are added to the fatty acid (0.1 mg) in a stoppered test-tube, and the mixture is heated at 80°C for 30 min. The solution is cooled, and it is filtered carefully before an aliquot is injected onto the HPLC column."

Both of the above methods can be used to prepare substituted phenacyl esters by replacing phenacyl bromide with the appropriate reagent. *p*-Bromophenacyl and naphthacyl esters can be analysed with greater sensitivity, but by-product peaks can occasionally be troublesome, overlapping with the analytes. Isomerization of *cis*-double bonds to the *trans*-configuration was observed in unsaturated phenacyl esters, when these were exposed to UV light during purification on a TLC plate [889]. A modified procedure in which hydrolysis of lipids and phenacyl ester formation were carried out sequentially in a single vessel has been described [416].

p-Bromophenacyl esters have been hydrolyzed back to the free fatty acids by stirring with zinc dust in chloroform–acetic acid (2:1 by volume) for 2 hours [683].

C. Separation of Fatty Acids by HPLC with Silver Ion Complexation

In silver ion (or argentation) chromatography, the facility with which the double bonds in the alkyl chains of fatty acids form polar complexes reversibly with silver compounds is exploited as a method of separation. Unsaturated compounds are fractionated according to the number, configuration (*cis* or *trans*) and on occasion the position of their double bonds. General reviews of this and related procedures have been published [258, 548]. In much of the published work with this technique, silver nitrate has been incorporated into silica gel layers on TLC plates, but in principle almost any chromatographic procedure can be adapted to the purpose. HPLC has been no exception. It has been used in three ways, i.e. with silica gel adsorbents impregnated with silver nitrate, with silver ions bound to ion-exchange resins, and in the reversed-phase mode with silver ions in the mobile phase (acetonitrile is an excellent solvent for silver nitrate). As yet, no single method can be said to be a clear favourite and preferable to the others, although the ion-exchange procedures appear to the setting the pace. To gain general acceptance, any method which employs silver ion complexation must work efficiently with real samples, not just with model compounds; it must be proved over a reasonable period of time, as silver ions can be lost from columns in some circumstances. Ideally, it should be adaptable to commercial prepacked columns, since only a few analysts have access to equipments for packing columns with special adsorbents.

In the light of the success of silver ion TLC, it was inevitable that analysts would make use of silica gel impregnated with silver nitrate in HPLC columns. Generally, the adsorbent has been prepared by making it into a slurry with a solution of silver nitrate in a suitable solvent, before taking the whole to dryness on a rotary evaporator or by other means. It appears that the precise manner in which this is done can have an important bearing on the nature of the subsequent separations. In one of the first published papers on the subject, Corasil™ II silica gel was coated with 0.8% silver nitrate and 1.75% ethylene glycol; the methyl ester derivatives of stearic, elaidic and oleic acids were then separated in this order by elution with hexane–heptane (1:1 by volume), and with refractive index detection [529]. Heath and co-workers [303, 304] used a variety of silica gel adsorbents, differing in make and particle size, with increasing loadings of silver nitrate, and packed by various methods, to demonstrate remarkable separations, both on an analytical and a preparative scale, of geometrical isomers of mono- and dienoic alkenol acetates. The optimum resolution was obtained with 5 μm silica gels and a 5% loading of silver nitrate, and with benzene as the mobile phase (refractive index detection). The use of silica gel particles with relatively high pore-sizes also helped. Unfortunately, silver nitrate leached slowly from the columns, and it was necessary to wash fractions of interest with an aqueous sodium chloride solution to remove it. Scholfield [733] prepared a column of silica gel, impregnated with silver nitrate, in the same way, and also obtained excellent separations of the methyl ester derivatives of geometrical isomers of unsaturated fatty acids (with benzene as the mobile phase). However, he noted that the resolution and the stability of the base-line deteriorated after about 50 samples had been run through the column. LiChrosorb™ Si 60, impregnated with 20% by weight of silver nitrate from an aqueous slurry, has been used to separate oxygenated unsaturated fatty acids (see Section F.1 below) [295].

Silver halides were found to be deposited fairly uniformly on silica gel adsorbents, but did not form sufficiently strong complexes with olefins to be of separatory value [11]. This also appeared to be true of silver acetate and sulphate (Christie and Stefanov, unpublished work).

More consistent separations of both positional and geometrical isomers of unsaturated fatty acids (methyl ester derivatives) were obtained on Spherisorb™ S5W silica gel with a 5% loading of silver nitrate, prepared from an aqueous slurry [60]. Figure 7.2 illustrates the nature of the separations attainable with a test mixture. A surprisingly good resolution of the complex range of isomers present in a margarine sample was also obtained (a 20% loading of silver nitrate gave the best results). In this instance, hexane–tetrahydrofuran (100:0.6 by volume) at a flow-rate of 2 ml/min was the mobile phase, with spectrophotometric detection at 205 nm. Although column performance deteriorated after about 20 analyses, it could be restored by flushing the column with pure tetrahydrofuran. The *trans* (*E*) isomers eluted ahead of the *cis* (*Z*) isomers, as in other forms of silver ion chromatography. A column prepared in a very similar way was used to separate the polyunsaturated fatty acids (3 to 6 double bonds) of cod liver

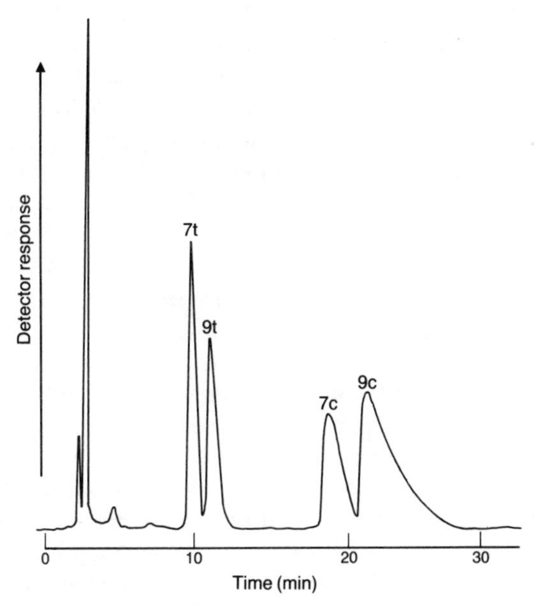

FIG. 7.2. HPLC separation of methyl octadecenoates (geometrical and positional isomers) on a column of silica gel impregnated with 5% by weight of silver nitrate, and eluted with hexane–tetrahydrofuran (99:1 by volume) with spectrophotometric detection at 205 nm [60]. (Reproduced by kind permission of the authors and of *Chromatographia*, and redrawn from the original paper.)

oil in the form of the methyl ester derivatives on a milligram scale, by elution with hexane–acetonitrile (100:0.4 by volume) [616]. Again, difficulties were experienced with silver ions leaching from columns prepared in this way, and this was confirmed by others [581]. Although this could be prevented by saturating the mobile phase with silver nitrate by means of a precolumn, there could be deleterious effects on the detector in the long term.

While only a limited range of unsaturated fatty acid isomers were separated in a systematic way in these and other studies, it appears probable that there would be a sinusoidal relationship between the retention volume and the number of carbon atoms between the double bond and the carboxyl group in positional isomers of the same chain-length, by analogy to studies with TLC adsorbents impregnated with silver nitrate [129, 261, 550]. In addition, a fatty acid derivative with an acetylenic bond (or an allenic double bond system) should elute just ahead of a double bond of the *cis*-configuration; thus, methyl stearolate (octade-9-ynoate) was found to migrate just before methyl oleate [549], while methyl crepenynate (octadeo-*cis*-9-en-12-ynoate) was slightly less polar than methyl linoleate [262] on layers of silica gel impregnated with silver nitrate. A *cis, trans*-conjugated double bond system resembled a single *cis*-double bond in similar circumstances [131]. It is worth noting that compounds not separable with aliphatic mobile phases can sometimes be resolved with aromatic solvents and *vice versa*.

More recently, it was claimed that a standard prepacked column of silica gel could be impregnated satisfactorily by pumping a solution of silver nitrate in acetonitrile through the column followed immediately by hexane, which presumably caused the salt to precipitate [547]. Although this method could potentially assist those analysts who are unable to pack their own adsorbents, the author would be concerned that some of the silver salt might precipitate prematurely in the solvent delivery system.

One disadvantage of columns prepared by any of the above methods appears to be that they do not have a long working life (see also Chapter 8), mainly because the silver ions are gradually leached from the adsorbents. Some analysts have therefore attempted to overcome the problem by using cation-exchange resins as support materials, since much more energy is required to break an ionic bond to a silver ion. To a considerable extent, they have succeeded although there appears to be uncertainty as to the best exchange resin to use for the purpose.

Research workers at the Northern Regional Laboratories (USDA) in Peoria have published a number of papers over a period of more than 20 years in which macroreticular cation-exchange resins, with sulphonic acid residues as the functional groups, have been loaded with silver ions for the separation of the methyl ester derivatives of unsaturated fatty acids. Silver ions did not leach to a significant extent, so columns could be re-used many times, while isocratic elution conditions (and refractive index detection) could be utilized for a relatively wide range of unsaturated compounds. Initially an experimental resin, which was not available commercially, was employed in a conventional column chromatography procedure. Later, Amberlite™ XE-284 was used for the purpose, but was ground in a mortar to a particle size more appropriate to HPLC technology [735]. This resin has a very high surface area, and therefore a large capacity for silver ions, but as a consequence rather polar solvents were required to elute the esters. With an analytical column (1 × 65 cm) and methanol at 0.98 ml/min as the mobile phase, base-line separations of methyl stearate, methyl elaidate and methyl oleate were obtained in about 2 hr, while almost as good a separation of a gram of these esters was obtained with a preparative column (2.2 × 55 cm). Similarly, geometrical isomers of methyl linoleate were well separated, but an elution time of more than 4 hr was needed to elute methyl linoleate itself.

Improved separations of geometrical isomers were subsequently attained with Amberlyst™ XN 1010 (again ground to a suitable particle size) as the support for the silver ions, but even longer elution times (up to 16 hr) were required for methyl linoleate, with methanol at a flow-rate of 3 ml/min as the eluent [203]. It was necessary to use methanol-hex-1-ene (9:1 by volume) as the mobile phase in order to recover methyl linolenate from the column. Fatty acid derivatives containing acetylenic bonds were also eluted safely from such columns, stearolate with oleate and crepenynate with linoleate. In addition, methyl esters of unsaturated hydroxy fatty acids have been resolved [686]. As retention times measured in hours were not practicable for many purposes, resins which were only partially loaded with silver ions were employed next, and they were certainly effective in permitting more rapid separations in addition to giving sharper peaks, both on an analytical and preparative scale [6,7,9]. The difficulties of preparing adsorbents reproducibly in this way have probably deterred other workers from taking this path, however. A better approach was to use ion-exchange resins fully-loaded with silver ions and to employ eluents containing acetonitrile, since this also complexes with silver ions and displaces unsaturated solutes; by programming from 3 to 10% acetonitrile into methanol or acetone as the mobile phase (with spectrophotometric detection at 215 nm), methyl arachidonate could be eluted as a sharp peak in under 2 hr, with sample sizes of up to 130 mg. One example (with isocratic elution) is illustrated in Fig. 7.3 [179,475]. Similarly, fractions enriched in particular polyunsaturated fatty acids were prepared in this way from fish oils (both as the esters and the free acids) [8].

Acetonitrile-acetone was better for the elution of free acids, as residual free sulphonic acid groups on the stationary phase catalyzed methylation in the presence of methanol. In addition, when the conjugated diene, *cis*-3,*cis*-5-tetradecadienoic acid, was purified in this way, some geometrical isomerization occurred; this did not happen apparently, when a column of silica gel impregnated with silver nitrate was employed for the purpose [180].

Other workers utilized macroporous cation-exchange resins, which were

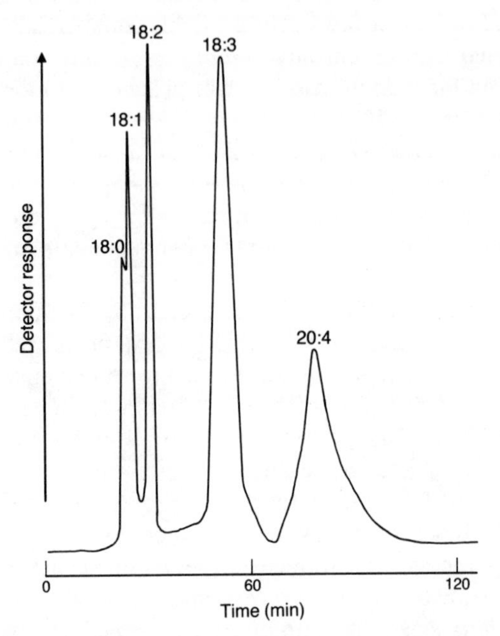

FIG. 7.3. Preparative-scale separation of methyl stearate, oleate, linoleate, linolenate and arachidonate (130 mg in total) by HPLC on a macroreticular ion-exchange resin impregnated with silver nitrate, eluted with acetone–acetonitrile (4:1 by volume) and with refractive index detection [179]. (Reproduced by kind permission of the authors and of the *Journal of the American Oil Chemists' Society*, and redrawn from the original paper.)

available commercially with a size distribution (30 to 70 μm) suitable for HPLC purposes, i.e. Bio Rad™ AG MP-50 [863] and Vydac™ "strong cation-exchange resin" [525], for the separation of geometrical isomers of insect attractants and of prostaglandins respectively.

In a different approach, a microparticulate silica was converted to an aluminosilicate, so that it could hold silver ions by ionic bonds [472]. Positional and geometrical isomers of octadecenoic acids, in the form of the *p*-bromophenacyl esters, were then resolved on columns packed with this material by elution with solvents of relatively-low polarity, such as 0.01% acetonitrile in chloroform–hexane (1:13 by volume). For example, a partial separation of the derivatives of *trans*-vaccenate and elaidate and a complete separation of *cis*-vaccenate and oleate were demonstrated. In one attempt to repeat this work, good separation was achieved, but the peak shapes were poor and double peaks were sometimes seen [581]. These faults were ascribed to destruction of the silica support, but could also have been a consequence of poor packing of the column.

The advantages of long column-life and negligible bleed of silver compounds attainable with silver-loaded ion-exchange resins would undoubtedly attract more analysts to the technique if cation-exchangers based on microparticulate silica gel with chemically-bonded sulphonic acid groups could be used, especially if prepacked columns were adapted to the purpose. There are encouraging signs that some of these objectives may indeed be possible. For example, columns (4.6 × 250 mm) custom-packed with silver-loaded Nucleosil™ 10 SA were employed for the resolution of insect sex attractants and of methyl esters of fatty acids with up to three *cis* or *trans* double bonds in only 10 min, with methanol as the mobile phase at 2 ml/min and at 40°C, and with refractive index detection [337]. Over 1000 samples were analysed, on the microgram to milligram scale, without signs of deterioration of the column. Although the authors were unable to reduce the retention times of components by adding other solvents to the eluent, they apparently did not try to use acetonitrile, which would certainly have assisted appreciably.

A prepacked column of Partisil™ SCX (10 μm particles) was loaded with silver nitrate by pumping 80 ml of 1 M aqueous silver nitrate through the column, followed by water (until no more silver ion was eluted), ethanol and solvents of decreasing polarity [526]. By eluting with various acetonitrile–dioxane and/or acetonitrile–chloroform mixtures, excellent resolution was achieved of a variety of prostaglandins in the form of the *p*-bromophenacyl derivatives. The column deteriorated somewhat if left unused for long periods while containing polar solvents, but was stable in continued use with the same solvents. Dioxane tended to cause some deterioration of the column, possibly because of reaction between peroxides and the silver ions. A very similar technique, but with a prepacked column of RSil™ CAT (5 μm particles) loaded *in situ* with silver nitrate, has been used for the separation of eicosenoids, including arachidonic acid, and prostaglandins in the free acid form [676]. In this instance, the mobile phase consisted of various mixtures of chloroform–methanol–acetic acid and chloroform–acetonitrile–acetic acid. The author (*J. High Res. Cromatogr. Cromatogr. Commun.*, **10**, 148–150 (1987)) has obtained some excellent separations of methyl ester derivatives of unsaturated fatty acids in preliminary experiments with a column of this type. These last columns would appear to have great potential for the separation of fatty acid derivatives. If they could also be used for the fractionation of molecular species of intact lipids (see Chapter 8), they would represent a considerable advance.

There is one further form of silver ion chromatography, i.e. the incorporation of silver salts into the mobile phase in conjunction with a stationary phase of the ODS type. In one of the first separations of this kind, methyl oleate was separated from methyl elaidate on a column of LiChrosorb™ RP8, eluted with isopropanol–1.5% aqueous silver nitrate (5:4 by volume) [736]; in this instance, the *cis*-isomer eluted ahead of the corresponding *trans*-isomer. A similar approach was used to separate the *p*-bromophenacyl ester derivatives of these and related fatty acids for UV detection [115], while others [714] employed dansyl-ethanolamine derivatives in the same way for fluorometric detection. Later, a wider variety of unsaturated fatty acids was separated as the methyl ester derivatives on a column packed with an ODS phase, and with methanol–water (5:1 by volume), containing 0.01 M silver perchlorate, as the mobile phase [856]. To this author, these separations appear only a little better than could have been achieved by reversed-phase chromatography alone, and the cost of silver salts would mitigate against a more frequent use of this technique, which appears only to have been tried with model mixtures. Whether the silver ions in the eluent could corrode the HPLC equipment might also be a cause for concern. In one of these reports, no deleterious effects were seen over a 2-year period [856]; in another, silver metal was found to be deposited in the tubes and detector cell, and it was necessary to remove it periodically by flushing with dilute nitric acid [736].

D. Separation of Fatty Acid Derivatives by HPLC in the Reversed-phase Mode

1. Introduction

HPLC in the reversed-phase mode is of considerable value for the separation of derivatives of those fatty acids encountered most often in animal and plant tissues, i.e. C_{14} to C_{22} fatty acids with an even number of carbon atoms and zero to six double bonds (separations of some of the less common fatty acids are discussed in Section E below). In more than 90% of the published work, ODS-bonded phases have been used, and this will permit some simplification of the discussion below. However, C_8 phases are increasingly being found to be of some utility. It should not be assumed that stationary phases from different manufacturers, that are nominally equivalent, will give identical results. Those readers who are familiar with the analysis of fatty acids by means of GLC on polar liquid phase will find one immediate and major difference, when they first encounter this form of HPLC, i.e. unsaturated fatty acids are eluted substantially ahead of the saturated fatty acid with the same chain-length, each double bond reducing the retention time by the equivalent of approximately two carbon atoms. Thus oleic acid derivatives tend to elute in a similar region to the corresponding palmitic acid derivatives. As these are almost invariably major components of plant and animal tissues, it is essential in assessing separation conditions for natural samples that these fatty acids especially should be adequately resolved. One further set of four compounds that can be troublesome consists of 14:0, 16:1, 18:2 and 20:4 fatty acids; any HPLC system intended for use with samples of animal origin should be able to distinguish these. The resolving power of columns packed with modern microparticulate phases is such that there need be little overlap of the main components of interest in natural samples, if the mobile phase is selected with care. On the other hand, because of the nature of the separation, different fatty acids are easily confused, and it is necessary to be especially vigilant to ensure that components separated in this way are identified correctly. This should not be a problem with lipids containing a few fatty acids only, as in seed oils, but can be troublesome with polyunsaturated fatty acids of animal phospholipids, for example. Most analysts have recourse to the use of model mixtures of pure standards in developing separations and as an aid to the identification of unknowns, and such mixtures are also invaluable for investigating the theoretical basis of a separation. However, published procedures are always more convincing when they are also applied to real samples.

Those analysts who are experience in the analysis of fatty acid derivatives by GLC will be familiar with the concept of identifying components by their "equivalent chain-lengths" or "carbon numbers" (see Section G.1 below for a brief discussion). Such parameters tend to be of less value in HPLC, where the capacity factor is more relevant (see Chapter 2).

In the following sections, separations are discussed in terms of the various types of derivative of fatty acids, that have been employed for the purposes of reducing polarity and aiding detection, since this either determines or is determined by the nature of the elution and detection systems utilized. It can also have some effect on the resolution achievable. On the other hand, although the nature of the derivative employed may require some modification of the mobile phase to bring off a particular component in a given time, it has relatively little effect on the order of elution components in any given mixture.

2. *Free fatty acids*

Fatty acids can be separated in the unesterified form with relative ease on columns of ODS stationary phases, although it has invariably been necessary to add acetic acid or phosphoric acid to the mobile phase as an ion suppressant. A number of interesting reports of applications to natural samples or to model mixtures have been published [40, 42, 59, 244, 320, 419, 510, 615, 780], and a few of these appear to be of special significance and are discussed further here. Various detectors have been employed, and spectrophotometric detection at low wavelengths in the UV spectrum has been of most value, although two novel detection systems, i.e. nephelometry and electrokinesis, have also been tried.

Acetonitrile–aqueous phosphoric acid (about 30 mM; pH 2) mixtures were used by Aveldano *et al.* [40] for the separation of free fatty acids, on a column (4.6 × (250 + 150) mm) of Zorbax™ ODS and with spectrophotometric detection at 192 nm (for unsaturated acids) or at 205 nm (for saturated fatty acids). Members of any series of fatty acids having the same chain-length and different degrees of unsaturation were always found to be separable with this system. With decreasing acetonitrile concentrations, the capacity factor for any component increased logarithmically, although the slopes for different series of fatty acids varied slightly. Similarly, series of fatty acids of the same degree of unsaturation but with different chain-lengths were also clearly separated. On the other hand, the longer the aliphatic chain, the greater was the increase in the capacity factor with decreasing concentrations of acetonitrile in the mobile phase. Thus as the solvent strength increased, the retention times of the shorter-chain fatty acids decreased faster than for longer-chain components. A consequence of this was that the order of elution of fatty acids in samples was changed according to the relative proportion of acetonitrile in the mobile phase. For example, the 16:1 (n – 9) fatty acid eluted before 20:4 (n – 6) with less than 60% of acetonitrile in the mobile phase, and after it with more than 80% acetonitrile present.

In an equally systematic study with methanol–water–phosphoric acid mixtures as the mobile phase, very similar conclusions were drawn [320].

If the separated components are to be collected for analysis by other methods, it is probably advisable to use a mobile phase containing acetic rather than phosphoric acid. Thus, some useful separations have been attained with tetrahydrofuran–acetonitrile–water–acetic acid (3:67:30:0.1 by volume) [42], and with methanol–water–acetic acid (89:11:0.2 by volume) [59] as mobile phases.

3. Methyl and other alkyl esters

Methyl esters are by far the most widely used derivatives in the analysis of fatty acids by GLC, and in addition their chromatographic properties have been studied by a variety of complementary techniques, including HPLC. These derivatives may therefore be the most suitable, if an appropriate detector is available, when it is intended to subject fractions emerging from an HPLC column to analysis later by other techniques. Again, ODS phases, ideally as 5 μm particles nowadays, have been preferred for the HPLC separations with acetonitrile–water or methanol–water mixtures as the mobile phase. There does not appear to have been a systematic comparison of the relative merits of these solvents with methyl ester derivatives and the same column, but acetonitrile tends to have a greater effect than methanol in reducing the retention times of unsaturated components. On the other hand, this is not necessarily advantageous as it can cause components with very different structures to elute in the same region.

Perhaps the most systematic study of the separation of methyl ester derivatives with real samples, in addition to model mixtures, was that of Aveldano *et al.* [40]. Figure 7.4 illustrates the separation of the methyl ester derivatives prepared from the phospholipids of mouse brain on a column (4.6 × 250 mm) of Zorbax™ ODS, maintained at 35°C; the mobile phase was acetonitrile–water (7:3 by volume) at 2 ml/min, changed to acetonitrile alone over 20 min towards the end of the separation. Experienced analysts have learned to identify fatty acids separated by GLC almost instinctively, but this has not been true of reversed-phase HPLC, where as can be seen from

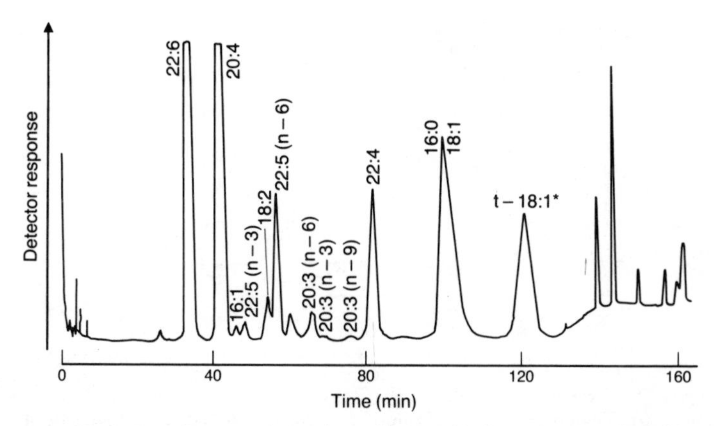

FIG. 7.4. Separation of methyl ester derivatives of fatty acids from the phospholipids of mouse brain by reversed-phase HPLC (0.15 mg scale) with spectrophotometric detection at 192 nm [40]. The elution conditions are described in the text. Methyl elaidate was added as an internal standard. (Reproduced by kind permission of the authors and of the *Journal of Lipids Research*, and redrawn from the original paper.)

the figure the relationship between retention time and structure is not immediately apparent. As described earlier for unesterified fatty acids, the order of elution of specific components was greatly dependent on the relative proportions of acetonitrile and water in the mobile phase. Note that methyl oleate and its *trans* equivalent were very well resolved.

As detection was by spectrophotometry at 192 nm, the response was much greater to the polyunsaturated components, which tended to elute first, than to saturated or monoenoic components. It was therefore concluded that this method of detection was of little value for purposes of quantification, but had considerable merit if used for the isolation of individual components for identification by other means, or for determination of radioactivity. The procedure was subsequently used to study the metabolism of isotopically-labelled arachidonic acid in rat brain *in vivo* [570].

In an equally systematic study, but with model systems, methanol–water mixtures were employed as the mobile phase [321].

Others have applied similar separation conditions to the isolation of polyunsaturated fatty acids from fish oils [616] or from rat liver microsomes [810]. Relatively simple mixtures of the methyl ester derivatives of fatty acids, including those from fungal and animal tissues in addition to model mixtures, have been separated on a number of occasions both on an analytical or small scale [45, 141, 312, 615, 731, 750, 862], and with up to gram quantities with suitable equipment [56, 220, 732]. For example, methyl crepenynate was purified in 100 mg aliquots on a column (10 × 250 mm) packed with an ODS phase, and eluted with methanol–water (19:1 by volume); the required compound eluted ahead of methyl linoleate as shown in Fig. 7.5 [220]. Smaller quantities of esters (up to as much as 5 mg) have been isolated on standard

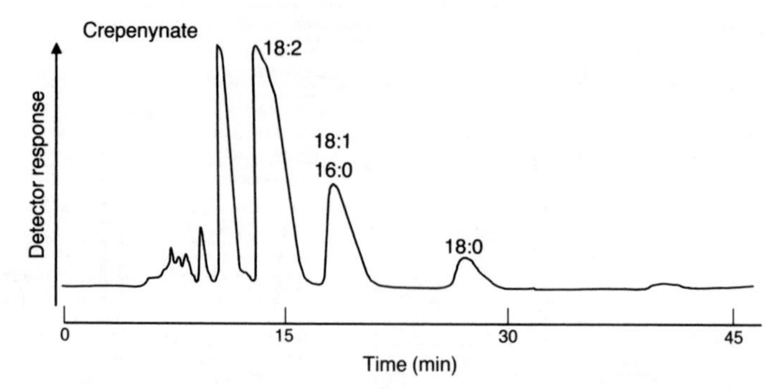

FIG. 7.5. Separation of methyl crepenynate from other esters of the seed oil, *Ixiolaena brevicompta*, by reversed-phase HPLC on a preparative scale (100 mg) with refractive index detection [220]. The column (10 × 250 mm) contained an ODS phase and was eluted with methanol at a flow-rate of 2.5 ml/min. (Reproduced by kind permission of the author and of the *Journal of Chromatography*, and redrawn from the original paper.)

analytical ODS columns. Obviously, some loss of resolution and peak symmetry must be expected in these circumstances. Esters of short-chain fatty acids, such as those in milk fat, have been separated [142], as have those of much longer chain-length than are normally encountered (up to C_{32}) [40, 251, 690]. For example, the methyl ester derivatives of fatty acids up to C_{30} in chain-length with five double bonds were isolated from rat testis on an ODS column, eluted with gradients of acetonitrile into methanol–water mixtures [251]. Others preferred different bonded phases, such as C_1 or C_8 phases, especially for longer-chain fatty acid derivatives [40, 690]. While the capacity to utilize gradients in the mobile phase can shorten analysis times appreciably, some excellent separations have been achieved with isocratic elution conditions over a wide range of chain-lengths [141, 142, 810, 811]. Increasing the flow-rate after the 16:0 and 18:1 fatty acid derivatives have eluted, for example, can hasten the elution of 18:0, which is often the last component to emerge with many natural samples.

One important application of reversed-phase HPLC may be for the separation of geometrical and positional isomers of long-chain monounsaturated fatty acids. The retention times of many of the isomeric methyl octadecenoates relative to that of methyl palmitate are illustrated in Fig. 7.6 [811]; the elution conditions were a column (4.6 × 200 mm) of Nucleosil™ C_{18} with methanol–water (89:11 by volume) at a flow-rate of 1.2 ml/min as the mobile phase. *cis*-Isomers always eluted before the corresponding *trans* isomers (also reported earlier with a simpler range of isomers [862]), and a double bond in the centre of the molecule had a greater effect on the retention time than if it were at either extremity of the aliphatic chain. Isomers with double bonds of the same configuration in adjacent positions would only be separable if they were present at either end of the molecule. In contrast, it is apparent from the figure that fatty acids with a mixture of *cis*

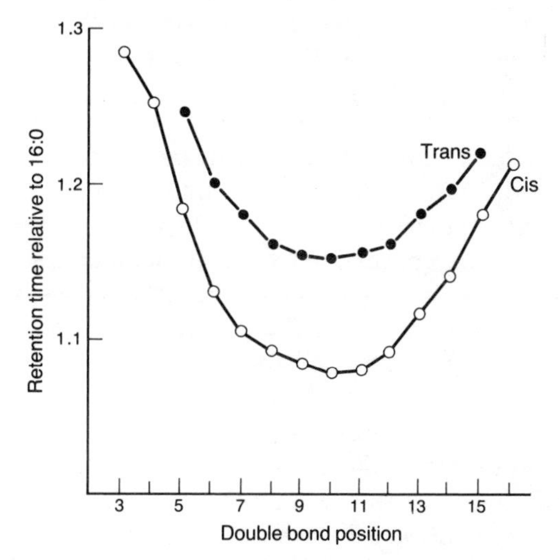

FIG. 7.6. Retention times relative to methyl palmitate of different positional and geometrical isomers of methyl octadecenoate on HPLC in the reversed-phase mode [811]. The double bond positions are designated from the carboxyl group. The chromatographic conditions are described in the text. (Reproduced by kind permission of the author and of *Lipids*, and redrawn from the original paper.)

and *trans* double bonds, that are approximately central, could probably be resolved into the two classes. If there were a wider spread of double bond positions, it is unlikely that complete resolution would be possible, although the technique is probably as good as any other for the purpose. An application to partially hydrogenated fish oils was described [811], and quantification of geometrical isomers at each chain-length was possible by making use of a refractive index detector, which operated on the interference principle and was highly sensitive. However, some preliminary separation by some form of silver ion chromatography would certainly be necessary in most circumstances to remove certain esters, which would otherwise co-chromatograph with those of interest.

In much of the work described above, UV detection at low wavelengths or differential refractometry were used, and in general there has been very little attempt to use such HPLC techniques for quantification purposes. The mass (or light-scattering) detector has also been applied to methyl esters of fatty acids, but it did not appear to be particularly suitable for such analyses, as some evaporation occurred in the detector [698, 793].

4. Derivatives suitable for UV detection

By preparing fatty acid derivatives which have a pronounced and distinctive absorption maximum in the UV region of the spectrum, it is possible to detect components separated by HPLC, and to quantify them spectrophotometrically with great sensitivity. Some specificity is also added to the analysis, as impurities, which are not derivatized, do not often interfere. As detection is dependent only on the properties of the alcohol moiety, generally an aromatic grouping of some kind [388], the response is directly proportional to the molar amount of the fatty acid. Innumerable types of derivative has been prepared for the purpose, but none appears to have any clear advantage in separation terms over the others. The choice may be governed by a need for higher than normal sensitivity if the sample size is particularly small, or it may simply be a question of the convenience or availability of a particular derivatization reagent. The nature of the derivative, because of its hydrophobicity, may have an influence on the composition of the mobile phase required to bring off a specific component in a given time, however. As described above (Section B.3), phenacyl esters have probably been the derivatives used most frequently, and they can certainly be recommended to to anyone attempting to use the technique for the first time. It appears that as little as 10 ng of a phenacyl ester can be measured accurately by this technique. Most work has been carried out on ODS stationary phases, but some complementary separations have been described with "fatty acid", C_8 and C_{30} phases (see below). In addition, some preliminary separations of *p*-nitrobenzyl esters of fatty acids on capillary columns of fused-silica packed with 5 and 10 μm ODS phases have been described; 90,000 theoretical plates were generated in a column 1 m in length [323]. Although such columns are not commercially available at present, this work is possibly a pointer to the future.

Most analysts have used UV detection at 254 nm in the analysis of phenacyl esters, probably because this is suited to inexpensive fixed-wavelengths detectors. However, it has been shown that an appreciable increase in sensitivity is possible if the detector can be operated at 242 nm, which is the absorption maximum for such compounds [889]. The use of phenacyl derivatives for the determination of fatty acid derivatives has been reviewed elsewhere [474].

One of the first attempts at a comprehensive separation of phenacyl ester derivatives was by Borch [88], who used a column (6.4 × 900 mm) of *micro*Bondapak™ C_{18}, eluted in a stepwise fashion with acetonitrile–water mixtures (from 67:33 to 97:3 by volume) at 2 ml/min. A representative separation of a model mixture of phenacyl esters is shown in Fig. 7.7. Geometrical isomers of unsaturated esters were well separated, as were some positional isomers, including the derivatives of oleic (*cis*-9) and petroselinic (*cis*-6) acids. The factors governing the order of elution are similar to those discussed earlier for methyl esters. Summarized briefly, *cis* isomers eluted before the corresponding *trans* isomers, *cis*-monoenoic esters eluted just after the saturated ester with two fewer carbon atoms, further double bonds in an ester had successively smaller effects, and a double bond in the centre of the chain reduced the retention volume more than one at either extremity of the

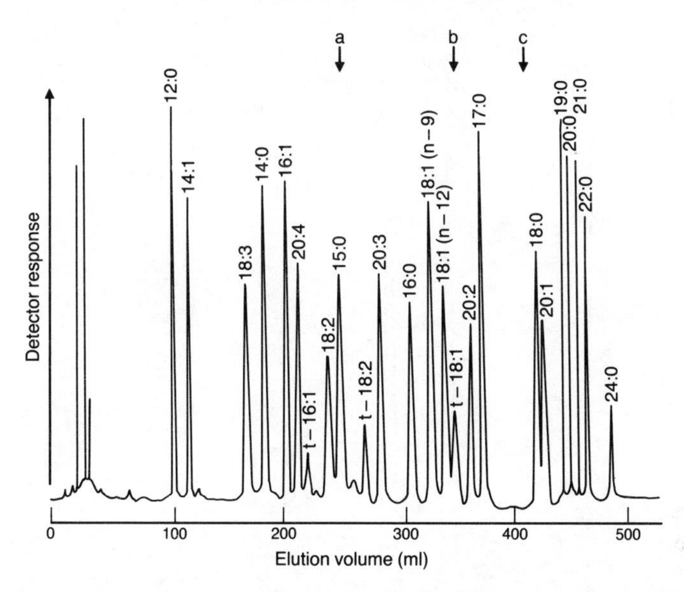

FIG. 7.7. Separation of a standard mixture of fatty acids in the form of the phenacyl derivatives by reversed-phase HPLC with spectrophotometric detection at 254 nm [88]. The column (6.4 × 900 nm) was packed with *micro*Bondapak™ C-18, and was eluted with acetonitrile–water in the proportions 76:33 (by volume) initially, changed to 74:26 at "a", to 4:1 at "b", and to 97:3 at "c", at a flow-rate of 2 ml/min. (Reproduced by kind permission of the author and of *Analytical Chemistry*, and redrawn from the original paper.)

molecule. The apparent complexity of this pattern should be contrasted with that obtained by capillary GLC, as illustrated in Fig. 7.11 below, in which components of different chain-lengths emerged in discrete groups, within each of which the elution sequence was readily predicted.

More recently, chromatograms very similar to that in Fig. 7.7 have been published by others [524, 889], but they were obtained on an ODS phase in columns with dimensions closer to current general practice. Wood and Lee [889], for example, employed a column (4.5 × 250 mm) containing an ODS phase (5 μm particles), with a linear gradient of acetonitrile and water (from 80:20 to 90:10 by volume) at a flow-rate of 2 ml/min as the mobile phase. Of the common fatty acids of animal tissues, only the resolution of palmitoleic and arachidonic acids was troublesome, although by varying the relative proportions of acetonitrile and water the problem could generally be solved. Not all natural samples contain such a wide range of fatty acids, and it proved a relatively simple matter to separate the components of various seed oils in the form of either the phenacyl or naphthacyl esters. In this instance, isocratic elution with acetonitrile–water (4:1 by volume) was sufficient for phenacyl esters; with naphthacyl esters, the same solvents in the proportions 87:13 respectively gave similar results.

In addition, it was possible to determine the fatty acid compositions of

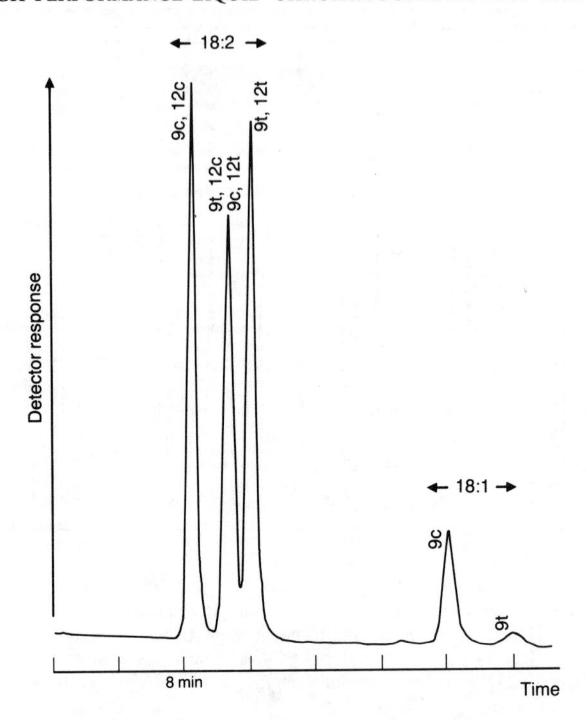

FIG. 7.8. Separation of geometrical isomers of C_{16} mono- and dienoic fatty acids in the form of the phenacyl derivatives by reversed-phase HPLC with spectrophotometric detection at 242 nm [889]. The column (4.5 × 250 mm) was packed with an IBM™ ODS (5 μm) phase, and was eluted with acetonitrile–water (4:1 by volume) at a flow-rate of 2 ml/min. (Reproduced by kind permission of the authors and of the *Journal of Chromatography*, and redrawn from the original paper.)

the seed oils quantitatively in term of the molar proportions of each constituent. Results obtained with standard mixtures confirmed that the accuracy was comparable to that attainable by means of GLC.

As discussed above for methyl esters, it was possible to separate a variety of geometrical isomers of unsaturated fatty acids as shown in Fig. 7.8, where the four configurational isomers of linoleic acid were resolved into three fractions, for example [889]. Later systematic studies [886] with a large number of positional and geometrical isomers of octandecenoic acid, as the phenacyl esters, confirmed the conclusions described above (Section D.3) for methyl esters, i.e. that *cis*-isomers generally eluted before *trans*-isomers, and that the greatest effects in reducing the retention volumes were observed when the double bonds were located approximately centrally in the alkyl chain. However, one major anomaly was noted, i.e. that the *trans*-2 isomer eluted ahead of the *cis*-2 isomer, a result which may be explained by variable interactions between the pi bonds and the carbonyl oxygen because of their relative geometries. Although some of the separations of isomers with double bonds near the carboxyl group were impressive, it is doubtful whether isomers in which the double bonds were approximately central, as in many samples of practical interest, would be separable. Qualitatively similar results were obtained with isomeric phenacyl octadecynoates, which were eluted a considerable distance ahead of the corresponding *cis*-octadecenoates.

Phenacyl ester derivatives have been employed in the analysis by means of reversed-phase HPLC of the fatty acids of butter fat and related samples [204, 416, 687], of free and bound fatty acids in river water [342], of archeological samples [625], of the cellular lipids of the microorganism, *Vibrio parahaemolyticus* [524], and of the phosphoglycerides of amniotic fluid [726].

Naphthacyl ester derivatives appear to have been used with standard mixtures only [159, 374, 889]. Electron-impact and chemical-ionization mass spectra of a few derivatives were obtained by introduction of the sample into the mass spectrometer via a solid probe, but they were of little value for structure elucidation [159].

Substituted phenacyl derivatives, which tend to have higher molar extinction coefficients, can be detected with much higher sensitivity than is possible with the unsubstituted compounds. For example, the limit of detection of the *p*-bromophenacyl ester of arachidonic acid was reported to be 10 pg [374]. On the other hand, the nature of the substituent group in the phenacyl ring appeared to make little difference to the quality of the separation that could be achieved with a comprehensive range of fatty acid standards [374]. Separations on columns containing *micro*Bondapak™ C_{18} or *micro*Bondapak™ "fatty acid phase" (somewhat more polar) were rather different and complemented each other; for example, the palmitoleic and arachidonic acid derivatives were separated rather easily on the latter phase. In other work [818], it was shown that separations attainable on a "C_{30}" phase were also distinctive in that there was an even greater effect of a double bond on elution volume than with an ODS phase. *p*-Bromophenacyl esters have been employed for the separation or analysis of fatty acids of biological interest [270, 440, 638, 704], of grain and feed extracts [837], of the minor C_{20} to C_{30} fatty acids in the serum of patients with adrenoleukodystrophy [22], and of the C_{30} to C_{56} fatty acids from *Mycobacterium tuberculosis* (with dioxane–acetonitrile mixtures as the mobile phase) [683].

p-Methoxyphenacyl esters were used in some particularly comprehensive studies of a wide range of fatty acids of bacterial origin [101, 102, 530]. In addition to unsaturated fatty acids, components containing cyclopropane rings and hydroxyl groups (analysed as the trifluoroacetate derivatives) were present in the samples. The fatty acids of a few seed oils have been quantified in the form of the phenylazophenacyl esters, which were detectable at 330 nm [853].

Others have separated fatty acids with UV detection in the form of isatinyl methyl esters [257], anthrylmethyl esters (also used with fluorescence detection) [54,62], *p*-nitrobenzyl esters [323,431], benzyl esters [667], pentafluorobenzyl esters [580,581], naphthylamides [348], isopropylidene hydrazides [10], 2-nitrophenylhydrazides (also detectable in the visible region of the spectrum at 400 nm) [536–538], and methoxy- and nitro-anilides [325, 851].

5. Derivatives suitable for fluorescence detection

The use of derivatives of fatty acids that can be detected by fluorescence spectroscopy has enabled a further substantial increase in the sensitivity of HPLC analysis, such that amounts can now be quantified which would not be possible with GLC on capillary columns or with GC-MS even. The first and most widely-used derivatives of this kind were anthrylmethyl esters, prepared simply by reaction of the free fatty acids with 9-diazomethylanthracene in an inert solvent [54, 588] (although this has not always give acceptable results in some hands, and alternative methods have been described [61,62,439]). Optimum sensitivity was obtained by fluorescent excitation at 360 nm and determining the emission at 440 nm. (UV detection at 256 nm could also be used.) Although the improvement in detectability was somewhat dependent on the chromatographic conditions, it was at least 10-fold better than with UV detection, and was certainly in the low picogram range. The nature of the mobile phase can have an appreciable effect on the fluorescence quantum yield and it is generally advisable, when all such derivatives are employed, to make use of an internal standard, such as the appropriate derivative of heptadecanoic acid, and to avoid gradient elution conditions (unless great care is taken in calibration). The use of fluorescence detection in HPLC has been reviewed [477,488]. Some applications to the analysis of free fatty acid in serum and other biological samples are discussed in Chapter 5.

The nature of the separations attained with anthrylmethyl esters on ODS phases has been similar to that reported above for most other derivatives, except that some modification to the mobile phase has been necessary to counter the effect of the bulky alcohol moiety. For example, methanol–water (47:53 by volume) at 60°C was used with an ODS phase for the separation of the fatty acids of plasma lipids, as shown in Fig. 7.9 [762]. As alternative mobile phases, acetonitrile alone was used with some standard mixtures [54] and for the analysis of fatty acids with very long chain-lengths in patients with adrenoleukodystrophy [433], while others made use of methanol–acetonitrile–1, 4-dioxane–water (16:1:1:2 by volume) for a similar purpose [296]. Acetonitrile–isopropanol–water (90:9:1 by volume) was utilized in the analysis of free arachidonic acid in tissues [572], and a similar system but containing methanol was used for the albumin-bound free fatty acids from plasma [240]. Methanol–water (9:1 by volume) was employed with a column

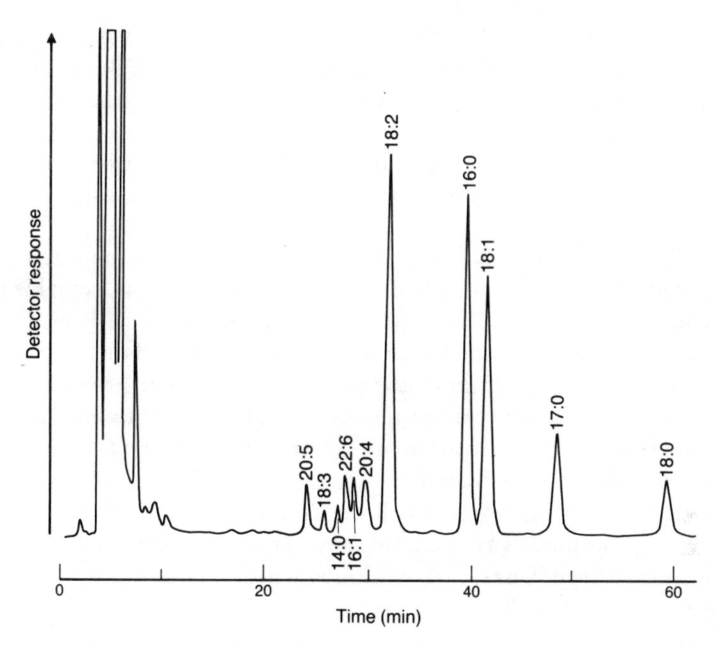

FIG. 7.9. Separation of the fatty acids from serum lipids in the form of the anthrylmethyl derivatives by reversed-phase HPLC and fluorescence detection, with excitation at 365 nm and emisssion at 412 nm [762]. The column (two 4.6 × 150 mm) contained Zorbax™ ODS phase, and was eluted with methanol–water (47:53 by volume) at a flow-rate of 1 ml/min. (Reproduced by kind permission of the authors and of *Clinica Chimica Acta*, and redrawn from the original paper.)

(4.6 × 150 mm) of a Zorbax™ C_8 phase for the analysis of the polyunsaturated fatty acids in fish and blood samples [347]. In addition, the value of a C_8 phase was confirmed by Baty and Pazouki [61], who obtained a particularly good resolution of the critical group, 20:4, 14:0, 16:1 and 18:2, with a complex acetonitrile–water gradient as the mobile phase.

Anthrylmethyl derivatives gave distinct molecular ions in HPLC-MS, that permitted the detection of minor components by ion-monitoring techniques [62].

A number of workers have made use of coumarin derivatives, including 4-methyl-7-methoxycoumarin [301, 473, 855, 913], 4-methyl-6, 7-dimethoxycoumarin [211] and 4-methyl-7-acetoxycoumarin [835] in analyses of fatty acids by reversed-phase HPLC with fluorescence detection. The derivatives were prepared by methods analogous to those employed for phenacyl esters. Because of the relatively small size of the alcohol moiety, coumarin esters have tended to give somewhat better resolution than other fluorescent derivatives, and samples derived from plasma lipids and other sources have been eluted from columns of ODS phases with solvents such as methanol–

water in the proportions of 85:15 [473] or 95:5 [211] by volume, and with mixture of acetonitrile and water. In one study [835], the fluorescence was determined in a post-column reaction chamber following hydrolysis of the ester bond. In this instance, the mobile phase was a gradient of methanol–water (9:1 by volume) into methanol–acetonitrile–water (7:7:5 by volume), while the column containing an ODS phase was maintained at 40°C.

As well as those derivatives described above, 9-aminophenanthrene [349], 2-naphthacyl [187], phenanthrimidazole [491], dansyl-semipiperazide [904], and methyl-6, 7-dimethoxy-1-methyl-2(1*H*)-quinoxalinonone [898, 899, 900] derivatives have been used in the analysis of fatty acids by HPLC with fluorescence detection. Of these, the last holds particular promise in terms of sensitivity, as it permitted the analysis of fatty acids in amounts as low as 0.3 femtomole. Quantification was simplified, as isocratic elution with acetonitrile–water (78:28 by volume) as the mobile phase, at a flow-rate of 2 ml/min, was possible, from a column containing a C_8 stationary phase [899]. The procedure was used to detect small changes with diet in the composition of the polyunsaturated components of the free fatty acid fraction of plasma.

E. Normal-phase Separations of Fatty Acid Derivatives by HPLC

Normal-phase partition HPLC has been little used for the analysis of lipids and their derivatives, but a few separations have been described which have indicated that there is some potential for further development. For example, the 4-methyl-7-methoxycoumarin derivatives of some short-chain fatty acids were separated on a column of silica gel eluted with a gradient of acetonitrile–dichloromethane (2:3 by volume) into isooctane [913]; components were eluted in reverse order of chain-length. Oleic and linoleic acids are separated in this order on a column of silica gel, buffered with citric acid–sodium citrate (pH 3), with hexane–diethyl ether (4:1 by volume) as the mobile phase [743]. Others [580, 581] described similar results and also separated pentafluorobenzyl derivatives of fatty acids of the same chain-length (C_{18}), but differing in degree of unsaturation, on a column of silica gel, eluted with dichloromethane–hexane half saturated with water (3:17 by volume). The separation was monitored spectrophotometrically at 263 nm. In this instance, the saturated derivative eluted first, followed by the monoene, then the diene and finally the triene. *Trans*-isomers eluted ahead of the corresponding *cis*-compounds with this system. Subsequently, a method for the preparation of such derivatives and an application to the analysis of the fatty acids in seed oils were described [579]. Synthetic *cis*- and *trans*-3-tetradecen-5-ynoic acids have been separated on a preparative scale using a procedure of this kind [180].

F. HPLC Separations of Some Less Common Fatty Acids

1. Hydroperoxides and other oxygenated fatty acids

HPLC is well suited to the separation and analysis of compounds with comparatively-unstable functional groups, such as hydroperoxides, since chromatography is carried out at or near ambient temperature under anaerobic conditions. In contrast in GLC methods which require elevated temperatures to effect separations, it is generally necessary to convert the functional group to a stable non-polar derivatized form prior to chromatography. The availability of HPLC methods for the separation of hydroperoxides has brought about a considerable increase in our knowledge of chemical and enzymatic oxidation processes in relation to lipids.

By means of HPLC, it is possible either to separate hydroperoxides as such or as the more stable hydroxides, following reduction. The first approach has generally been favoured for the isolation of particular components for structural analyses or for studies of the further reactions of such compounds. For example, fatty acids with a hydroperoxide group in either position 9 or 13, derived from the enzymatic oxidation of linoleic acid with soyabean and potato lipoxygenases, were separated with relative ease by Chan and Prescott [116] by adsorption chromatography of the methyl esters on a column of Partisil™ silica gel, eluted with hexane–ethanol (99.5:0.5 by volume). As such compounds contaïned a conjugated double bond system, they were detected spectrophotometrically at 235 nm. Subsequently, it was shown that isomeric hydroperoxides, prepared by chemical autoxidation of linoleic acid, could be separated both according to the positions of the hydroperoxide groups and to the configurations of the double bonds by this or very similar methods as shown in Fig. 7.10(A) [113, 117, 628, 823, 829]. An analogous method was used for the separation of hydroperoxides derived from phenyl linoleate [294]. In other work in which linoleate hydroperoxides were subjected to HPLC in the form of the free acids, in order to avoid the preparation of any form of derivative and any possibility of artefact formation, better resolution was obtained with heptane–acetic acid (97:7:2.3 by volume) [441], or with other solvents containing acetic acid [823, 850, 892, 901] as the mobile phase.

Similar methods were used to separate hydroperoxides obtained by oxidation of oleic acid [114], and of linolenic and arachidonic acids [226, 671], or their derivatives; with the latter, hexane–isopropanol mixtures were used as the eluent. As an alternative, others separated the hydroperoxides (and related derivatives) from linolenic acid by reversed-phase HPLC on a column (4.6 × 220 mm) of Supelco™ LC-8, with a mobile phase consisting of acetonitrile–tetrahydrofuran-0.1% phosphoric acid (50.4:21.6:28 by volume) [400]. Di-peroxides formed from methyl linolenate were also separated both on an analytical and on a preparative scale by HPLC in the adsorption mode [162]. More recently, hexane–dichloromethane–ethyl acetate (7:4:1 by

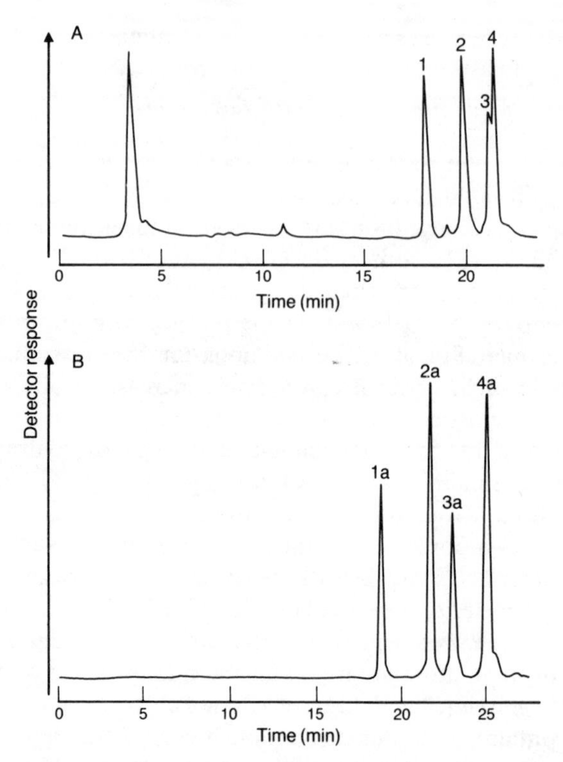

FIG. 7.10. A. An HPLC separation of hydroperoxide isomers derived from methyl linoleate on a column of silica gel with spectrophotometric detection at 234 nm [113]. Details of the chromatographic conditions are given in the text. 1, methyl 13-hydroperoxy-*cis*-9, *trans*-11-octadecadienoate; 2, methyl 13-hydroperoxy-*trans*-9, *trans*-11-octadecadienoate; 3, methyl 9-hydroperoxy-*trans*-10, *cis*-12-octadecadienoate; 4, methyl 9-hydroperoxy-*trans*-10, *trans*-12-octadecadienoate.

B. An HPLC separation of hydroxy isomers derived from the above hydroperoxides on a column of silica gel with spectrophotometric detection at 234 nm; 1a to 4a represent the hydroxides derived from hydroperoxides 1 to 4 respectively. (Reproduced by kind permission of the authors and of *Biochimica Biophysica Acta*, and redrawn from the original paper.)

volume) was used as the mobile phase to elute hydroperoxides of ester derivatives of cyclic fatty acids from columns of silica gel [41]. Both silica gel and reversed-phase chromatography were used to isolate hydroxy, keto, epoxyhydroxy and trihydroxy fatty acids, formed by reaction of linoleic acid hydroperoxide with hematin [190]. Peroxides and secondary oxidation products, separated by HPLC, have been characterized by chemical-ionization MS [222].

It proved possible to separate positional isomers of linoleate hydroperoxides by gel-permeation chromatography on a column (4 × 600 mm) packed with a porous polymer gel, TSK-Gel™ LS-140, and eluted with hexane–ethanol (7:3 or 9:1 by volume) [30–32], although the resolution was not as good as that attained by adsorption chromatography. With HPLC in the reversed-phase mode and a column packed with an ODS phase, it was possible to separate geometrical isomers only [113]. More components appeared to be separated (as the free acids) on a column of *micro*Bondapak™ "fatty acid phase", eluted with tetrahydrofuran–acetonitrile–water (5:7:9 by volume) [399]. Yet others separated hydroperoxides, derived from linoleic acid and freed from phospholipids by phospholipase A hydrolysis, on a column containing an amine-bonded phase [843].

If the hydroperoxides themselves were not required for any reason, better resolution for analytical purposes has been obtained by reducing to the corresponding hydroxides with sodium borohydride in methanol, prior to chromatography on silica gel as before, with the results shown in Fig. 7.10(B) [113]. Similar methods were used for hydroperoxides derived from free linoleic acid and its phospholipid esters [672, 675, 823]. Hydroxylated derivatives of arachidonic acid, derived from peroxidation, were isolated in this way for characterization by other means [87], as were those prepared from the fatty acids of membrane phospholipids [341]. This approach was used for the separation of oxidation products from the fatty acids of soyabean oil, except that the esters were hydrogenated catalytically to the fully saturated compounds [577]. By HPLC with a combination of silica gel and ODS phases, ten different hydroxy compounds were obtained from the autoxidation of docosahexaenoic acid [846]. The isomeric methyl 9-hydroxy- to 13-hydroxy-stearates were all separable by adsorption chromatography, as were the 15- and 16-hydroxy isomers, and only the 13- and 14-hydroxy isomers were poorly resolved. Similar separations were obtained with the oxidation products of phenyl oleate and linoleate [293]. A further advantage of the hydroxy derivatives is that they are more readily identified by MS than are the original hydroperoxides.

Monohydroxy derivatives of arachidonic acid (as the phenacyl ester derivatives) have been isolated by HPLC in the reversed-phase mode [879], and related derivatives of a docosapentaenoic acid were obtained by adsorption chromatography [107]. These papers are cited as representatives only of a large number of papers on the separation of hydroxy-eicosenoids and prostaglandins by means of HPLC, a topic reviewed comprehensively by Hamilton and Karol [272].

One further approach to the separation of linoleate hydroperoxides, following methylation and borohydride reduction to the hydroxy esters, consisted in HPLC separation on a column of silica gel impregnated with silver nitrate, with 0.4% ethanol in hexane as the mobile phase [295]. In this instance, isomers containing a *trans*, *trans*-double bond system eluted before those with *cis*, *trans*-double bonds.

A number of oxygenated fatty acids other than peroxides of synthetic or natural origin, have been separated by HPLC. For example, *p*-nitrobenzyl ester derivatives of 22 different synthetic monohydroxy fatty acids (C_{16} to

C_{22}), including a number of positional isomers, were separated by HPLC in both the adsorption and reversed-phase modes, with detection at 265 nm [50]. On a column (10 × 250 mm) of silica gel (Spherisorb™ S5W), with hexane-isopropanol (1000:4 by volume) as the mobile phase, the 6-, 7-, 9-, 10, 11- and 12-hydroxystearate derivatives were all well resolved, although they overlapped with components of different chain-lengths when these were present. In the reversed-phase mode (ODS phase) with a gradient of 30 to 0% water in acetonitrile as the mobile phase, the resolution of positional isomers was poorer, but components of different chain-lengths were well separated. Following conversion to the *tert*-butyldimethylsilyl derivatives of the hydroxyl groups, the separation achieved was essentially by chain-length only [48]. Various synthetic diastereoisomeric 9, 10, 12-trihydroxystearates have been resolved by HPLC on a column of silica gel [161]. In contrast, reversed-phase chromatography (ODS stationary phase), with acetonitrile as the mobile phase and refractive index detection, were used to separate methyl ester derivatives of synthetic epoxy fatty acids, prepared from oleic, linoleic and linolenic acids; the greater the number of epoxyl groups, the faster the compound was eluted [643]. Nitroxide derivatives of fatty acids, prepared by radical-catalyzed autoxidation of linoleic acid, have been separated by HPLC [207].

2- and 3-hydroxy fatty acids of bacterial origin were initially analysed in the form of the phenacyl esters (of the carboxyl group), with the hydroxyl group underivatized [102]. On separation by HPLC in the reversed-phase mode, the 2- and 3-hydroxy compounds eluted in positions equivalent to 2.7 and 3.32 carbon atoms respectively ahead of the corresponding normal fatty acids, but the important C_{10} to C_{14} components were obscured by by-products of the derivatization procedure. However, following trifluoro-acetylation, the resulting derivative of the 2-hydroxy fatty acid eluted just after the corresponding normal fatty acid, while the 3-hydroxy derivative eluted just before the normal fatty acid. Other workers separated (*omega*)- and (*omega*-1)-hydroxy fatty acids, derived from microsomal oxidation of fatty acids *in vitro*, by HPLC in the adsorption mode [152]. A procedure for the rapid screening of oxygenated fatty acids in plants and natural waters, involving HPLC in the reversed-phase mode, has also been described [703].

2. *Polymerized fatty acids*

During the autoxidation of unsaturated fatty acids, catalyzed by free radical reactions and especially at high temperatures, a number of complex polymeric products are formed, including linear dimeric, mono- and dicylic dimeric, and higher oligomeric fatty acids, in addition to the starting materials and oxidized monomers. These can have adverse physiological effects when consumed, so methods for their analysis are of some importance. Many different techniques have been employed for the purpose, but HPLC has recently come to the fore. For example, HPLC in the gel-permeation mode with a column of Styragel™ 500, with tetrahydrofuran as the mobile phase and with refractive index detection, was used to fractionate the methyl ester derivatives of the fatty acids from polymerized triacylglycerols according to their molecular weights in the regions 200 to 400 (monomers), 500 to 700 (dimers) and above 800 (trimers mainly) [373]. Each fraction was then characterized by MS. Similar fractions were also obtained on a column of LiChrosorb™ NH_2 with methanol containing 0.05% acetic acid as the mobile phase [914], and by adsorption chromatography on a column of Supercosil™ LC-Si, eluted with cyclohexane–isopropanol–acetic acid (99.3:0.5:0.2 by volume) [847]. With the latter, the Tracor Instruments transport-flame ionization detector was used to monitor the separation, and to quantify the components.

Dimer fractions from methyl linoleate hydroperoxides were obtained by low-pressure gel-permeation chromatography, and were then further fractionated by HPLC on a column of silica gel with hexane containing 0.7 or 2.5% isopropanol as the mobile phase [540].

3. *Some miscellaneous unusual fatty acids*

Analyses of some less common fatty acids such as short-chain [142, 204, 416, 687, 904], e.g. those of milk fat, and long-chain ($> C_{22}$) fatty acids, including mycolic acids [22, 40, 251, 683, 690, 818], by HPLC procedures were described in Section D above. Again, the technique is probably best suited to samples not easily analysed by GLC, and one important application has been to the separation and quantification of fatty acids containing cyclopropene rings, which decompose at elevated temperatures. Good separations of the methyl ester derivatives of malvalic and sterculic acids from the monoenoic esters of the same chain-length were obtained on a column (4.6 × 250 mm) of LiChrosorb™ RP-8, eluted with acetonitrile–water (17:3 by volume) [72, 232]; methyl sterculate eluted just after methyl oleate under these conditions. With careful calibration, spectrophotometry at 206 nm could be used both for detection and quantification with natural seed oils containing these compounds. Very similar chromatographic conditions but with refractive index detection were employed by others for the same purpose [493]. Such procedures were probably not sufficiently sensitive for the analysis of cyclopropene fatty acids in such commercially important oils as cotton seed oil. However, Wood [887] showed that this was indeed possible if the phenacyl derivatives and UV detection at 242 nm were utilized. Fatty acids containing cyclopropane rings have also been separated in this way [72, 102, 232, 493, 581].

Other fatty acids containing unusual functional groups, that have been isolated or analysed by HPLC, include pyrene-substituted fatty acids [338], partially-deuterated (from non-deuterated) palmitic acid [155], and C_9 to C_{12} dibasic fatty acids [624].

G. Alternatives to HPLC for the Analysis of Fatty Acids

1. Gas–liquid chromatography

It could take several chapters to describe alternatives to HPLC for the separation and analysis of fatty acids in detail, and the topic is treated here only briefly in order to widen the perspective of the discussion. Further information can be obtained in the author's book [133], and in many other review articles cited below.

As stated in the introduction to this chapter, GLC would in most circumstances be the method of choice for the analysis of fatty acids in the form of the methyl ester derivatives (prepared as described in Section B.2) Until recently packed columns (of the order of 4 mm i.d. × 2 m in length), containing polar polyester liquid phases on inert supports have been preferred for the pupose, but there has been a strong swing towards the use of open-tubular or capillary columns constructed from fused silica over the last 2 or 3 years. Indeed the resolution and speed of analysis attainable with the latter are such that it is not difficult to foresee that all new gas chromatographs will soon be supplied with facilities for capillary columns as the standard. The analysis of fatty acid methyl esters by GLC on columns of fused silica or glass has recently been reviewed [3, 366, 367, 452]. In some circumstances, it may be necessary to subject fatty acids in the free form to GLC analysis and the required techniques have also been reviewed [450].

The liquid phases in use for the GLC analysis of methyl ester derivatives are either non-polar silicones and hydrocarbons or polar polyesters. With the former, unsaturated components are eluted before the saturated fatty acids with the same chain-length, while the reverse is seen with the latter. Polyesters can be classified according to their degree of polarity, from those of relatively low polarity, e.g. PEGA (polyethylene glycol adipate), BDS (butanediol succinate) and Silar ™ 5CP, to those of somewhat higher polarity, e.g. DEGS (diethyleneglycol succinate) and EGSS-X, and to very-highly polar phases, sold under various trade designations, e.g. Silar 10C, Silar 9CP, SP-2340 and OV-275. The last group has been increasingly preferred for use in packed columns, as these phases are capable of better resolution than is attainable with those of lower polarity, especially of geometrical isomers of unsaturated components. Usually, the column packing materials consist of acid-washed and silanized support materials, of a uniform fine grade (100 to 120 mesh), coated with 10 to 15% of the polyester.

The absolute retention times of fatty acid derivatives on a GLC column have very little value for identification purposes, as they are dependent on a variety of instrumental parameters which are not amenable to standardization, and on the nature and age of the packing material. It is of course possible to use authentic standards as a guide to identification, but a relatively few fatty acids only are available from commercial suppliers, in comparison to the wide spectrum found in natural lipids of animal and bacterial origin especially. A secondary reference standard consisting of a natural fatty acid mixture of known composition can then be helpful. For example, cod liver oil has been employed in this way to aid in the identification of marine oils [4], and the lipids of bovine and porcine testes have been utilized similarly with animal tissue lipids [329, 330]. In GLC analyses in general, Kovat's retention indices have been of great importance as an objective means of recording retention data, and analogous parameters, i.e. *equivalent chain-length* (ECL) [539] or *carbon numbers* [890], have proved of great value to lipid analysts. They are calculated by reference to the straight line obtained by plotting the logarithms of the retention times of a series of saturated straight-chain fatty acid methyl esters against the number of carbon atoms in the aliphatic chains. The retention times of the fatty acids in a natural mixture are measured under identical operating conditions, and the ECL values are read directly from the graph. Some values obtained with packed columns by the author [133] with some popular stationary phases, that varied appreciably in polarity are listed in Table 7.1 as a guide. ECL values

TABLE 7.1
Equivalent chain length (ECL) values of methyl ester derivatives of some unsaturated fatty acids on packed columns of EGSS-Y, EGSS-X and Silar 10C

	ECL value		
Fatty acid	EGSS-Y	EGSS-X	Silar 10C
16:1(n-9)	16.62	16.57	17.08
16:2(n-4)	17.45	17.65	
18:1(n-9)	18.52	18.53	19.00
18:2(n-6)	19.20	19.42	20.07
18:3(n-6)	19.67	20.00	21.10
18:3(n-3)	20.02	20.40	21.28
18:4(n-3)	20.52	21.05	22.12
20:1(n-9)	20.45	20.50	20.90
20:2(n-6)	21.15	21.40	
20:3(n-9)	21.33	21.63	22.22
20:3(n-6)	21.53	21.77	22.72
20:3(n-3)	21.60	21.95	23.10
20:4(n-6)	22.00	22.43	23.53
20:4(n-3)	22.47	23.00	24.13
20:5(n-3)	22.80	23.50	24.77
22:1(n-9)	22.35	22.43	22.80
22:4(n-6)	24.00	24.45	25.70
22:5(n-6)	23.85	24.57	25.96
22:5(n-3)	24.80	25.53	26.83
22:6(n-3)	25.20	26.18	27.40

have some obvious physical meaning, and they have some value in inter-laboratory comparisons, although this should not be overestimated as some changes will occur as columns age. With the widely-used EGSS-X stationary phase, a single double bond increases the apparent chain-length of the fatty acid by the equivalent of about 0.5 of a carbon atom. In addition, the position of the double bond system influences the retention time of components, and in general a fatty acid of the (n-9) series is eluted before one with the same number of double bonds but of the (n-6) series, while this is in turn eluted before one of the (n-3) series. The effect of double bond position on ECL values has been well documented in a review by Jamieson [369]. It should be noted that this means of identification can only be used under isothermal operating conditions.

The first capillary columns were constructed of stainless steel; they were costly, had a short working life and were of little use for the quantitative analysis of polyunsaturated fatty acids, because of irreversible adsorption on to the walls of the tubing. Modern capillary columns of fused silica do not suffer from such problems, and with care they will last for one to two years, while affording high resolution and low adsorptivity. Glass capillary columns are surprisingly robust also, but some skill is required to straighten the ends before they are installed into gas chromatographs. Although it is tempting for anyone new to the technique to purchase columns of 50 m in length in order to obtain the maximum resolution possible, excellent results can be obtained with columns of 10 or 25 m with most samples. One advantage of shorter columns indeed is that analysis times can be appreciably reduced, so that more samples can be analysed in a given time. For analytical purposes, the internal diameter of the column should be 0.2 to 0.3 mm, though new columns of 0.1 mm appear promising. The author is not convinced of the value of so-called "wide-bore" columns.

Many different injection systems, e.g. "on-column" or "split/splitless", are available commerically for use with capillary columns. While there does not appear to have been any systematic study of the relative merits of these in lipid analyses, there may be advantages in the use of "on-column" systems with lipid derivatives at the extremes of the volatility range, such as in the chromatography of short-chain esters or in contrast of intact triacylglycerol molecules.

The high efficiency inherent in the use of capillary columns has meant that the resolving power of the phase itself is less important than with packed columns; polyester phases of low polarity then have certain advantages in that there tends to be little overlap of fatty acids of different chain-lengths. For example on column of EGSS-X, as can be seen from Table 7.1, 18:3(n-3) and 20:1(n-9) fatty acids overlap, as do 20:4(n-6) and 22:1(n-9). The author has favoured columns of fused silica coated with Silar™ 5CP (or polyesters of equivalent polarity) for the analysis of fatty acids of animal origin, as this appeared to give satisfactory resolution of all the important polyunsaturated

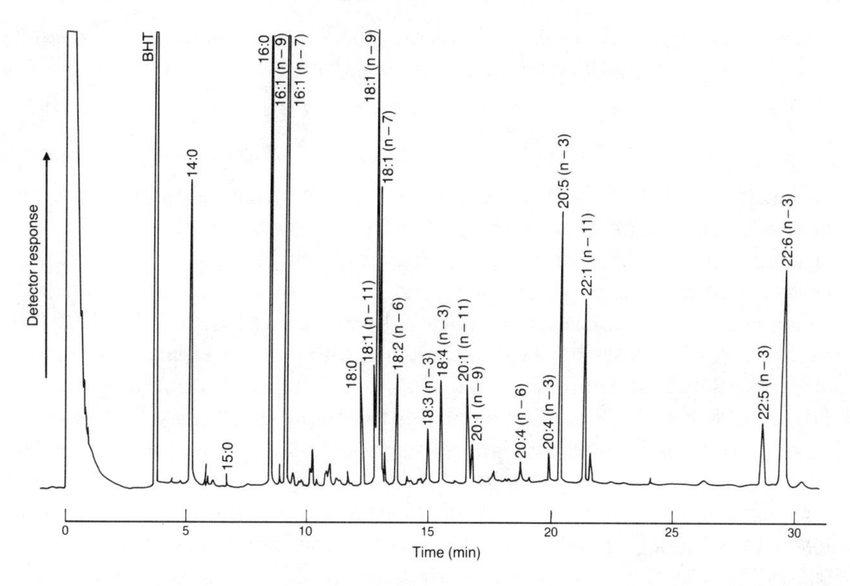

FIG. 7.11. Separation of the methyl ester derivatives of the fatty acids of cod liver oil by means of GLC with a fused-silica capillary column (0.25 mm i.d. × 25 m in length) coated with Silar™ 5CP. Hydrogen was the carrier gas, and temperature-programming was carried out from when the solvent emerged; the oven temperature was held at 140°C for 3 min, then raised at 4°/min to 195°C, at which it was maintained for a further 20 min. The Carlo Erba gas chromatograph was fitted with a "split/splitless" injection system set at a ratio of about 100:1.

fatty acids with no interference by components differing in chain-length by two carbon atoms. Figure 7.11 illustrates a typical separation, i.e of the fatty acids of a commercial fish oil. For all practical purposes, there is no overlap of components of different chain-lengths, and positional isomers of unsaturated fatty acids are well resolved. It has also been established that satisfactory analyses of many esters can be obtained with fused-silica capillaries coated with non-polar phases [788]; such columns are long-lived and relatively easy to prepare. Unfortunately, the essential fatty acids, 18:2 and 18:3, tend not to be resolved.

One valuable application of capillary GLC with the most polar phases, such as OV-275 or SP-2340, has been for the separation of *cis* and *trans* isomers of unsaturated fatty acids, especially monoenes [5, 188, 242, 305, 365], although good results have also been obtained with longer than usual packed columns. *Trans* isomers are eluted ahead of the corresponding *cis* compounds. With complex mixtures such as partially hydrogenated fats, that contain a wide range of positional isomers which may be resolved into a number of additional peaks, interpretation of the results obtained may be difficult initially, but it can be mastered with experience.

While manual methods can give satisfactory results in the quantification of peaks from packed columns, it is essential to use electronic integration for analyses with capillary columns.

2. *Thin-layer chromatography*

Adsorption TLC [133] has been much used for the isolation of fatty acids containing polar substituent groups, especially oxygenated moieties, and the chromatographic behaviour of positional isomers of hydroxy-, acetoxy- and keto-stearates in particular has been studied in some detail [551].

Chromatography on layers of silica gel impregnated with 10 to 20% (by weight) of silver nitrate has been of immense value in the separation of fatty acids according to the number, configuration and to some extent the positions of the double bonds. (The principle of this technique is discussed in Section C above.) For example, it is reportedly possible to separate methyl ester derivatives of fatty acids with zero to six double bonds on silver nitrate TLC plate with a double development in hexane–diethyl ether–acetic acid (94:4:2 by volume), provided that the relative humidity is maintained below 50% [195]. On the other hand, most workers have preferred to separate methyl esters of fatty acids with zero to two double bonds on one silver nitrate-impregnated plate, with hexane–diethyl ether (9:1 by volume) as developing solvent, and those with three to six double bonds on a further plate, with hexane–diethyl ether (2:3 by volume) as the eluent [133]. Spots are visualized under a UV lamp after spraying with a solution of 2′,7′-dichlorofluorescein in 95% methanol (0.1% by weight). Esters with zero to two double bonds can be recovered from the adsorbent by elution with diethyl ether, while those with three to six double bonds may require elution with a solvent as polar as chloroform–methanol (9:1 by volume) for quantitative recovery. To remove excess silver ions and the spray reagent, the lipids should be taken up in hexane–diethyl ether (1:1 by volume) and washed first with sodium chloride solution containing dilute ammonia (pH 9.0).

Silver nitrate TLC and GLC used sequentially make a powerful technique for the identification and analysis of fatty acids in natural mixtures. Fatty acids containing one *trans* double bond, for example, can be quantified by separation first on a silver nitrate plate (they migrate between the saturated and *cis* monoene components), followed by elution together with the saturated components from the adsorbent. By analysing before and after this separation by means of GLC, the relative amounts of the *trans*- and *cis*-monoene components can be determined [148].

3. *Identification of unsaturated fatty acids*

The first step in the identification of a pure fatty acid isolated by silver nitrate TLC (which will have given an indication of the number of double bonds), or a combination of this technique with HPLC or other procedures, is the determination of its chain-length. This can be accomplished by hydrogenation for comparison of the retention time of the product on a GLC column with that of authentic saturated standards. A simple method to accomplish hydrogenation has been described [133]. Next the fatty acid can be oxidized at the double bond under controlled conditions either with ozone, which is a mild reagent with few side reactions although special equipment is required to generate it, or with a permanganate–periodate solution, which is a little more vigorous but makes uses of readily available chemicals. The scission products can again be identified by their retention times relative to authentic standards when subjected to GLC analysis. These methods have thoroughly reviewed in other publications [133, 678, 734].

Spectroscopic methods, including IR, UV and NMR spectrometry, have an important place in identifying and sometimes in locating specific functional groups, including double bonds, in fatty acyl chains. In addition, mass spectrometry, when used in conjunction with GLC, has proved an immensely powerful weapon in the hands of natural product chemists including lipid analysts. Unfortunately, methyl ester derivatives of unsaturated fatty acids *per se* have proved to be unsuitable for identification by MS, since the double bonds migrate when ionized. To overcome this problem, it has proved possible to "fix" the double bonds, generally by preparing some form of oxygenated derivative, which is then converted to a form suitable for gas chromatography. For example, double bonds have been oxidized to vicinal diols and then converted to trimethylsilyl ether derivatives, which were sufficiently volatile and gave excellent mass spectra with fragmentation patterns that were readily interpretable in terms of the positions of the original double bonds [33, 106].

A simpler approach has been to convert the fatty acids at the carboxyl group to various amide derivatives, which appear to stabilize the double bonds during ionization. Pyrrolidides were first used for the purpose, and gave spectra in which the diagnostic ions were small but usable with monoenes and dienes, although the results were less certain with polyenes [26]. More recently, Harvey showed that much better spectra could be obtained with picolinyl esters, both of unsaturated fatty acids and of those with other functional groups, such as hydroxyl groups or cyclopropane rings [287–289]. Of equal importance was the observation that picolinyl ester derivatives of unsaturated fatty acids in natural samples could be satisfactorily separated by GLC on fused-silica capillary columns, coated with non-polar methyl-silicone phases; the identity of each component was then established by MS [140, 881]. A procedure which appears to offer some of the best features of amidation of the carboxyl groups with derivatization of the double bonds has been described [27, 409, 429], i.e. in which the double bonds of pyrrolides were deuterated; the ions diagnostic for the positions of the original double bonds in such saturated derivatives were often readily identifiable. It is the author's impression that picolinyl esters would be better suited to such a method. In addition, an HPLC system has recently been developed for the fractionation of picolinyl esters, and this assisted in the isolation of minor components for subsequent analysis by GC-MS (Christie and Stefanov, *J. Chromatogr.*, **392**, 259–265 (1987)). Procedures for the location of double bonds by MS in unsaturated fatty acids have been reviewed very fully [533, 730].

CHAPTER 8

The Separation of Molecular Species of Glycerolipids

A. The Nature of the Problem

Each lipid class in a tissue exists in nature as a complex mixture of related components in which the composition of the aliphatic residues varies from one molecule to the next. Sometimes, as in cholesterol esters, only the single fatty acid component will change. On the other hand in triacylglycerols, each of the three positions in the molecules may contain a different fatty acid. Both the long-chain base and fatty acid constituents of sphingolipids can be variable, and can exist in distinctive combinations. For a complete structural analysis of a lipid, it is therefore necessary to separate it into *molecular species*, i.e. into groups of molecules with single specific alkyl or acyl moieties (fatty acids, alcohols, ethers, etc.) in all the relevant portions of the molecule. With lipids that contain only one aliphatic residue, this can frequently be accomplished without difficulty. When there are two aliphatic residues, the task is much more difficult, but is not impossible. While triacylglycerols can be subjected to some considerable molecular simplification, it is not yet feasible technically to obtain single species, especially if differing enantiomeric forms are considered. Determination of the composition of molecular species is not merely an academic challenge. Specific molecular species of phospholipids are involved in important biological reactions, and the molecular compositions of edible fats and oils can govern their physical and nutritive properties, and their commercial value.

Under ideal circumstances, the analyst would wish to separate a lipid into individual molecular species in its native form, in order that the biosynthesis or metabolism of every part of the molecule could potentially be studied, or so that the physical properties of each species could be assessed in relation to those of the whole. With polar complex lipids, however, the analysis can be simplified in a technical sense by converting the compounds to non-polar forms by derivatization or by partial hydrolysis techniques. The approach of the analyst will depend on the nature and amount of information required.

Determination of the positional distributions of fatty acids in glycerolipid molecules is usually achieved primarily by means of specific enzymatic hydrolysis methods with subsequent chromatographic separation and analysis of the products, and not simply by chromatography *per se* [133]. Such procedures are not discussed in detail here therefore. It is possible that this will change when chiral-phase chromatography is brought to bear on the problem, as appears likely to occur in the near future.

The chromatographic procedures utilized for the separation of molecular species of glycerolipids resemble in kind those used for the separation of fatty acid derivatives, and described in the previous chapter. With most lipids, the separations achieved depend on the combined physical properties of all the aliphatic residues in each molecule. Considering triacylglycerols for illustrative purposes, reversed-phase HPLC will separate molecules according to the combined chain-lengths of the fatty acids, with the retention times being reduced by the equivalent of approximately two carbon atoms for each double bond in the three fatty acid constituents. Silver ion chromatography will separate those molecules containing three saturated fatty acids from those with one monoenoic and two saturated fatty acids, and these are in turn separated from further fractions with an increasing degree of unsaturation. Adsorption chromatography can be used to separate molecular species containing three normal fatty acids from those containing two normal fatty acids and one with a polar substituent, such as a hydroxyl group. Often no single method will give the required degree of fractionation, but if two of the above separation modes are used in sequence, a high degree of molecular simplification may be possible. Procedures for the separation of molecular species of lipids in general have been comprehensively reviewed elsewhere [133,452].

It is worth noting that in the determination of molecular species, it is normally advisable to calculate results in terms of molar rather than weight percentages.

B. HPLC Separation of Triacylglycerol Species

1. Silver ion chromatography

The principles of silver ion chromatography are discussed in Chapter 7 above (Section C) in relation to the separation of fatty acid derivatives. Silver ion complexation chromatography, used in conjunction with TLC, revolutionized the study of triacylglycerol structures during the 1960s (reviewed by Litchfield [490]). With relatively simple equipment, it was possible to obtain distinct molecular fractions, separated on the basis of a single well-defined property, i.e. degree of unsaturation. When fractions obtained in this way were subsequently separated by reversed-phase TLC or by high-temperature GLC, additional information was obtained on the chain-length distributions of the fatty acid constituents. It was inevitable that attempts would be made to transplant the technique of silver ion chromatography to HPLC, but only a handful of papers have appeared on the subject to date, testifying to the technical difficulties that have been encountered.

Indeed, it is obvious that the problems have yet to be fully mastered and that further developmental work is necessary. A stable reliable HPLC system, making use of silver ion complexation, would be of enormous value to lipid analysts.

The first apparently successful separation of triacylglycerols by HPLC with silver complexation made use of a reversed-phase column (Nucleosil™ $5C_{18}$) and elution with methanol–isopropanol (3:1 by volume) containing 5×10^{-2} M $AgClO_4$ [856]. The silver ions reduced the retention volumes of the unsaturated components relative to those on reversed-phase HPLC alone. Excellent resolutions were obtained both with standards and with triacylglycerols from olive oil, but the elution pattern was complicated by chain-length separations and no attempt was made at quantification. Some problems were experienced with silver depositing in optical flow cells. There have been two further attempts to use a method of this kind, in which acetonitrile–acetone (2:1 by volume) and acetonitrile–tetrahydrofuran–dichloromethane (3:1:1 by volume), containing 0.2 M silver nitrate [660], and acetonitrile–acetone (13:17 by volume), containing 2% silver nitrate [66], were used as the mobile phases. Some change in molecular selectivity was again observed, but it appeared insufficient to this author to be worth the risk of corroding the HPLC equipment. Significant recent improvements in the resolution attainable with more conventional reversed-phase chromatography of triacylglycerols have also tended to lessen the value of these methods (see the next section of this chapter).

A more conventional approach to silver ion HPLC consisted in impregnating silica gel (Partisil™ 5) with 10% by weight of silver nitrate by evaporation from a solution in acetonitrile [776], or better from methanol–water (9:1 by volume) [275]. The column (4 × 250 mm, for analytical purposes) was packed with the adsorbent in the form of a slurry with carbon tetrachloride. A separation of a triacylglycerol mixture on such a column, maintained at 6.8°C, eluted with benzene at 1 ml/min and with refractive index detection, is shown in Fig. 8.1 [776]. Mono-, di- and triunsaturated fractions, including some positional isomers, were obtained. Methyl santalbate (octadec-9-yne-11-*trans*-enoate) was added as an internal standard. Over the range studied, resolution increased as the silver ion load was decreased and was optimum at 5%; the 10% loading selected was a compromise to improve the speed of analysis. The resolution was also greatly improved by lowering the temperature of the column, as this increased the stability of the complexes with the silver ions. Good results were obtained in quantitative analyses, no difference in the response of the detector being found with degree of unsaturation (see Section B.3.i below, however). Even better qualitative results had been reported in brief by a French group in the previous year [545]. They used a column of a porous pellicular silica gel (Spherosil™ XOA 600) impregnated with 10% by weight of silver nitrate, and eluted with toluene–hexane–ethyl acetate (30:20:1 by volume); a Pye Unicam LCM2

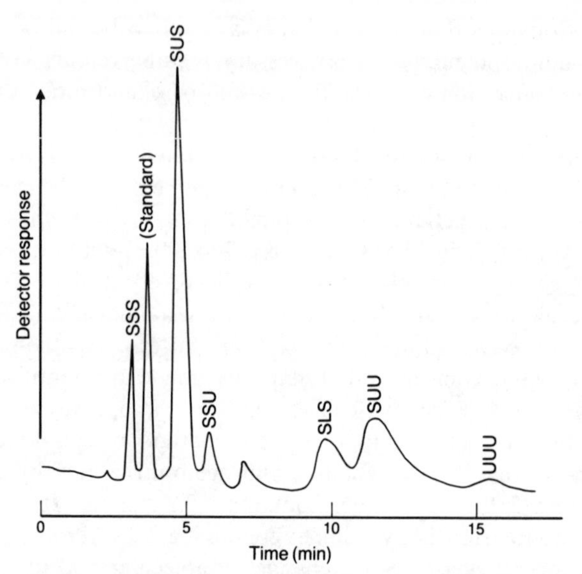

FIG. 8.1. Separation of triacylglycerols on a column (4 × 250 mm) of silica gel (Partisil™ 5), impregnated with silver nitrate (10%). The mobile phase was benzene at a flow-rate of 1 ml/min, and refractive index detection was used [776]. Abbreviations: S = saturated, U = monounsaturated and L = linoleoyl residues. The internal standard was methyl santalbate (octadec-9-yne-11-*trans*-enoate). (Reproduced by kind permission of the authors and of the *Journal of Chromatography*, and redrawn from the original paper.)

transport-flame ionization detector was used. Although these papers appear superficially to be a great step forward, the range of unsaturated components separated was rather limited, and the experience of others with columns prepared in this way has been that silver ions tend to elute from the columns in continued use, so that the resolution deteriorates (see Chapter 7).

An attempt to use a macroreticular ion-exchange resin loaded with silver ions for triacylglycerol separations was not successful, partly because solvents of high polarity were necessary for elution of components, causing solubility problems, and partly because a small proportion of residual sulphonic acid groups catalyzed some transesterification of the solute [6]. The author (*J. High Res. Cromatogr. Cromatogr. Commun.*, **10**, 148–150 (1987)) has, however, had some preliminary success with a column of a similar type, but with aprotic solvents as the mobile phase.

2. *Reversed-phase HPLC*

i. Theoretical considerations

The great resolving power of HPLC in the reversed-phase mode has been utilized to particularly good effect for the separation of molecular species of triacylglycerols. The topic has been reviewed briefly elsewhere [15, 273, 639,

659]. The principle of the separation process, as applied to fatty acid derivatives, is described in Chapter 7. In short, components elute in ascending order of chain-length, while a double bond in the chain reduces the retention time by the equivalent of about 2 carbon atoms. When applied to intact triacylglycerols, the separation is in ascending order of the total number of carbon atoms in the aliphatic chains of the three fatty acids, with a double bond in any of the fatty acids reducing the retention time to that of a component with two fewer carbon atoms. The relative retention time of a given component has been defined in terms of an "equivalent carbon number" (ECN) or "partition number" value, defined as the actual number of carbon atoms in the aliphatic residues (CN) less twice the number of double bonds (n) per molecule (the carbons of the glycerol moiety are not counted for this purpose), i.e.

$$ECN = CN - 2n$$

Two components having the same ECN value are said to be "critical pairs". For example, triacylglycerol species containing the fatty acid combinations 16:0–16:0–16:0, 16:0–16:0–18:1, 16:0–18:1–18:1 and 18:1–18:1–18:1 have the same ECN value and tend to elute close together. (The positions of the fatty acids within the triacylglycerol molecules have no effect on the nature of the separation in this instance.) The ECN concept was useful in the early days of the technique, when the resolving power was relatively limited. For example, it defined closely the composition of the fractions obtained when milk fat was separated on a column of a lipophilic hydroxyalkoxypropyl Sephadex™ [487]. On the other hand, the formula is now only of utility as a rough rule of thumb, by way of a guide to what may elute in a given area of a chromatogram, since the greatly increased resolving power of modern HPLC phases means that the factor for each double bond has to be defined much more precisely. Also, this factor can be no longer be treated as a constant, as a second double bond in a molecule has a slightly different effect from the first. Accordingly, more complex formulae are necessary to define the order of elution of triacylglycerols from modern reversed-phase columns, which in essence means from ODS stationary phases, as these have been used almost exclusively for the purpose. (A normal-phase partition separation of triacylglycerols on silica gel has been described, but it is doubtful whether this procedure has any practical significance [661].)

The quality of the separations obtained is greatly dependent on the composition of the mobile phase. It has been found to vary somewhat with the brand of packing material, in addition to such factors as the column dimensions and particle sizes, the column temperature and the injection solvent. These parameters affect the resolution of specific components, but not the general nature of the separations, and they are discussed at length below. The same detector and quantification problems arise in triacylglycerol analysis as in many other aspects of the HPLC of lipids. For convenience here, the nature of the separations of triacylglycerol molecular species that have been achieved are discussed first, before the merits of the various types of detector for quantification purposes are assessed.

Natural triacylglycerols may consist of an enormous number of different molecular species. For example, if only five different fatty acids are present, there can be 75 different molecular species (not counting enantiomers). Identification of those species separated by reversed-phase HPLC can therefore be troublesome. If non-destructive detectors are used, it is possible to collect the component triacylglycerols as they elute from the column, and then to identify them by GLC analysis of their constituent fatty acids. Fractions have also been quantified in this way by adding an internal standard at an early stage in the procedure [658, 752]. This approach is tedious and time-consuming, but may be essential with samples from new sources. Mass spectrometry can also be a considerable aid to identification, when available to the analyst (see Section B.4 below). Neither of these methods will be suitable for all components, especially when species overlap or are present in very small amounts. Some form of the ECN concept can then be of value.

Under steady state conditions, i.e. with a given column and isocratic elution at a constant temperature and flow-rate, it would be expected from theoretical considerations that there should be a rectilinear relationship between the logarithm of the capacity factor (k), or retention time or volume, and the carbon number of each member of a homologous series of triacylglycerols containing saturated fatty acids only. This was first shown to be true in practice by Plattner [658], and it has since been confirmed by many others [177, 201, 202, 245, 317, 648, 663, 794]. Similarly, it was shown that there was a rectilinear relationship between the logarithm of retention volume and the total number of double bonds in a series of C_{54} triacylglycerols [658]. It would also be expected that there would be linear relationships between the logarithms of the retention volumes and carbon numbers of homologous series of triacylglycerols containing particular unsaturated fatty acids. As it was not easy to obtain suitable standards to test this hypothesis, it was at first assumed to be true before eventually being proven [663, 794]. It was also shown that both the position and configuration of a double bond in a fatty acid constituent of a triacylglycerol had an effect on retention volume.

From such evidence, El-Hamdy and Perkins [202] suggested that triacylglycerol species might be identified from a *theroretical carbon number* (TCN), defined as—

$$\mathrm{TCN} = \mathrm{ECN} - \sum_{3}^{1} U_i$$

where U_i was a factor determined experimentally for different fatty acids (zero for saturated fatty acids, and roughly 0.2 for elaidic, 0.6 to 0.65 for oleic, and 0.7 to 0.8 for linoleic acid residues). Thus, the TCN value for a triacylglycerol containing 18:1–18:1–18:1 (triolein) was calculated as—

$$\mathrm{TCN} = (3 \times 16) - (3 \times 0.6) = 46.2$$

while for 18:1–18:1–16:0, 18:1–16:0–16:0 and 16:0–16:0–16:0, which have the same ECN value, the TCN values were 46.8, 47.4 and 48.0 respectively. The values of U_i will vary with the elution conditions and have to be determined independently by each analyst for his own conditions, but the concept of the TCN has been found to be of value by many workers. A list of TCN and related data for some important triacylglycerol species is contained in Table 8.1. The numerical values can be seen to have some immediate relevance, unlike relative retention times, for example.

As an alternative, Goiffon *et al.* [245, 246] presented a scheme for identifying triacylglycerol species in which the principle of additivity of solution free energies of saturated and unsaturated acyl residues (up to C_{18}) was utilized. The results were presented as a plot of log k (or retention volume) against the number of double bonds in the triacylglycerol. In practical terms, identifications were found to be most readily accomplished by plotting the number of double bonds in the molecule against the logarithm of the retention time for each component relative to that of triolein, expressed as log α. Parallel straight lines were obtained for all the homologous series. For a given triacylglycerol species, log α was the sum of the equivalent values for each of the three constituent fatty acids, the latter being equal to one third of log α

TABLE 8.1
Theoretical carbon number (TCN) values of some unsaturated triacylglycerol species [202, 648]

				TCN	
Triacylglycerol	ECN	CN	No. of double bonds	[202]	[648]
18:2–18:2–18:2	42	54	6	39.6	
18:2–18:2–18:1	44	54	5		41.8
18:3–18:1–18:1	44	54	5		42.1
18:2–18:2–16:0	44	52	4		42.3
18:3–18:1–16:0	44	52	4		42.7
14:0–14:0–18:1	44	46	1	43.4	
18:2–18:1–18:1	46	54	4		43.9
18:1–18:1–16:1	46	52	3		44.3
18:2–18:1–16:0	46	52	3		44.5
18:1–16:1–16:0	46	50	2		44.9
16:0–16:0–18:2	46	50	2	45.2	45.2
18:1–18:1–18:1*	48	54	3	46.2	45.9
16:0–18:1–18:1	48	52	2	46.8	46.6
16:0–18:1–16:0	48	50	1	47.4	47.3
18:0–18:1–18:1	50	54	2	48.8	48.5
18:0–16:0–18:1	50	52	1	49.4	49.3
18:0–18:0–18:1	52	54	1	51.4	51.3

*Triolein (trielaidin had an ECN value of 47.4 [202]).

of the corresponding simple triacylglycerol. The retention time of any triacylglycerol could be calculated from such data, or from the graphical relationships. However, others [794] pointed out that it was less easy to identify an unknown in this way, and have suggested an alternative method to solve the problem. For those more mathematically inclined, Takahashi *et al.* have proposed an identification technique based on a matrix model, but utilizing essentially the same principles [815, 816].

In one further approach, Podlaha and Toregard employed the term "ECN" rather more in the sense of the *equivalent chain-length* principle first enunciated for fatty acids (see Chapter 7) [663], and if used more extensively would at least have the merit of being consistent with established procedures elsewhere. For a given homologous series in this instance, the carbon numbers were found to be related in a rectilinear manner to the ECN values. A range of parallel straight lines were again obtained for series of homologues differing in degree of unsaturation. Sempore and Bezard [752] confirmed the relationships found by Goiffon, Podlaha and coworkers, and demonstrated how the two approaches could be harmonized through the graphical determination of both $\log \alpha$ and the ECN for every component from the same data. Under their chromatographic conditions for each species—

$$\mathrm{ECN} = 17.54 \log \alpha + 45.60.$$

With a different elution system, the numerical values of the constants would be expected to vary, but the principle should hold.

All of these methods present the analyst with lengthy preparatory work, not only in standardizing the chromatographic conditions, but also in running a variety of natural samples and standards to obtain retention data. Graphical methods can be tedious and can obviously only be used if sufficient data points are available. Perrin and Naudet [642] have therefore simplified the task by tabulating retention data relative to that of triolein for 120 different triacylglycerol species commonly encountered in seed oils. The absolute values listed probably cannot be reproduced directly by other workers, but the data is certainly important for predicting relative orders of elution. As an example, Table 8.2 contains some of this information for components of the same ECN value (= 44).

It is not possible to separate positional isomers within a particular molecular species by HPLC in the reversed-phase mode under normal conditions, but this has been accomplished after bromination (see Section B.5 below).

ii. Columns and packing materials.

Virtually all the work on the reversed-phase separation of triacylglycerols has been carried out with ODS phases, and there appears to be only one brief report of an attempted use of a C_8 phase, which gave very poor results

TABLE 8.2

Retention times (RT) relative to triolein, carbon numbers (CN) and number of double bonds (DB) of triacylglycerols of equivalent carbon number = 44, when separated by reversed-phase HPLC [642]

Triacylglycerol	RT	CN	DB	Triacylglycerol	RT	CN	DB
18:1–18:2–18:2	0.549	54	5	14:0–16:0–18:2	0.684	48	2
16:1–18:1–18:2	0.571	52	4	16:1–18:0–18:3	0.691	52	4
18:1–18:1–18:3	0.583	54	5	12:0–16:0–18:1	0.695	46	1
16:1–16:1–18:1	0.593	50	3	14:0–14:0–18:1	0.702	46	1
16:0–18:2–18:2	0.605	52	4	16:0–16:0–18:3	0.709	50	3
14:0–18:1–18:2	0.621	50	3	14:0–16:0–16:1	0.712	46	1
16:0–16:1–18:2	0.629	50	3	12:0–18:0–18:2	0.718	48	2
12:0–18:1–18:1	0.631	48	2	12:0–16:1–18:0	0.747	46	1
16:0–18:1–18:3	0.643	52	4	14:0–18:0–18:3	0.751	50	3
14:0–16:1–18:1	0.646	48	2	12:0–16:0–16:0	0.766	44	0
16:0–16:1–16:1	0.654	48	2	14:0–14:0–16:0	0.774	44	0
18:0–18:2–18:3	0.660	54	5	12:0–14:0–18:0	0.812	44	0

[246]. Most commercial brands of ODS phase have been used by one analyst or another for the purpose, and while it is evident from direct comparisons that some are better than others [202, 215, 648, 660], the reasons for this are not clear. It is possible that the total carbon content of the alkyl groups in the phase is a factor, and it has been demonstrated that phases with a higher content of alkyl groups (usually designated as "ODS2"), from the same manufacturer, retain triacylglycerols more strongly [201]. Such is the variability and uncertainty, that it is not possible to recommend any particular brand.

Initially, particle sizes of 10 μm were employed in the stationary phase, but most of the recent published work was carried out with 5 μm particles. The smaller the particle size, the better the resolution that would be anticipated from a column of constant dimensions, and a few excellent separations have now been achieved with 3 μm phases [193, 215, 648, 650, 752, 763].

With 5 μm particles, the standard column dimensions have generally been in the range of 4 to 5 mm internal diameter by 250 to 300 mm in length, and some very good separations of triacylglycerol species have been achieved with columns of this kind. Within limits, the longer the column, the better the resolution attainable, and some authors have coupled two or more columns together (up to 100 cm in total length) to improve the separations [215, 646, 663, 664, 697, 698, 794, 815]. Similarly with 3 μm phases, which are usually supplied in columns of 10 cm length, it has proved advantageous to use longer columns or to couple as many as three together to improve resolution [193, 215, 752, 763]. Of course, the time taken for a given separation is increased proportionately when this is done.

iii. Temperature and injection solvents.

Most analysts have carried out reversed-phase separations of triacylglycerols with their columns at ambient temperature. Over the temperature range 14.5 to 25.5°C, Jensen [372] found that the lower the temperature, the longer the retention time and the better the separation, a result confirmed later by others [235]. However, sharper peaks for high melting triacylglycerols, such as tristearin, were obtained at the higher temperature. Indeed the improvement in resolution obtained at lower temperatures was so small, that some workers have preferred to work at higher than ambient temperatures (30 to 45°C), to speed up the analysis and to improve the separations of relatively saturated fats [177, 215, 752, 767].

Depending on the sensitivity of their detection system, analysts working with natural samples containing many different molecular species have generally tended to inject 1 to 5 mg of lipid in a small volume of solvent on to the reversed-phase column. It is self evident that the smaller the volume injected, the less chance there is of band spreading and loss of resolution. It is probably less obvious that the nature of the injection solvent can also have an effect. The problem was studied by Tsimidou and Macrae [833, 834], who concluded that the ideal injection solvent should be identical in composition to the mobile phase, but that when the solubility of the sample was a problem, it should be injected in acetone (not suitable for saturated triacylglycerols), or in a mixture of acetone–tetrahydrofuran of similar polarity to the mobile phase. Tetrahydrofuran alone could be used only if the volume injected was no more than 5 μl, while chloroform alone was not recommended. Unfortunately, others [767] came to rather different conclusions, possibly because they used different mobile phases. In the absence of a definitive explanation, analysts should be aware of the problem and may have to experiment with their own systems.

iv. The mobile phase.

Perhaps the single most important factor in the separation of triacylglycerols in the choice of the mobile phase. In the early work on the problem, methanol–water [637] and methanol–chloroform [858, 859] mixtures were utilized, but it soon became apparent that solvent combinations based on acetonitrile gave much better resolution. The latter are now used almost universally, although propionitrile is preferred by some analysts. An element of subjectivity must be brought to bear here in discussing eluents, since there are so many other variables involved in comparing published separations from different laboratories, not least the skill of the analyst concerned. In order to obtain the optimum separations of triacylglycerols, it is necessary to add some other solvent to the acetonitrile to increase the solubility of the solute, to change the polarity and to modify the selectivity. It is well established that the relative proportion of acetonitrile to the modifier solvent has a marked effect on the elution time of a given triacylglycerol species, i.e. the lower the polarity of the mobile phase, the lower the retention volume [201]. There has been a considerable debate in the literature as to which solvent mixture is best. No consensus has emerged, partly because there appears to be no objective criterion that can be used to assess relative merits, and partly because the nature of the columns and other equipment used in different laboratories may impose some restraint. The detection system available to the analyst may also be a limitation. With refractive index detectors, isocratic elution must be used; with UV spectrophotometric detection, solvents such as acetone or chloroform especially are opaque in the regions of interest, although other solvents may still gave excellent results and gradient elution may be possible. In general as would perhaps be expected, gradient elution conditions tend to give sharper peaks over the whole range of components present in a natural fat, than is possible under isocratic conditions.

The flow-rate of the mobile phase for optimized separations does not appear to have been studied in a systematic manner in these particular circumstances, but from theoretical considerations would be expected to give the best results at 0.5 to 1.5 ml/min with a column of standard dimensions.

Acetonitrile–acetone mixtures have been used more often than any other solvent combination as the mobile phase in the reversed-phase separation of triacylglycerol species, in an isocratic manner with refractive index detection and in gradients with the mass detector. It has certainly permitted some fine separations, especially with seed oils, and an application to palm oil is illustrated in Fig. 8.2 [177]. (Further data on these samples were given in a subsequent paper [178].) Here, the mobile phase was acetone–acetonitrile (62.5:37.5 by volume) at 30°C. Five main groups of peaks, distinguishable by their ECN values, were recognized. Group 1 (ECN = 44) consisted of the combinations 18:1–18:2–18:2, 16:0–18:2–18:2 with 14:0–18:1–18:2, and 14:0–18:2–16:0 with 14:0–18:1–14:0. (Note that the acyl groups are listed in an arbitrary order and do not represent any specific positions within the triacylglycerol molecules.) The order of elution was consistent with the data in Table 8.1 and 8.2, and this could be used to predict the order of other species that might conceivably be present. (The original paper [642], from which the data in the table was taken, should be consulted for the remaining partition groups.) Group 2 (ECN = 46) consisted of 18:1–18:1–18:2, 16:0–18:1–18:2, 16:0–18:2–16:0 with 14:0–18:1–16:0, and 14:0–16:0–16:0. Group 3 (ECN = 48) is a particularly important one in confectionery fats and comprised predominantly 18:1–18:1–18:1, 16:0–18:1–18:1, 16:0–18:1–16:0 and 16:0–16:0–16:0. In group 4 (ECN = 50), only three main components could be recognized, i.e. 18:0–18:1–18:1, 16:0–18:1–18:0, and 16:0–16:0–18:0, while the fifth group (ECN = 52) containing 18:0–18:1–18:0 and 16:0–18:0–18:0.

Many other workers have used acetonitrile–acetone, in slightly different proportions depending on the nature of the sample and of the column and

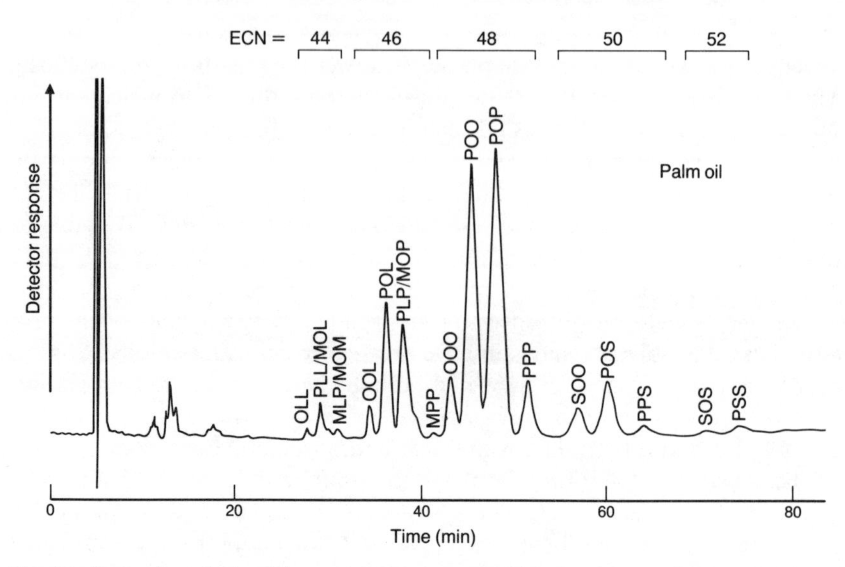

FIG. 8.2. Separation of triacylglycerols from palm oil by reversed-phase HPLC on a column (8 × 100 mm) of *micro*Bondapak™C_{18} (5 μm) and a column (4 × 250 mm) of LiChrosorb™ RP-18 (5 μm) in series, and maintained at 30°C [177]. The mobile phase was acetone–acetonitrile (62.5:37.5 by volume) at a flow-rate of 1.1 ml/min, and refractive index detection was employed. The abbreviations refer to fatty acyl residues: M, 14:0, P, 16:0; S, 18:0; A, 20:0; Pa, 16:1; O, 18:1; L, 18:2; Ln, 18:3. The ECN values are listed above the appropriate groups of peaks. (Reproduced by kind permission of the author and of *Revue Francaise des Corps Gras*, and redrawn from the original paper.)

its packing material, for the separation or isolation of the triacylglycerol species in palm oil [177, 202, 646, 795], human plasma and lipoprotein fractions [641], rape seed oil [794, 795], soybean oil [202, 246, 494, 615, 642, 658, 660, 833], loba oil [646], cocoa butter [418, 615, 646], olive oil [193, 202, 633, 642, 794], corn (maize) oil [193, 202, 642, 658, 833], peanut oil and fractions obtained from this by silver ion TLC [69, 193, 752, 794], coconut oil [317, 658], sunflower oil [193, 658, 794, 833], the seed oil of *Argania spinosa* [210], linseed oil [658, 794], safflower oil [658, 833], palm kernel oil [317], cod liver oil [795], beef tallow [177], lard [177], horse fat [794], standard mixtures [315] and avocado oil [494]. It is perhaps invidious to pick out one of these as of particular interest, but Dong and DiCesare [193] have obtained some remarkably good separations of seed oils over relatively short time periods by elution with acetonitrile–acetone (3:7 by volume) at 1 ml/min from two columns (4.6 × 100 mm each) in series, containing a 3 μm ODS phase.

The triacylglycerols in butter fat, which contains a large number of different fatty acids, have presented a particular challenge to analysts [177, 697, 698, 794, 795]. Figure 8.3 represents a separation of bovine milk fat obtained in the author's own laboratory, with a gradient of acetone into acetonitrile, an

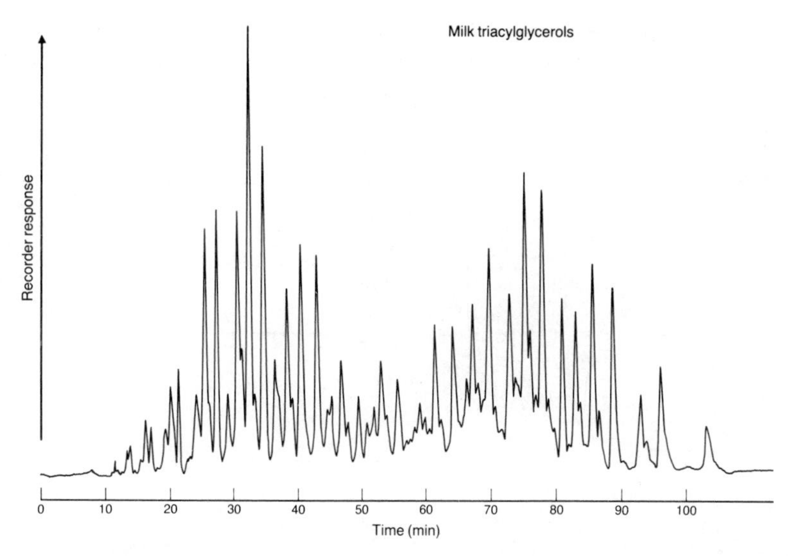

FIG. 8.3. Separation of triacylglycerols from cows' milk by reversed-phase HPLC on a column (5 × 250 mm) of Spherisorb™ ODS2 (5 μm) with mass detection (Applied Chromatography Systems, Macclesfield, Cheshire). The mobile phase was acetone–acetonitrile (1:1 by volume) over the first 15 min, and then was changed by a linear gradient to acetone 100% over a further 105 min; the flow-rate was 0.5 ml/min.

ODS2 phase and mass detection (unpublished work). It also graphically illustrates one of the problems of reversed-phase separations of triacylglycerol species, in that such is the complexity that no simple theoretical scheme would be of assistance in identifying peaks. Here, mass spectrometry coupled to HPLC would appear to be the only solution, and a few of these components have been identified elsewhere by this means from bovine and goat milk fat [453, 454].

The main difficulty associated with acetonitrile–acetone as the mobile phase, other than its unsuitability with UV detection, is that trisaturated species of high molecular weight, above C_{48}, tend to be insoluble and can crystallize out in the column. This was first pointed out by Lie Ken Jie [486], and has since been confirmed by many others. One solution has been to raise the column temperature to ensure that all of the components remain in solution, and this may be favoured where the analyst has the appropriate equipment available to him [177, 372, 752, 794]. Acetonitrile in combination with other solvents can permit elution of saturated triacylglycerols with less difficulty, and Fig. 8.4 illustrates a separation of hydrogenated milk fat, with a gradient of 1,2-dichloroethane into acetonitrile, on the same column as in the previous figure (Author, unpublished work). Not only were all components with odd and even carbon atoms separated, but fractions containing branched-chain fatty acids could be discerned. With the same sample and

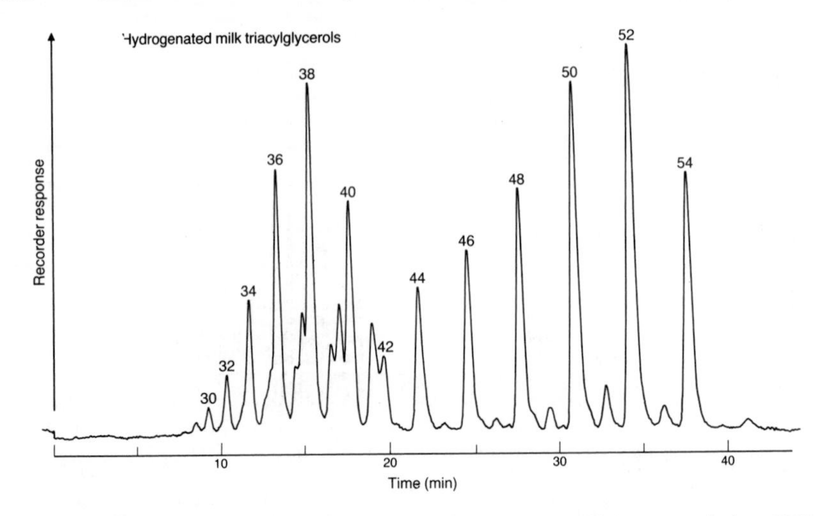

FIG. 8.4. Separation of hydrogenated triacylglycerols from cows' milk by reversed-phase HPLC on a column (5 × 250 mm) of Spherisorb™ ODS2 (5 μm), with mass detection (Applied Chromatography Systems, Macclesfield, Cheshire). The mobile phase was 1,2-dichloroethane-acetonitrile (1:1 by volume) for 5 min, and was changed by a linear gradient to 1,2-dichloroethane-acetonitrile (4:1 by volume) over a further 55 min; the flow-rate was 0.5 ml/min. The numbers above the peaks refer to the total numbers of carbon atoms in the fatty acid constituents, and in this instance are also the ECN values.

the elution scheme of Fig. 8.3, there was a better separation in the early region of the chromatogram, but components of about C_{48} and above were badly shaped and C_{54} did not appear to elute at all.

Dichloromethane–acetonitrile has been used to good effect as a mobile phase, especially by Privett's group, with transport-flame ionization detection [648, 650]. The eluent was tested first with model mixtures, and so good was the resolution that it was possible to distinguish molecular species containing petroselinoyl (i.e. with the *cis* double bond in position 6) from those with oleoyl (position 9) residues. It also gave good results with soybean oil, cocoa butter and olive oil, and a separation of the last of these is shown in Fig. 8.5. In this instance, gradient elution was employed. In addition to the major components, many minor species could be recognized in each partition group. In work by others, a comparison of a number of solvent combinations was made, in which the novel expedient of normalization of the compositions of the solvents to give elution of the last component of interest in olive oil in a constant time was utilized; a mobile phase of dichloromethane–acetonitrile was reported to give the best resolution [633]. Refractive index detection was used in this instance.

Chloroform was employed with acetonitrile as the mobile phase for reversed-phase separations with mass detection by one group of workers [794]. It appeared to give particularly good results with saturated triacyl-

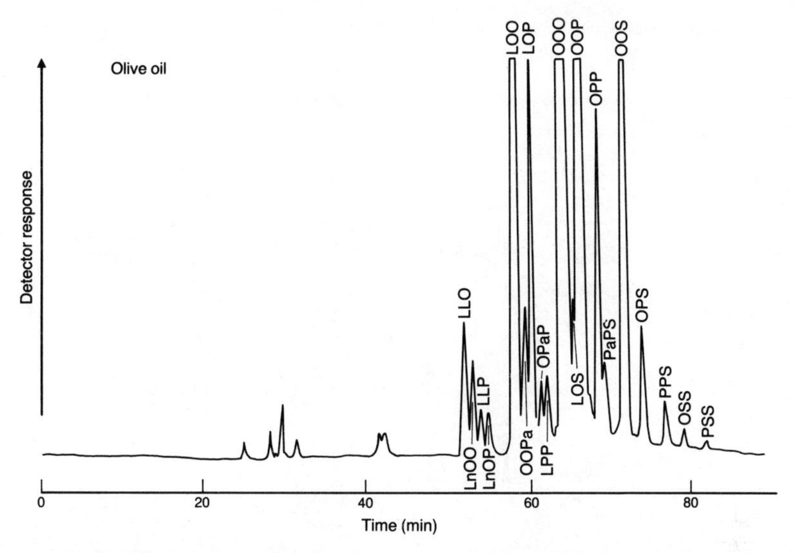

FIG. 8.5. Separation of olive oil triacylglycerols by reversed-phase HPLC on a column (4.6 × 250 mm) of Zorbax™ 5 μm ODS (15% carbon content) in series with a similar column but with a 6% carbon loading; transport-flame ionization detection was employed [648]. The mobile phase was a gradient of acetonitrile–dichloromethane from 7:3 to 4:6 by volume over 120 min, at a flow-rate of 0.8 ml/min. See Fig. 8.2 for a list of abbreviations. (Reproduced by kind permission of the authors and of *Lipids*, and redrawn from the original paper.)

glycerol standards, and even tribehenate gave a symmetrical peak when it was eluted from a column of Lichrosphere™ RP-18 (5 μm) with acetonitrile–chloroform (51:49 by volume). The same solvent combination in a gradient elution scheme gave excellent resolution of standard triacylglycerol mixtures and of olive oil. In addition, it was found to give good results with lard and tallow [644].

Among other solvents tried in admixture with acetonitrile for triacylglycerol separations, with refractive index detection generally, were diethyl ether [737], isopropanol [633], methanol [317, 633, 767], ethanol [438, 698, 767] and isopropanol–ethanol [401]; none appeared to be as good as the solvents discussed above. Some crude separations of natural fats with *n*-propanol, ethanol or methanol as mobile phase have been described [235].

By a careful choice of solvents, it has been possible to use UV detection in the reversed-phase separation of triacylglycerols. Shukla *et al.* [763] showed that by using a wavelength of 220 nm, i.e. away from the region in which isolated double bonds absorb, sufficiently sensitive detection and good quantification could be obtained (see Section B.3.ii below). They employed tetrahydrofuran–acetonitrile (73:27 by volume) as the mobile phase with two columns of a 3 μm ODS phase in series, and obtained the chromatogram of cocoa butter shown in Fig. 8.6. Each of the main groups with the same ECN value were resolved from the others, as were molecular species within

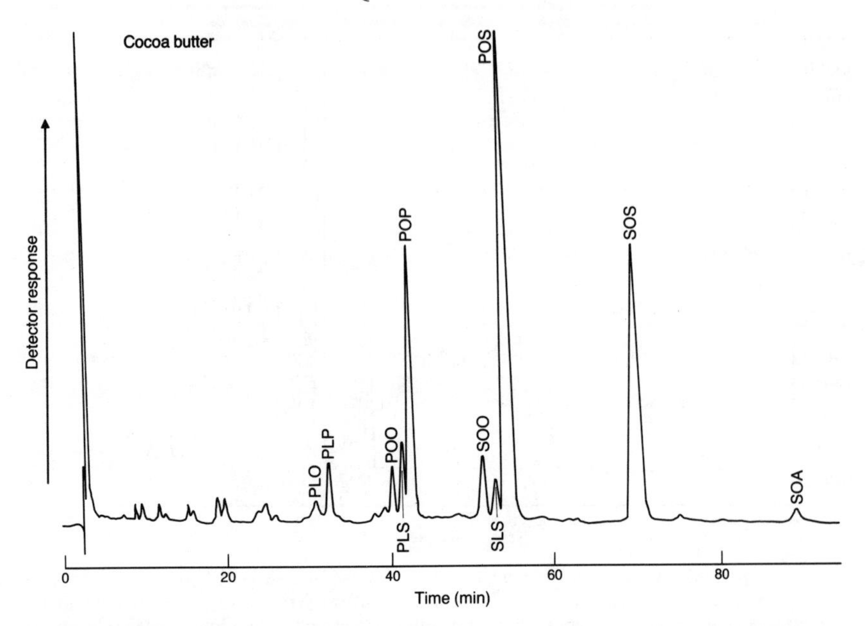

FIG. 8.6. Separation of triacylglycerols from cocoa butter by reversed-phase HPLC, using two columns (4.5 × 150 mm) in series of Spherisorb™ S30-ODS2 (3 μm particles); UV detection was at 220 nm [763]. The mobile phase was acetonitrile–tetrahydrofuran (73:27 by volume) at a flow-rate of 1 ml/min. See Fig. 8.2 for a list of abbreviations. (Reproduced by kind permission of the authors and of *Fette Seifen Anstrichmittel*, and redrawn from the original paper.)

each group. Indeed the resolution obtained is at least as good as any published to date. Others separated olive oil with a similar system [193]. Tetrahydrofuran–acetonitrile had earlier been shown to be a good combination for solubilizing high melting triacylglycerol species to facilitate analysis [372]. It may, however, be necessary to purify the tetrahydrofuran by distillation from lithium aluminium hydride to remove UV-absorbing oxidized material.

An acetonitrile–isopropanol–hexane mixture has also been used as mobile phase with UV detection [314], as later was acetonitrile–ethanol–hexane [313], which appeared to give good separations of trisaturated triacylglycerol species. Subsequently, a linear gradient of acetonitrile (100%) to acetonitrile–ethanol–hexane (2:2:1 by volume) was used to effect separations of standard mixtures, soybean oil, coconut oil and butter fat, but with mass detection [316].

Several research groups have found propionitrile on its own to be an excellent mobile phase for the elution of triacylglycerol molecular species from reversed-phase columns, and indeed some would claim that it may be the best available. It has been employed with most detection systems. Schulte [737] first made use of it for the fractionation of cocoa butter, and it was later used for this oil [235], and for the separation of palm oil [235, 663], soyabean oil, sunflower oil [215, 235], olive and wheat germ oils [215], and corn, cotton seed, hazelnut, rapeseed (high and low erucic acid), and safflower oils [235]. In Kuksis' laboratory, a gradient of 30 to 90% propionitrile in acetonitrile was used to effect separations of triacylglycerols from peanut, corn, and soyabean oils, lard, bovine and goat milk fat, human plasma and rat liver [453, 454, 456, 512, 565]. Mass spectrometry was utilized to detect and identify components in this research (and incidentally to correct some misidentifications made in earlier work from other laboratories). This work is described in greater detail below (Section B.4).

Propionitrile does have a number of disadvantages. In particular, it is very costly and extremely toxic. Cost may be less of a problem if the pure solvent is utilized as the mobile phase, since it may later be recovered, provided that a non-destructive detection system is used. It should be purified by distillation from phosphorus pentoxide to eliminate alkaline impurities, which can accumulate on storage and cause rapid deterioration of ODS phases [737].

3. *Detectors and quantification*

i. Refractive index detection.

Differential refractometry was used in many of the isocratic elution schemes described above simply to detect fractions as they eluted from the columns (see Chapter 2 for a general discussion). In a number of instances, it was also used for quantification purposes. Most workers have assumed that differences in the chain-length or degree of unsaturation of the fatty acyl moieties would have little or no effect on the refractive indices of molecular species of triacylglycerols, and have equated the areas of peaks on the recorder chart directly to mass [313, 315, 646, 664]. In practice in at least one study [69], it was found that the results obtained by this simple method, with fractions containing components of the same degree of unsaturation, were in good agreement with those obtained by a procedure expected to give a high degree of precision (GLC analysis of the fatty acid constituents in the presence of an internal standard). Similarly, it gave excellent results with triacylglycerols containing epoxy fatty acids [662].

On the other hand, a few systematic studies have demonstrated that careful calibration with pure standards can improve the accuracy of quantitative analyses. All such work has confirmed that the response of the refractive index detector is related in a rectilinear manner to the amount of solute in the eluent. In addition, with a range of saturated standards, the response of this detector was found to vary with the chain-length of the fatty acid components, but in an unpredictable and essentially uninterpretable manner [486]. Others [642, 752] showed that the refractive indices of triacylglycerols were functions both of the number of carbon atoms and of the number of double bonds. Correction factors were determined for the methyl ester derivatives of the pure fatty acids, encountered in the seed oils of interest, from their physical and chromatographic properties, and these were used in a simple formula to calculate response factors (relative to triolein) for different triacylglycerol species. Similar conclusions were obtained from a study of the response of the RI detector to pure triacylglycerols containing various unsaturated fatty acids [215].

ii. Ultraviolet detection.

Although UV detection at 205 nm has been used a great deal for the detection of lipid classes separated by HPLC, it is of much less value in analyses of molecular species. The response of the detector is mainly to isolated double bonds in this region of the spectrum, and molecular species must almost by definition differ widely in degree of unsaturation. Herslof [314], however, pointed out that a UV detector operated at 215 nm, where ester bonds (but not double bonds) exhibit a weak absorbance, was no less sensitive than most refractive index detectors. Peak area measurements were then a good measure of the amount of each triacylglycerol solute, especially when isocratic elution conditions were used [313, 314]. Shukla *et al.* [763] studied the operation of a UV detector at 220 nm in the separation of triacylglycerols in somewhat more detail (see Fig. 8.6). They found no more than 5% differences in the molar extinction coefficients for particular molecular species, and obtained acceptable results in quantitative analyses of natural oils. Of course, the range of solvents that could be used at this wavelength was somewhat limited, and oxygen dissolved in tetrahydrofuran, for example, exhibited an unacceptable absorbance in the range 212 to 220 nm, if it were not removed by purging with helium [763]. When a wide range of compositions was utilized in a solvent gradient with UV detection, the base-line drift was found to be much too great to be tolerated [698].

iii. Infrared detection.

Although infrared detection at 5.75 μm (for ester bonds) has been used in qualitative analyses of molecular species of triacylglycerols, it does not appear to have been used quantitatively for the purpose [619, 620, 660]. This is a pity, as the IR detector may have many good qualities in this particular application. It certainly appeared to give excellent results in a separation of estolide triacylglycerols by HPLC in the adsorption mode (see Section B.5 below) [634].

iv. Transport-flame ionization detection.

This type of detection system would appear to be the ideal one for separations of molecular species of triacylglycerols. With the detector developed by Privett and coworkers, it was shown that the response was rectilinear with respect to mass for individual components of a standard mixture, varying in unsaturation from zero to nine double bonds [650]. The dynamic range extended to about 200 μg, and minor components could be quantified with relative ease. In addition, no base-line drift was obtained during gradient elution. It remains to be determined whether the commercial detector of this type, i.e. that manufactured by Tracor Instruments of Texas, can give equally good results.

v. Mass detection.

The mass detector has been shown to be as sensitive as any other to have been used in triacylglycerol separations, and exhibited no base-line drift, even with lengthy gradients [316, 697, 698, 794, 833, 834].

For quantitative work with the equipment from Applied Chromatography Systems (Macclesfield, UK), Robinson and Macrae found that it was essential for all the instrumental parameters, such as atomizer inlet pressure, evaporator temperature, mobile phase and flow-rate, to be standardized prior to analysis and quantification, since changes in any of these could affect the response [697]. The detection limits were of the order of 1 μg for most triacylglycerol standards. As expected, the response with respect to mass was found to be sigmoidal for all triacylglycerol standards, but was approximately linear in the range 10 to 25 μg for saturated compounds (limited by solubility problems with the mobile phase selected for the work), and in the range 10 to 60 μg for unsaturated compounds. The logarithm of the detector response was rectilinear with respect to the logarithm of sample mass, but the slopes of the lines for different standards were not parallel. With natural samples, it would therefore be necessary to calibrate the detector with as wide a range of standards as possible to optimize the quantification. Detector reproducibility was high only under conditions of isocratic elution and not with gradient elution. With the same type of detector, Herslof and Kindmark [316] obtained an identical detection limit, but the reproducibility of the analysis of molecular species of soybean oil triacylglycerols was found to be quite acceptable, even with gradient elution; relative standard deviations of 2% were obtained for the main components. One possible explanation for the better results obtained in the latter work was that the data were expressed in terms of the relative proportions (percentages) of the different components, and not in terms of the absolute amounts of each.

With the laser light-scattering detector of their own construction, Stolyhwo *et al.* [136, 644, 794, 795] reported that the response was again not rectilinear, but was proportional to a power of the concentration of the sample. However, the nature of the triacylglycerol species made very little difference to the response, calibration was therefore a rather simple matter, and good reproducibility was obtainable with natural samples.

vi. Chemical detection.

A detection system, involving post-column chemical reactions, has been described in which the triacylglycerols emerging from a column of an ODS phase were hydrolyzed with potassium hydroxide; the glycerol produced was oxidized and converted to a derivative, which could be detected spectrophotometrically at 461 nm [438]. As little as 0.1 nmole of trilaurin could be detected, and the response was rectilinear from 0.3 to 60 nmole.

4. Mass spectrometry

All the published work to date on the analysis and identification of molecular species of triacylglycerols by HPLC coupled to MS has come from Kuksis' laboratory in Toronto [453, 454, 456, 512, 565]. A Hewlett-Packard mass spectrometer, equipped with a direct liquid-inlet interface, was used in the chemical-ionization mode with the solvents of the mobile phase, i.e. acetonitrile and propionitrile, serving as the reagent gas. Because of interference from impurities in the solvents, ions below $m/z = 200$ were ignored. With chemical-ionization MS under these conditions, a limited fragmentation only was observed, but with increased sensitivity and with the diagnostic ions in sufficient abundance. The total ion current chromatograms obtained from the system confirmed that the nature of the separation and the resolution obtained were as good as those from most other laboratories, where different detection techniques were employed. The chromatographic conditions and the samples analysed are discussed above (Section B.2.iv).

The advantages of MS in the chemical-ionization mode with ammonia as reagent gas have been known for some time, from studies with pure standards or with fractions separated by gas chromatography [558, 559, 739]. Similarly, from the chemical-ionization spectra of the individual triacylglycerol fractions obtained by reversed-phase HPLC, it was found to be possible to identify all of the major molecular species together with those minor components which were adequately resolved, in a number of oils and fats of commercial importance [453, 512]. The protonated molecular ions (MH^+) for each peak clearly gave the molecular weight and assisted greatly with the identification. (There were also minor peaks at $m/z = (M+41)^+$ and $(M+55)^+$, corresponding to addition of acetonitrile and propionitrile respectively.) Most diagnostic information was obtained for ions representing a cleavage and random loss of one fatty acyl moiety, i.e. $(MH—RCOOH)^+$, since this permitted identification of the 1,2- and 1,3-diacylglycerol fragments, and thence allowed the structure of the parent molecule to be deduced unequivocally. Thus, the mass spectrum of dipalmitoyllinoleoylglycerol exhibited the MH^+ ion at $m/z = 831$, together with $(MH—RCOOH)^+$ ions at $m/z = 551$ and 575, which corresponded to dipalmitoyl- and palmitoyllinoleoylglycerols respectively; any other triacylglycerol of the same molecular weight would exhibit $(MH—RCOOH)^+$ ions with different m/z values. As an example, the

TABLE 8.3
The important diagnostic ions in the chemical-ionization mass spectra of molecular species of triacylglycerols from corn oil [512]

Triacylglycerol	MH^+	$(MH—RCOOH)^+$
18:2–18:2–18:3	877	599, 597
18:2–18:2–18:2	879	599
18:1–18:2–18:3	879	601, 597
16:0–18:2–18:3	853	575, 597
18:1–18:2–18:2	881	601, 599
16:0–18:2–18:2	855	575, 599
18:1–18:1–18:2	883	603, 601
18:0–18:2–18:2	883	603, 599
16:0–18:1–18:2	857	577, 601
16:0–16:0–18:2	831	551, 575
18:1–18:1–18:1	885	603
18:0–18:1–18:2	885	605, 601
16:0–18:1–18:1	859	577, 603
16:0–18:1–16:0	833	577, 551
18:0–18:1–18:0	889	605, 607
16:0–18:0–18:1	861	579, 605

important diagnostic ions in the triacylglycerol species of corn oil are listed in Table 8.3. With more complex oils than corn oil, such as peanut oil, there was a considerable overlap of fractions of different chain-length, but a preliminary simplification accomplished by silver ion TLC enabled positive identification of components separated later by HPLC. While it was possible to distinguish to some extent between a given fatty acid in a primary as opposed to the secondary position of the glycerol moiety, it was not of course possible to differentiate enantiomers.

Quantification by means of the total ion current of components separated by this HPLC-MS system presented a number of difficulties [565]. For example, the yields of the quasi-molecular and fragment ions were found to vary with the chain-length, degree of unsaturation and positional distribution of fatty acyl groups in the molecules. As pure mixed acid standards, and structural isomers especially, were not available, it was necessary to calculate calibration factors by a comparison of peak area measurements obtained in separations of natural oils and fats with their known compositions (from independent chromatographic analyses). With positional isomers, the $(MH—RCOOH)^+$ ions varied in yield over a 1- to 3-fold range in intensity, while the nature of the fatty acid affected the ion current over a 1.25-fold range. In spite of the fact that these factors were obtained with a gradient of reagent gases, they were found to applicable in a general manner. When the system was calibrated in this way, it was possible both to identify and quantify molecular species in natural fats and oils.

5. *Triacylglycerols containing unusual or modified fatty acids*

Triacylglycerols containing fatty acids with oxygenated substituent groups have been separated by HPLC in both the reversed-phase and adsorption mode. For example, seed oils containing the epoxy fatty acid, vernolic acid, were separated into fractions consisting of trivernolin, divernoyllinolein, divernoylolein and divernoylstearin by reversed-phase chromatography on an ODS column with acetonitrile–acetone (2:1 by volume) as the mobile phase, and with refractive index detection [662]. The same procedure was used to separate the triacylglycerols containing oxygenated fatty acids in castor oil and in the oil from *Linum mucronatum* seeds [659]. Adsorption chromatography on a column of Partisil™ PXS 10/25 PAC, with hexane–tetrahydrofuran (9:1 by volume) as the mobile phase, was used to separate normal from estolide triacylglycerols in the seed oil, *Sapium sebiferum* [634]. In this instance, IR spectrophotometry at 1750 cm^{-1} was used both to detect and quantify each of the components. Synthetic estolide triacylglycerols, containing four to six acyl groups, were separated by reversed-phase HPLC on an ODS column, with acetonitrile–acetone (2:1 by volume) as the mobile phase; components of higher molecular weight were retained more strongly [635].

After chemical autoxidation of trilinolein, a monohydroperoxide fraction was isolated for HPLC analysis [618]. On separation on a column of silica gel with hexane–isopropanol (99:1 by volume), six fractions were discerned, while in the reversed-phase mode with acetonitrile–isopropanol (3:1 by volume) as the mobile phase, three fractions were obtained. Each component was found to contain positional and configurational hydroperoxide isomers of the kind discussed in Chapter 7 (Section F.1). In addition, a number of triacylglycerols with related fatty acids were isolated from autoxidized vegetable oils. HPLC with columns of silica gel has been used to separate and quantify unchanged from oxidized triacylglycerols, either with transport-flame ionization [17] or mass [644] detection. HPLC methods for the analysis of polymerized triacylglycerols are discussed in Chapter 5 (Section B.3).

Seed oils of the Flacourtiaceae (chaulmoogric oils) are known to contain fatty acids containing cyclopentenyl groups, i.e. hydnocarpic, chaulmoogric and gorlic acids. Two seed oils with such constituents were successfully resolved into the main molecular species by HPLC in the reversed-phase mode by Shukla and Spener [764]. The equipment and elution conditions were very similar to those discussed above in relation to Fig. 8.6.

One of the important problems in triacylglycerol analysis in commercial laboratories is to determine the proportion of the monounsaturated species with the monoenoic component in position *sn*-2, as this fraction has important physical properties when used in confectionery products. This cannot be accomplished by conventional reversed-phase HPLC, but it has proved possible after bromination of the triacylglycerols. In one laboratory [235], brominated cocoa butter and "cocoa butter equivalents" were fractionated on an ODS phase with acetonitrile–isooctane–isopropanol (5:2:3 by volume) as the mobile phase, and with UV detection at 215 nm. "Unsaturated" components with more than one double bond eluted well ahead of the monounsaturated fractions, although in this instance, the resolution was apparently not sufficient to separate positional isomers. This was, however, accomplished in another laboratory [664] by HPLC on an ODS phase with propionitrile at a flow-rate of 0.5 ml/min as the mobile phase, and with refractive index detection. The species with the brominated fatty acid in the primary positions eluted just ahead of the isomer with this component in the secondary position, and the relative proportions of the two were determined in a number of natural fats in this way.

C. HPLC Separation of Species of Mono- and Diacylglycerols and Related Compounds

1. Conventional separations of mono- and diacylglycerols

A series of synthetic saturated 3-acyl-*sn*-glycerols were separated by reversed-phase HPLC on an ODS phase with a gradient of tetrahydrofuran–water–isopropanol–acetonitrile (from 7.6:23:5:64.4 to 3.2:0:70:26.8 respectively by volume) as the mobile phase, and with UV detection at 220 nm [434]. In similar work with a C_8 stationary phase, the resolution was somewhat inferior [517]. A limited range of commercial monoglyceride emulsifiers were separated in the same way by isocratic elution with acetonitrile as the mobile phase, and with UV detection at 204 nm [801].

Separations of glycerophospholipids in the form of diacylglycerol derivatives are discussed in some detail below (Section E). Diacylglycerols are important components of animal tissues in their own right, either as intermediates in the biosynthesis of phospholipids or in the role of cellular messengers, when formed as hydrolysis products of the glycerophosphoinositides. Diacylglycerol species in microsomes and mitochondria of normal and dystrophic human muscle [465] and in lung microsomes [713] were separated and analysed by HPLC in the reversed-phase mode in the form of the naphthylurethane derivatives. The method used was developed originally for the analysis of molecular species of phosphoglycerides and is discussed in greater detail below (Section E). Similarly, diacylglycerols from potato tuber microsomes were analysed in a comparable manner [389]. Acetate derivatives of natural diacylglycerols in a human cell line have also been fractionated by reversed-phase HPLC [721].

Diesters of chloropropane diol, which presumably have some biosynthetic relationship to the triacylglycerols, have been detected as minor constituents of goat milk in some circumstances. These components were isolated, and they were separated into molecular species by HPLC in the reversed-phase mode on an ODS phase, with a gradient of 30 to 90% propionitrile in acetonitrile as the mobile phase; in this instance, mass spectrometry was used to detect and to identify the fractions [454].

2. *Separation of enantiomers*

One remarkable recent achievement of HPLC has been the resolution of 1- and 3-monoacyl-*sn*-glycerols on a column (4 × 250 mm) containing a chiral stationary phase, consisting of *N*-(*S*)-2-(4-chlorophenyl) isovalerol-D-phenylglycine bonded chemically to an aminopropyl-silanized silica gel [813]. The monoacylglycerols were converted to the 3,5-dinitrophenylurethane derivatives, and they were eluted from the column with hexane–ethylene dichloride–ethanol (40:12:3 by volume) as the mobile phase, with UV detection at 254 nm. With this system, the 1-stearoyl-*sn*-glycerol derivative was eluted well ahead of the 3-enantiomer. The *sn*-2 isomer eluted before both. Subsequently, the work was extended to monoacylglycerols containing other fatty acids [355]. Even though partial separation of homologues occurred, the resolution obtained was probably adequate to cope with natural racemic mixtures of monoacylglycerols containing different fatty acids.

The complete resolution of enantiomeric diacylglycerol derivatives would be an even more useful achievement, as it could simplify appreciably the stereospecific analysis of triacylglycerols, a feat presently accomplished by a complex combination of chemical hydrolysis, synthesis and enzymatic hydrolysis techniques (reviewed elsewhere [133, 138]). Some complete separations of diastereoisomeric (*S*)-1-(1-naphthyl)ethyl carbamate derivatives of synthetic mono- and dialkylglycerols were achieved on a column (10 × 500 mm) of silica gel with hexane–ethyl acetate (85:15 by volume) as mobile phase, and with UV detection at 254 or 280 nm [528]. The 2,3-O-hexadecyl-*sn*-glycerol derivative eluted ahead of the 1,2-isomer, for example, but because of chain-length separations of homologous compounds, the resolution of mixtures was rather spoiled. Unfortunately, partial resolution only of a diastereoisomeric diacylglycerol derivatives was achieved. It would be of great interest to ascertain whether a further improvement could be obtained by using a chiral stationary phase.

3. *Separation of alkyl- and alkenylglycerols*

Ether-linked glycerolipids were reduced with Vitride reagent ($NaAlH_2(OCH_2CH_2OCH_3)_2$) in order to obtain the alkylglycerol and alk-1-enylglycerol moieties, which were benzoylated and purified by TLC [79]. These derivatives were then further resolved into the individual components by HPLC in the reversed-phase mode, on a column of Ultrasphere™ ODS with acetonitrile–isopropanol (84:16 by volume) as the mobile phase; UV detection as 228 nm was employed and pentadecyldibenzoylglycerol was added as an internal standard to assist in quantification.

Synthetic enantiomers of 1- and 3-alkyl-*sn*-glycerols, in the form of the di-3,5-dinitrophenylurethane derivatives, were resolved by HPLC on a column containing a chiral stationary phase, under similar conditions to the analogous monoacyl derivatives (see previous section) [814].

D. HPLC Separation of Species of Intact Glycerophospholipids

1. Phosphatidylcholine and related lipids

Methods for the separation of intact glycerophospholipids by means of HPLC in the reversed-phase mode have evolved rapidly in recent years, and indeed this may now be the preferred approach to analysis (reviewed elsewhere [137, 377, 503, 696]). Procedures for the analysis of intact lipids are inherently simpler in that fewer preparatory steps, where selective losses could potentially occur, are required. In addition, they enable biochemical studies of the turnover of any part of the molecules to be carried out, and the physical, chemical and biological properties of fractions isolated in this way can be determined directly. It has always been considered that nonpolar derivatives of phospholipids are capable of being resolved more cleanly than are the intact compounds, but with HPLC methods at least the difference is now much less marked than it was formerly.

With reversed-phase HPLC of molecular species of phospholipids, the nature of the separation is similar to that discussed above for triacylglycerols, in that it is dependent on the combined chain-lengths and degree of unsaturation of the fatty acyl or alkyl chains. The first published separation of molecular species of phosphatidylcholine in this way now appears relatively crude, and it was accomplished on a column of hydroxyalkoxypropyl Sephadex™ (C_{11} to C_{14}) at 41.6°C, with methanol–water (4:1 by volume) as the mobile phase, and with refractive index detection [36]. In this instance, only four fractions were obtained according to the partition number of the constituent fatty acids. In nearly all the subsequent work, ODS phases from various manufacturers have been used. For example, Waters "Fatty Acid Analysis" and C_{18} columns, with methanol–water–chloroform (70:19:10 by volume) as mobile phase and refractive index detection, were employed in separations of saturated standards and of egg phosphatidylcholine into six fractions [674]. It has become apparent that some ODS phases are much better than others for phospholipid separations, although the reason for this is not clear. For example, in a comparison of three different commercial phases, the reduced plate height in separations of molecular species of egg phosphatidylcholine was found to vary from 2.99 to 4.90 [787]. The alkyl content of the phase did not appear to be the most important factor in the separation. In this work, Nucleosil™ C_{18} gave the most consistent results, but a number of other analysts have obtained excellent separations of phospholipids with Ultrasphere™ ODS.

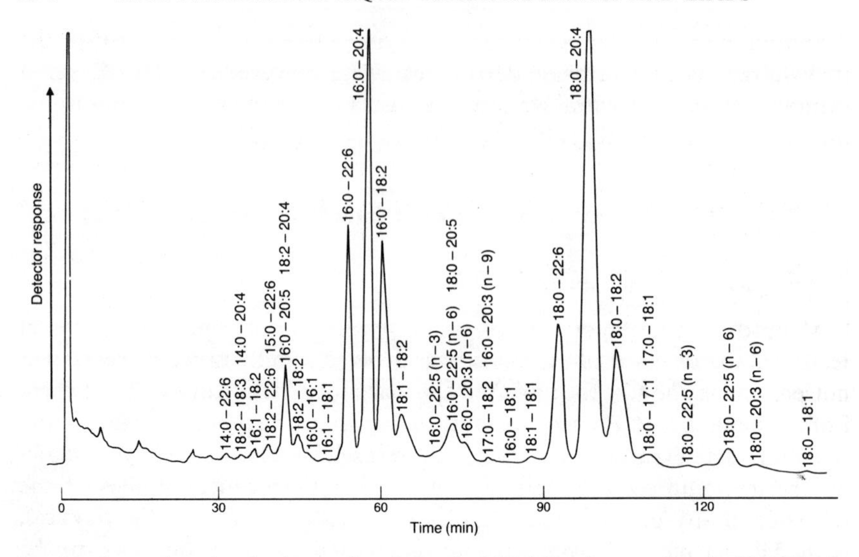

FIG. 8.7. Separation of molecular species of phosphatidylcholine from rat liver by HPLC in the reversed-phase mode [630]. A column (4.6 × 250 mm) of Ultrasphere™ ODS was used, with methanol–water–acetonitrile (90.5:7:2.5 by volume) containing 20 mM choline chloride as the mobile phase at a flow-rate of 2 ml/min, and with UV detection at 205 nm. (Reproduced by kind permission of the authors and of the *Journal of Lipid Research*, and redrawn from the original paper.)

A seminal paper on the subject of separations of phospholipid molecular species was published by Patton *et. al.* in 1982 [630]. They fractionated molecular species from different phospholipid classes on a column (4.6 × 250 mm) of Ultrasphere™ ODS, with 20 mM choline chloride in methanol–water–acetonitrile (90.5:7:2.5 by volume) as the mobile phase. UV spectrophotometric detection at 205 nm was used to locate the fractions, and these were collected for phosphorus assay and fatty acid analysis; 220 μg of sample was applied to the column in 50 to 100 μl of ethanol. In Fig. 8.7, the separation obtained for phosphatidylcholine from rat liver is illustrated. (Note that with UV detection at 205 nm, the response was highly dependent on the degree of unsaturation, and peak heights did not reflect the relative abundances of the components.) The separation could be considered as bimodal, with in essence those molecular species containing a 16:0 fatty acyl group eluting before those containing 18:0. As with triacylglycerols, the position of the acyl group within the molecule had no effect on separation in the reversed-phase mode, although the saturated components are known to be located predominantly in position *sn*-1 in this particular sample. All of the major components were clearly resolved, and a few only of the minor fractions contained two distinct species.

The problems of identification of molecular species were somewhat less than with triacylglycerols, since only two fatty acids were present, but the degree of unsaturation perhaps varied rather more than in most of the triacylglycerols studied to date. In phospholipid species containing a given saturated fatty acid with different unsaturated components, the retention time increased in the order 20:5, 16:1, 22:6, 20:4, 18:2, 22:5(n – 3), 22:5(n – 6), 20:3(n – 6), 20:3(n – 9) and 18:1. Thus even the position of the double bond had an effect on the elution order, with (n – 3) before (n – 6) then (n – 9). The relative retention times (RRT) for each species were calculated in relation to that of 16:0–22:6 (= 1.0), and they are listed in Table 8.4 for reference purposes. The same numerical values were obtained for corresponding species from different phospholipids. It was also possible to construct a graphical relationship from this data to predict the order of elution of molecular species not present in the samples. For the sake of consistency with previous work from other laboratories, it might with hindsight have been better if the retention data had been calculated in terms of some form of the carbon number concept (see below). Unfortunately, as data for disaturated species were not recorded, it is not possible to do this retrospectively. (The response of the UV detector at 205 nm towards fully saturated species would be poor.)

The method was used initially for separations of molecular species of rat liver phospholipids [630], and subsequently for phosphatidylcholine in plasma [629] and intestinal lymph [631] lipoproteins. Essentially the same elution conditions have now been used in a number of laboratories to effect similar separations of phosphatidylcholine species. For example, they were used to study phosphatidylcholine biosynthesis via lysophosphatidylcholine

TABLE 8.4

Relative retention times (RRT) for molecular species of phospholipids separated by HPLC in the reversed-phase mode under the conditions of Fig. 8.7 [630]

Molecular species	RRT[a]	Molecular species	RRT
14:0–22:6	0.579	16:0–22:5(n – 6)	1.355
14:0–20:4	0.621	18:0–20:5	1.355
18:2–18:3	0.621	16:0–20:3(n – 6)	1.432
16:1–18:2	0.687	16:0–20:3(n – 9)	1.470
18:2–22:6	0.759	17:0–18:2	1.470
15:0–22:6	0.759	16:0–18:1	1.549
16:0–20:5	0.791	18:1–18:1	1.644
18:2–20:4	0.791	18:0–22:6	1.738
18:2–18:2	0.834	18:0–20:4	1.841
16:0–16:1	0.895	18:0–18:2	1.941
16:1–18:1	0.972	17:0–18:1	2.066
16:0–22:6	1.000	18:0–17:1	2.066
16:0–20:4	1.059	18:0–22:5(n – 3)	2.251
16:0–18:2	1.117	18:0–22:5(n – 6)	2.355
18:2–18:1	1.156	18:0–20:3(n – 6)	2.495
16:0–22:5(n – 3)	1.317	18:0–18:1	2.679

[a]Calculated arbitrarily relative to 16:0–22:6 (= 1.0).

precursors in plasma [800] and in rabbit alveolar macrophages [699], to study the reacylation of the hydrolysis product of platelet activating factor in human neutrophils [125], to identify the phosphatidylcholine species in potato tubers and tobacco leaves [181, 182, 389], and to isolate choline glycerophosphalipid sub-fractions from the sarcoplasmic reticulum and mitochondria of heart and skeletal muscle [252]. With some modification, these conditions were used to separate the same species in egg and human bile [105]. In this instance, analyses of samples of known composition permitted the calculation of correction factors, so that UV detection could be used in quantitative analyses.

With plant lipids, the nature of the molecular species differed markedly from those found in the corresponding lipids of animal tissues, and under the same conditions, fractions containing the following fatty acid combinations eluted in the order 18:3–18:3, 18:3–18:2, 18:2–18:2 with 18:3–16:0, 18:2–16:0, 18:0–18:3 and 18:0–18:2 [181, 182, 389].

Good resolution of phosphatidylcholine species from egg, bovine brain and pig liver was obtained by HPLC on a column (4 × 300 mm) of Nucleosil™ C_{18} with methanol–1 mM (pH 7.4) potassium phosphate buffer (9.5:0.5 by volume) as the mobile phase, and with UV detection at 205 nm [778]. On the other hand, the nature of the fractions obtained was somewhat different from that shown in Fig. 8.7, in that there was some change in the order of elution of components. The retention times of individual species in this work were discussed in terms of the hydrophobicities of the aliphatic chains, relative to the disaturated components with equivalent combined chain-lengths. One double bond in a molecular species reduced the "hydrophobic carbon number" (HCN) by 1.8 units, a second reduced it by a further 1.4 units, and a third by a further 1.0 units; these factors varied a little when the double bonds were in both aliphatic chains. For reference purposes, the numerical values obtained are listed in Table 8.5. They could be used to calculate the relative retention times of components not present in the samples investigated. Although the data can only be applied directly to the elution system used in this particular study, the concept could be adapted to any reversed-phase separation.

Other workers obtained good separation of phosphatidylcholines from egg and rat liver by elution with methanol–water–acetonitrile (8:1:1 by volume) at a flow-rate of 1.5 ml/min from a column of Nucleosil™ C_{18} at 30°C [787]. Here the mass detector was compared qualitatively with UV detection at 214 nm. A column containing a polymeric styrene–divinylbenzene (PLRP-S™) as the stationary phase was eluted with acetonitrile–methanol–water (70:15:15 by volume), and with either mass or refractive index detection, to give six main fractions from rat liver phosphatidylcholines [145]. When the mass detector was employed, a stream splitter was inserted after the column to divert most of the eluent for collection. Ionic species were not incorporated into the mobile phase in the last two separations, a factor that

TABLE 8.5
The effects of double bonds in reducing the "hydrophobic carbon number" (HCN) values of molecular species of phosphatidylcholine [778]

	Reduction in HCN	
No. of double bonds	A[a]	B
1	1.8	
2	3.2	3.7
3	4.4	4.9
4	5.4	6.1
5	6.4	7.3
6	7.4	8.5
7		9.6

[a]A = when the double bonds are in one aliphatic chain only; B = when the double bonds are in both chains.

might be important in isolating fractions of high purity preparatively for physical chemical studies say. Other workers employed methanol–hexane–0.1 M ammonium acetate (71:5:7 by volume) as the mobile phase for the elution of species of egg phosphatidylcholine from a column (4.6 × 75 mm) of Ultrasphere™ ODS (3 μm particles) [417]. In this work, thermospray mass spectrometry was the detection system (see Section 4 below).

Individual molecular species of isotopically-labelled platelet activating factor (PAF) were separated by Patton's elution procedure in essence (described above) [364]. In related work, the conditions were modified slightly in that a column (4.6 × 150 mm) of Spherisorb™ ODS II (3 μm particles) at 35°C was eluted with 30 mM choline chloride in methanol–water–acetonitrile (79.5:18:2.5 by volume) for the first 41 minutes at 0.5 ml/min, then with 12 mM choline chloride in the same solvents in the ratio 90.5:7:2.5 respectively for a further 44 min [557]. In both of these studies in place of a conventional detection system, fractions were collected for determination of their radioactivity. In addition to detection in this way, it proved possible to monitor the elution of molecular species of PAF from a column containing a Hamilton PRP-1™ resin (more polar than ODS) by spectrophotometry at 210 nm, when a gradient of 40 to 70% of acetonitrile in water containing 1 mM methanesulphonic acid was the mobile phase [719].

Molecular species of lysophosphatidylcholine, including positional isomers, were separated by reversed-phase HPLC on a column of *micro*Bondapak™ C_{18}, eluted with methanol–water (85:15 by volume) as mobile phase, and with UV detection at 214 nm [584]. The 2-acyl isomer eluted before the 1-acyl compound. Not surprisingly, the Patton procedure described above has been adapted to this purpose, and synthetic lysophospholipids were fractionated with 20 mM choline chloride in methanol–water–acetonitrile (57:23:20 by volume) as the mobile phase with Ultrasphere™ ODS as the stationary phase, and with UV detection at 203 nm [166]. The procedure was also applied to lysophospholipids, including the 1-alkyl and 1-alkenyl forms, produced by phospholipase A_2-catalyzed hydrolysis of phosphatidylcholine from bovine heart and brain and from soyabeans [167]. In addition, the method of Patton and coworkers has been adopted to the isolation of molecular species of plasmenylcholine [516, 885].

Brief details have been published of a separation of phosphatidylcholine, from the microsomal fraction of rat brain, into molecular species with one to six double bonds in total by HPLC with silver ion complexation [377, 386]. The column contained Nucleosil™ 10 sulphonic acid Ag^+ (from Chrompak Ltd), it was maintained at 45°C, and was eluted with methanol at a flow-rate of 2 ml/min. Although not apparently noted here, it would be surprising if some hydrolysis or trans-methylation of the lipids did not take place on he column under such conditions (see Chapter 7, Section C).

Fractions enriched in plasmalogens were isolated from bovine heart phosphatidylcholine by a form of silver ion chromatography in which the silver was bound in a similar way in HPLC columns; a gradient of water into methanol was employed as the mobile phase, and plasmalogens species tended to elute after the diacyl components [516]. When the ion-exchange medium was converted to the mercury form, the plasmalogens were hydrolyzed completely on the column.

2. *Phosphatidylethanolamine and other common phosphoglycerides*

Phosphatidylethanolamine and phosphatidylinositol from rat liver were separated into molecular species under exactly the same conditions employed for phosphatidylcholine by Patton *et al.* [630], that were described in the previous section (and illustrated in Fig. 8.7). Indeed, fractions identical in composition were obtained, although the relative proportions were rather different, as expected. After modifying the mobile phase to 30 mM choline chloride in methanol–25 mM KH_2PO_4–acetonitrile–acetic acid (90.5:7:2.5:0.8 by volume), molecular species of phosphatidylserine were resolved. Others have fractionated species of phosphatidylethanolamine and phosphatidylglycerol of plant origin [181, 389] and phosphatidylethanolamine (including plasmalogens) of the sarcoplasmic reticulum and mitochondria of heart and skeletal muscle [252] by essentially the same method.

Phosphatidylethanolamine from rat liver has also been resolved on a column of Nucleosil-5™ C_{18} with methanol–water–acetonitrile (8:1:1 by volume) at 30°C as the mobile phase [787]. In this instance, mass detection was used.

In a different approach, egg phosphatidylethanolamine was converted chemically to the fluorescent dansyl derivative, and some molecular fractions were resolved on a column (4.5 × 250 mm) containing a C_8 stationary phase, and eluted with methanol–aqueous potassium phosphate (pH 7) (83:17 by volume) [715, 716]. The fluorescence detector was set at an excitation wavelength of 360 nm and an emission wavelength of 420 nm.

Molecular species of phosphatidylglycerol from plant chloroplasts were resolved by HPLC on a column (4.6 × 250 mm) containing Rainin Microsorb™ reversed-phase packing material, and with 1-ethylpropylamine–acetic acid–methanol–acetonitrile (0.3:0.5:34.7:64.5 by volume) as the mobile phase at a flow-rate of 0.8 ml/min [594, 777]. Of particular interest here was what appears to have been the first published application of the Tracor Instruments transport-flame ionization detector to lipid analysis. Excellent base-line stability was recorded in spite of the presence of ionic species in the mobile phase, and minor components present at a level of as little as 1.2 nmol could be quantified. Although the linearity of the detector response was not determined rigorously, it was confirmed that direct quantification of components by integration of peaks from the detector gave results which were comparable to those obtained by alternative methods. A stream splitter enabled fractions to be collected for determination of radioactivity or for GLC analysis of the fatty acid constituents. In the sample studied, species eluted in the order 16:1*t*–18:3, 16:1*t*–18:2 and 16:0–18:2. An alternative method has been described in which phosphatidylglycerol (after methylation of the acidic moiety with diazomethane) was fractionated under chromatographic conditions developed primarily for plant galactolipids (see Section F below) [415].

Diphosphatidylglycerol (cardiolipin) would not be expected to be the easiest lipid class to resolve into molecular species, as it contains 4 molecules of fatty acid per mole. Nonetheless, considerable simplification of this lipid from various rat tissues and from bovine mitochondria was achieved on a column (4.6 × 100 mm) of Microsorb™C_{18} (3 μm particles), eluted with gradients of acetonitrile–methanol–10 mM phosphate buffer in various proportions, and with UV detection at 208 nm [822]. Each of the fractions obtained was collected, and its fatty acid composition was determined by GLC analysis.

3. *Oxidized phospholipids*

Oxidation of the unsaturated fatty acid constituents of membrane phospholipids is an important biological process in living tissues, and it may occur adventitiously in foods, with deleterious effects on keeping quality. Procedures for the separation of the oxidized fatty acid contituents *per se* were described in the previous chapter (Section F.1), and those for oxidized triacylglycerols are described above (Section B.5). It has also proved possible to isolate or identify molecular species of intact phospholipids containing oxidized fatty acids. Generally, HPLC in the reversed-phase mode has been used, although adsorption chromatography gave better results with other lipid classes. For example, dilinoleoylphosphatidylcholine was subjected to autoxidation in air, and was separated into two oxygenated fractions in addition to the unchanged material by reversed-phase HPLC, the oxidized materials eluting first [673, 675]. A column containing *micro*Bondapak™ C_{18} phase was eluted with methanol–water–chloroform (10:1:1 by volume) as the mobile phase, and with refractive index detection. Slightly modified conditions were used to isolate the products of autoxidation of 1-stearoyl,2-arachidonoyl-phosphatidylcholine, with UV detection at 240 nm.

Somewhat better resolution of molecular species of autoxidized soyabean phosphatidylcholines (and of a synthetic dilinoleoyl species) was obtained on an ODS column with a gradient of methanol–water (91:9 to 95:5 by volume) as the mobile phase [164]. In this instance, UV detection at 206 nm was used to identify the normal molecular species, while at 234 nm, only the oxidized species were seen. Fractions in which either one or both of the fatty acids were oxidized were identified. Some commercially-available detectors are capable of monitoring both of these wavelengths simultaneously, and one such has been used in a study of the oxidation of 1-stearoyl,2-linoleoyl-phosphatidylcholine [843]. Here, fractions were eluted from a column (4.6 × 250 nm) of Ultrasphere™ ODS with a mobile phase consisting of a gradient of methanol–acetonitrile–water containing 20 mM choline chloride (from 90.5:2.5:7 to 90.5:8.8:0.7 by volume). Detection at these two wavelengths was also used with adsorption chromatography on a column of silica gel, in which the oxidized species lagged only slightly behind the unchanged forms, to determine the relative proportions of oxidized to unoxidized materials in the phosphatidylethanolamine and phosphatidylcholine from the livers of rats treated with carbon tetrachloride [826]. The mobile phase in this work was a gradient of water into acetonitrile.

4. Mass spectrometry

Thermospray HPLC-mass spectrometry has recently been applied to the detection and identification of molecular species of intact phospholipids separated in the reversed-phase mode [417]. The chromatographic conditions are described in Section D.1 above. The positive ion mass spectra of phosphatidylcholine and phosphatidylethanolamine species were obtained by thermospray with the filament on for more efficient ionization. In this way, a simple fragmentation pattern was obtained, which gave information both on the nature of the polar head group and of the composition of the fatty acyl groups. The protonated molecular ion was distinct and identifiable, and the base peak was the diacylglycerol ion, resulting from the loss of the phosphorus head group, which also gave characteristic ions. Glyceride ions containing each of the fatty acyl moieties were evident and were of diagnostic value, although positional isomers could not be distinguished. The procedure was applied both to synthetic compounds and to egg yolk phospholipids.

Molecular species have been identified and quantified by mass spectrometry of intact phospholipids, separated by adsorption chromatography, as discussed in Chapter 6 above (Section B.1) [381]. This work has also been reviewed in detail elsewhere [377, 503].

E. HPLC Separation of Glycerophospholipids After Partial Hydrolysis

When silver ion TLC and high-temperature GLC were the principal methods available to lipid analysts for the separation of molecular species of glycerophospholipids, it greatly simplified the task technically if the polar head group were removed and the resulting diacylglycerol moiety converted to a stable non-polar derivative. Of course, such methods could not be used if information were required on the rate of turnover of specific atoms in the head group, for example. On the other hand if this approach is adopted in conjunction with HPLC, all classes of glycerophospholipids can be converted to the same type of derivative, so they are analysed under exactly the same elution conditions. Improved resolution may often be obtained, with no requirement for inorganic ions in the mobile phase, and UV-absorbing or fluorescent derivatives may be employed, simplifying detection and quantification. In addition, other complementary chromatographic techniques can more easily be brought to bear for the further resolution of fractions separated by HPLC.

Perhaps the simplest method of deactivating the polar head group of glycerophospholipids is to remove the base moiety by hydrolysis with the enzyme phospholipase D, and then to react the phosphatidic acid formed with diazomethane to produce phosphatidic acid dimethyl esters. This approach was utilized for the separation of molecular species of phosphatidylcholine, but the initial published separations by means of reversed-phase HPLC showed some tailing of peaks [339, 340]. On the other hand, comparable results to those obtained by other methods were obtained when a column (4 × 250 mm) of LiChrosorb™ RP-18 was eluted isocratically with acetonitrile–isopropanol–methanol–water (50:27:18:5 by volume) at a flow-rate of 1.5 ml/min [573]. The separation was monitored by UV spectrophotometry at 205 nm, and fractions were collected for identification and quantification, after conversion to the methyl ester derivatives of the fatty acid constituents, by GLC with an added internal standard. This procedure was used subsequently for the separation of molecular species of the alkylacyl and diacyl forms of the choline glycerophospholipids of guinea pig polymorphonuclear leukocytes [568]. For the preliminary isolation of the ether and diacyl species, adsorption HPLC on a column of silica gel, maintained at 30°C, was employed with hexane-isopropanol (19:1 by volume) as the mobile phase at a flow-rate of 1 ml/min.

A rather similar method was used to obtain fractions consisting of derivatives of hepta-, hexa- and pentaacyl monophosphoryl lipid A from bacterial lipopolysaccharides [682, 819]. These were subsequently identified by laser desorption and fast atom bombardment MS.

Most workers have preferred to hydrolyze glycerophospholipids to diacylglycerols by means of the enzyme phospholipase C, and then to esterify the free hydroxyl group to reduce the polarity further. The main sources of the enzyme preparation are the microorganisms, *Clostridium welchii* and *Bacillus cereus*, and the properties and substrate specificities have been reviewed [96]. The enzyme from *C. welchii* is most suited to the hydrolysis of phosphatidylcholine, while that from *B. cereus* is best for phosphatidylethanolamine and for the acidic phospholipids. Procedures for specific phospholipids have been described in detail elsewhere [133]. The following method is given as an example, and is suited to the hydrolysis of phosphatidylcholine [689].

> "Phospholipase C from *C. welchii* (1 mg) in tris(hydroxymethyl)-methylamine buffer (0.5 M; pH 7.5; 2 ml) containing 2×10^{-3} M calcium chloride is added to phosphatidylcholine (5 mg) in diethyl ether (2 ml). After shaking the mixture for 3 hr, it is extracted three times with diethyl ether (4 ml portions). The solvent is dried over anhydrous sodium sulphate, before being evaporated in a stream of nitrogen at room temperature. The required 1,2-diacyl-*sn*-glycerols are obtained in pure form by preparative TLC on silica gel impregnated with 10% by weight of boric acid, with hexane-diethyl ether (1:1 by volume) as the mobile phase. After locating the appropriate band under UV light by spraying with dichlorofluorescein solution, it is eluted from the adsorbent with diethyl ether, and is then derivatized immediately."

Acetate derivatives of diacylglycerols are perhaps the simplest to prepare, and they have been much used in TLC and GLC separation procedures, while Nakagawa and Horrocks [569] have used them to particularly good effect with HPLC. For example, diradylacetylglycerols prepared from the ethanolamine glycerophospholipids of bovine brain were separated into the alkenylacyl, alkylacyl and diacyl forms by adsorption HPLC on silica gel. They were eluted from a column (3.9 × 300 mm) containing *micro*Porasil™, maintained at 36°C, with cyclopentane–hexane–methyl-*t*-butyl ether–acetic acid (73:24:3:0.03 by volume) as the mobile phase at a flow-rate of 2 ml/min; UV detection at 205 nm was employed. The nature of the separation obtained is shown in Fig. 8.8. Such resolution has been attained only with difficulty by other methods, and it has been almost impossible to achieve with intact phospholipids (phosphatidylethanolamine only has been partially separated into the diradyl forms (see Chapter 6)). For quantification purposes, the fractions were transmethylated in the presence of methyl heptadecanoate (as an internal standard) for GLC analysis.

The alkenylacyl-, alkylacyl- and diacylglycerol acetates, isolated in this

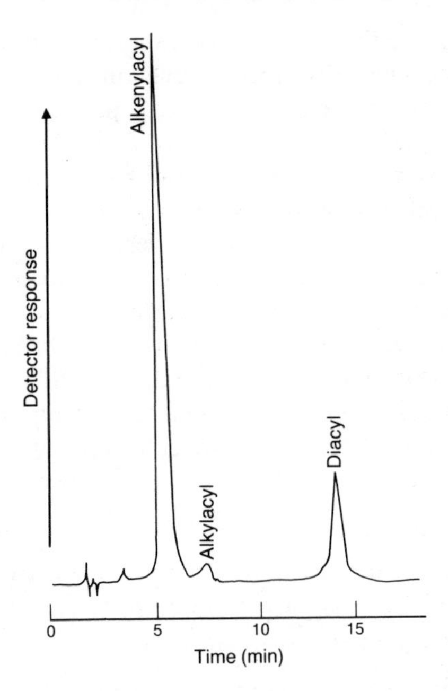

FIG. 8.8. Separation of alkenylacyl-, alkylacyl- and diacylglycerol acetates, derived from the ethanolamine glycerophosphatides of bovine brain [569] by HPLC in the adsorption mode. The chromatographic conditions are given in the text. (Reproduced by kind permission of the authors and of the *Journal of Lipid Research*, and redrawn from the original paper.)

way, were then each separated into molecular species by HPLC in the reversed-phase mode. For the alkenylacyl and alkylacyl derivatives, a column (4.6 × 250 mm) containing Zorbax™ ODS and maintained at 33°C was eluted with acetonitrile–isopropanol–methyl-*t*-butyl ether–water (63:28:7:2 by volume) at 0.5 ml/min; for the diacyl form, the same solvents were used but in the proportions 72:18:8:2 respectively. UV detection at 205 nm was again employed. Each of the diradyl forms was separated into 22 fractions, and indeed the recorder traces resembled those reported by others for intact phospholipids, although the resolution was perhaps slightly better here with the acetate derivatives. The fractions were collected once more for identification and determination of the alkyl and acyl moieties by GLC methods. The first component to elute consisted of 18:1 with 22:6; the retention times of all the remaining species relative to this one were calculated, and a graphical scheme was devised as a guide to the identification of minor components.

These procedures have also been used to separate the corresponding molecular species of each of the diradyl forms of the choline-, ethanolamine-, inositol- and serine-glycerophospholipids in rat alveolar macrophages [571], of phosphatidylcholine in rat liver microsomes [574], and of phosphatidic acid, phosphatidylinositol and native diacylglycerols in human phytohemagglutinin-stimulated T-cells [721]. In addition, they were employed to study the metabolism of isotopically-labelled lipids in rat brain *in vivo* [570].

Because disaturated molecular species of phosphatidylcholine, such as those predominating in lung, could not easily be detected and quantified spectrophotometrically at 205 nm, the use of refractive index detection for the purpose was explored and was found to give satisfactory results [357]. In this study, the diacylglycerol acetate derivatives were separated on a column of LiChrosorb™ RP18, eluted with acetonitrile–isopropanol–water (35:15:1 by volume) as the mobile phase. As expected, very different patterns were seen when refractive index and UV detection were compared. On the other hand, the proportions of the various molecular species obtained by integration of the signal from the RI detector corresponded very well with results obtained by acetylating the diacylglycerols with [^{3}H]-acetic anhydride, and collecting the fractions from the HPLC columns for liquid-scintillation counting. The limit of RI detection for the dipalmitoyl species was about 20 μg.

Instead of using diacylglycerol acetate derivatives, Kuksis and coworkers prepared the *tert*-butyldimethylsilyl ethers for separation and analysis by reversed-phase HPLC with mass spectrometric detection [654, 655]. A column containing a Supelco C_{18} phase was eluted with a gradient of 30 to 90% propionitrile in acetonitrile. The chemical-ionization mass spectra obtained were very similar to those described above for triacylglycerols (see Section B.4), in that a good protonated molecular ion and ions representing loss of each of the fatty acyl moieties were seen. This method was applied to rat liver and egg yolk phosphatidylcholines and phosphatidylethanolamines, including components containing deuterated fatty acids. Brief details of an application to the total lipids of plasma have also been published [453].

By converting diacylglycerols prepared from phospholipids to UV-absorbing derivatives, it has proved possible to use the high sensitivity and specificity of spectrophotometric detection in the analysis of molecular species. *p*-Nitrobenzoates were first used for the purpose, and they were resolved on a Brownlee RP18 column with isopropanol–acetonitrile (7:13 by volume) as the mobile phase and detection at 254 nm [57]. Monoacylglycerol *p*-nitrobenzoates, prepared by hydrolysis with the lipase of *Rhizopus arrhizus*, were separated similarly, but with acetonitrile as the mobile phase. Subsequently, *p*-methoxybenzoate derivatives, which had improved storage properties, were utilized under the same chromatographic conditions in a study of the molecular species of phosphatidylethanolamine, phosphatidylglycerol and diphosphatidylglycerol from *Escherichia coli* [58]. Direct quantification from the detector signal was possible, and components were identified, after collection, by GLC analysis of the fatty acid constituents and by a form of the equivalent chain-length concept. The detection limit was in the nanomoles range. The phosphatidylglycerol from plants has been fractionated in the

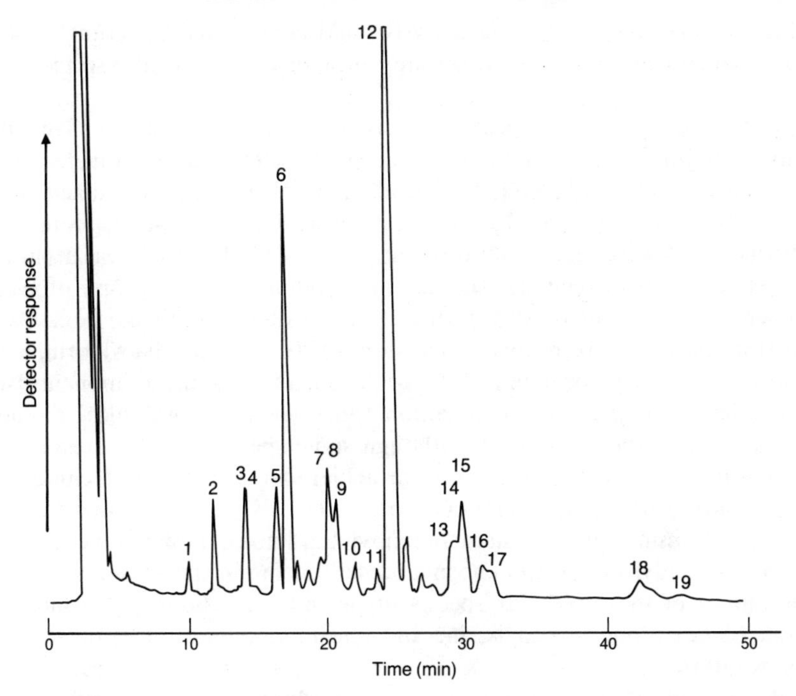

FIG. 8.9. Separation of molecular species of the dinitrobenzoate derivatives of diacylglycerols, derived from phosphatidylcholine of hearts of rats fed a diet enriched in corn oil, by HPLC in the reversed-phase mode [817]. The chromatographic conditions are given in the text. Abbreviations: 1, 16:1–22:6, 18:2–22:6; 2, 16:1–20:4(n – 6), 18:2–20:4(n – 6); 3, 18:1–22:6; 4, 16:0–22:6; 5, 18:1–20:4(n – 6), 16:0–22:5(n – 3); 6, 16:0–20:4(n – 6); 7, 18:1–16:1, 18:1–18:2; 8, 18:0–22:6; 9, 16:0–16:1, 16:0–18:2, 18:1–22:4(n – 6); 10, 16:0–22:4(n – 6): 11, 18:0–22:5(n – 3); 12, 18:0–20:4(n – 6); 13, 18:1–18:1; 14,15, 16:0–18:1, 18:0–18:2; 16, 16:0–16:0; 17, 18:0–22:4(n – 6); 18, 18:0–18:1; 19, 18:0–16:0. (Reproduced by kind permission of the authors and of *Lipids*, and redrawn from the original paper.)

form of the *p*-anisoyl diacylglycerols, under chromatographic conditions developed for related derivatives prepared from galactoglycerolipids (see Section F below) [415].

Dinitrobenzoate derivatives were employed by others [425, 817] to obtain some impressive separations of molecular species of phosphatidylcholine from rat heart, human platelets and hamster cell cultures, as illustrated in Fig. 8.9. With a column (4.6 × 250 mm) of Ultrasphere™ ODS and elution with acetonitrile–isopropanol (4:1 by volume) as the mobile phase, 29 distinct fractions were detected, identified and quantified. 12:0–12:0 or 18:0–18:0 species could be added as an internal standard if required. When methanol–isopropanol (19:1 by volume) was the mobile phase, only 17 fractions were seen but some not separated by the previous system were in fact resolved. Thus by collecting fractions containing more than one component from the first eluent, a more comprehensive analysis could be obtained by rerunning with the second eluent. In this way, as many as 36 distinct molecular species were obtained from each lipid class.

Simple diacylglycerol benzoate derivatives were found to have good chromatographic properties, and could be detected and quantified spectrophotometrically at 230 nm [81]. The alkenylacyl, alkylacyl and diacyl forms of the derivatives prepared from the ethanolamine glycerophospholipids of bovine brain were first resolved by means of TLC, then each was separated by HPLC into molecular fractions on a column (4.6 × 250 mm) of Ultrasphere™ ODS, eluted isocratically with various acetonitrile–isopropanol mixtures, i.e. in the respective proportions of 70:30 for the diacyl form, 65:35 for the alkenylacyl form and 63:37 for the alkylacyl form. Components were identified from the relative retention times using the graphical methods described elsewhere [20, 569]. Although some species were not resolved in this way, complete separations were achieved by using the technique in combination with silver ion TLC. These methods were later used to study the metabolism of unique diarachidonoyl and linoleoylarachidonoyl species of phosphatidylcholine and phosphatidylethanolamine in rat testes [80] and in studies of the molecular species of phosphatidylcholine, phosphatidylethanolamine, phosphatidylserine and phosphatidylinositol in rat erythrocytes [700].

In an interesting modification of this procedure, total serum phospholipid extracts were hydrolyzed by phospholipase C, benzoylated, and the resulting monoacylglycerol, diacylglycerol and ceramide benzoates were fractionated as above but with gradient elution, i.e. from 30 to 60% isopropanol in acetonitrile [68]. A rather complex recorder trace was obtained, but the components were identifiable by graphical methods. As mentioned above, Kuksis and coworkers have published some details of a similar method [453]. A related procedure in which high-temperature GLC rather than HPLC was employed has been used more often for this purpose as described in Chapter 5 (Section D).

Rather more sensitive detection (down to 10 pmoles at 290 nm) was possible with naphthylurethane derivatives of diacylglycerols [447]. Initially, they were separated on a column of LiChrosorb™ RP18, maintained at 60°C, with a gradient of 4 to 10% water in acetonitrile as the mobile phase, but later [713] 4 to 14% water in methanol was used. The nature of the separation appeared to be similar to that obtained with other derivatives, and the method has been applied to phosphatidic acid, phosphatidylcholine and diacylglycerols synthesized by rat microsomes [713], to the native diacylglycerols of human muscle tissue [465], and to phosphatidylglycerol from the microsomes, mitochondria and surfactant from rat lung [729].

Although fluorescence detection should be capable of even greater sensitivity, the only fluorescent derivatives to have been employed for the purpose have been the rather polar dansylethanolaminephosphates [715, 716].

F. HPLC Separation of Glyceroglycolipids

The galactosyldiacylglycerols of plant lipids, in both the intact form and as diacylglycerol derivatives, have been successfully separated into molecular species by reversed-phase HPLC by several research groups. In the first such separation to be published, mono- and digalactosyldiacylglycerols of spinach chloroplasts were eluted from a column of LiChrosorb™ RP18 with methanol–water (19:1 by volume) as the mobile phase, and with refractive index detection [902]. Five monogalactosyldiacylglycerol fractions were obtained, i.e. 18:3–16:3, 18:3–16:2, 18:3–18:3, 18:2–18:3 and 18:1–18:3; the digalactosyldiacylglycerols were slightly more complex, but this lipid also contained some 16:0 fatty acid. Similar chromatographic conditions were employed to fractionate mono- and digalactosyldiacylglycerols from chloroplasts of *Dunaliella salina*, i.e. with two columns (4.6 × 250 mm each) in series of Biophase-ODS™, with methanol–1 mM phosphate buffer (pH 7.4) (19:1 by volume) as the mobile phase, and with UV detection at 205 nm [496]. Subsequently, the elution conditions were modified somewhat so that the Tracor Instruments transport-flame ionization detector could be used (see Section D.2 above) [777]. The separation obtained with the digalactosyldiacylglycerols is illustrated in Fig. 8.10. Again, the nature of the response was very different from that obtained with UV detection, and acceptable quantitative analyses were obtained with as little as 75 nmoles of each lipid.

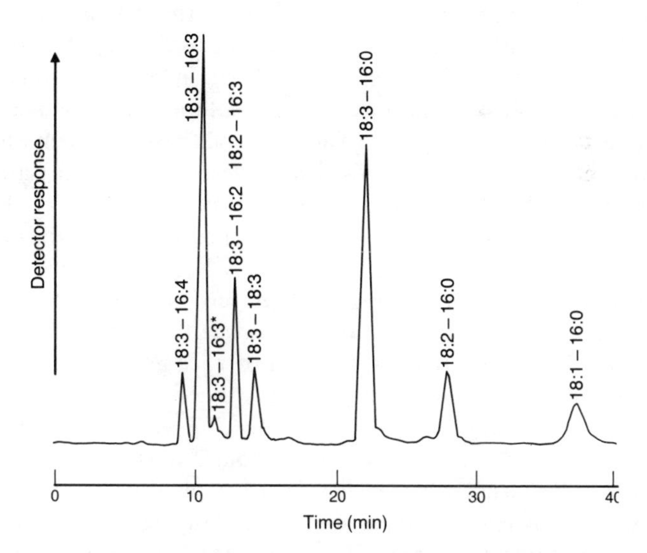

FIG. 8.10. Separation of molecular species of digalactosyldiacylglycerols from *Dunaliella salina* by HPLC in the reversed-phase mode [777]. A column (4.6 × 250 mm) of Rainin Microsorb™ was eluted with methanol–water (96:4 by volume) as the mobile phase at a flow rate of 0.8 ml/min. The Tracor Instruments (Austin, Texas, USA) transport-flame ionization detector was employed. (Reproduced by kind permission of the authors and of the *Journal of Chromatography*, and redrawn from the original paper.)

Others [181] obtained comparable resolution of the same lipid classes from potato tubers and tobacco leaves with the procedure developed by Patton *et al.* [630] for molecular species of phospholipids.

Sulphoquinovosyldiacylglycerols (after methylation of the acidic group with diazomethane) and mono- and digalactosyldiacylglycerols from leaves of *Aquilegia alpina* were fractionated on a column (4.6 × 125 mm) of Spherisorb™ C_6 with a gradient of 50 to 0% water in acetonitrile as the mobile phase, and with spectrophotometric detection at 200 nm [415]. Response factors were determined for each molecular species so that the technique could be used quantitatively. As in related work with phospholipids, graphical relationships between relative retention times and double bond numbers were of value in identifying fractions. Components containing *cis*- and *trans*-16:1 fatty acids could not be resolved by reversed-phase HPLC, but this was accomplished by silver ion TLC. As an alternative, the galactolipids were converted to the diacylglycerols by a chemical procedure [309]; the *p*-anisoyl derivatives were then prepared and fractionated under similar conditions, except that a gradient of 30 to 0% water in acetonitrile was the mobile phase with selective detection at 250 nm.

G. Alternatives to HPLC Procedures

Until recently, silver ion TLC and high-temperature GLC on packed columns were the methods of greatest value for the separation of molecular species of lipids. With the latter technique, fused-silica capillary columns are now widely preferred. Silver ion TLC still has much to commend it as a small-scale preparative procedure. In this section in order to broaden the perspective of the discussion, a brief account only is given of some established methods, which are still widely used in preference to HPLC, and of some recent work, which in the author's opinion sets new standards in analysis. The reader should look to other texts for comprehensive or historical reviews of the subject [133, 452].

The principle of silver ion chromatography is discussed in Chapter 7 (Section C). When used in conjunction with TLC, about 10% by weight of silver nitrate is incorporated into the layer of silica gel G when it is prepared. With triacylglycerols, components migrate in the order SSS > SSM > SMM > SSD > MMM > SMD > MMD > SDD > SST and so forth, where S, M, D and T denote saturated, monoenoic, dienoic and trienoic fatty acyl groups respectively. The usual practice has been to separate the least saturated fractions first with a mobile phase consisting of say hexane–diethyl ether (4:1 by volume), and then to separate the more unsaturated fractions with a more polar mobile phase, such as the same solvents in the ratio 2:3. With care, some resolution of isomeric species is possible; e.g. with the species S_2M, a fraction with the monoenoic fatty acids in position 2 is separable from that with the same component in one of the primary positions. Fractions have been quantified by eluting them from the adsorbent, washing to remove any silver salts and transmethylating in the presence of an internal standard for GLC analysis. Good results have also been obtained by brominating to eliminate the double bonds, and charring followed by photodensitometry [126].

Although some limited success has been achieved in separating intact species of phospholipids by silver ion TLC, most workers have preferred to hydrolyze with phospholipase C and to prepare the diacylglycerol acetates for analysis. With a mobile phase such as chloroform–methanol (99:1 by volume), species migrate in the order SS > SM > MM > SD > MD > DD > ST > MT > STe > MTe > DTe > SP > SH, where Te, P and H represent tetraenoic, pentaenoic and hexaenoic species respectively.

In many studies, information on the chain-length distributions of the fractions separated in this way, both with triacylglycerols and diacylglycerol acetates, has been obtained by high-temperature GLC on packed columns

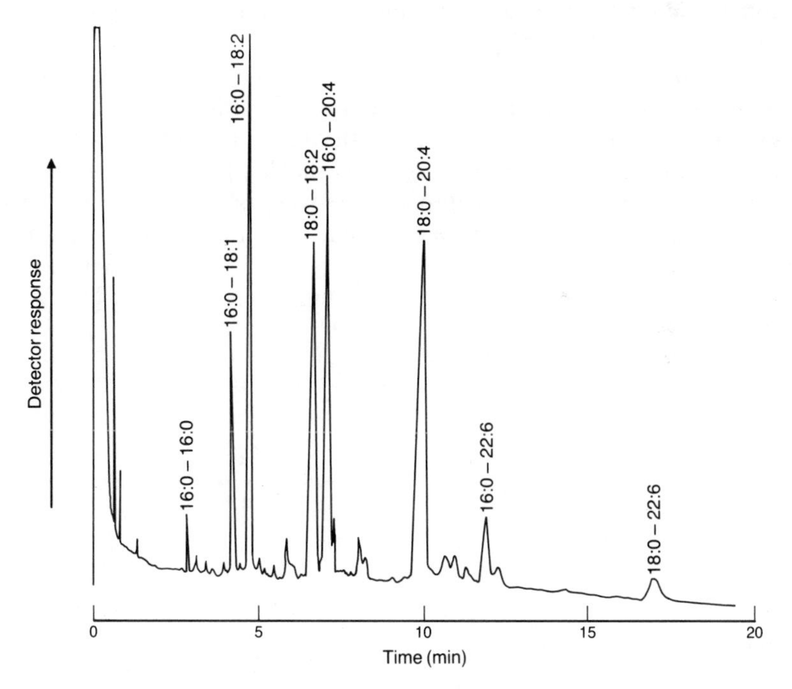

FIG. 8.11. Separation by high temperature GLC of the trimethylsilyl ether derivatives of diacylglycerols, prepared by phospholipase C hydrolysis from the phosphatidylcholines of rat liver [562]. The column was a 10 m × 0.25 mm glass capillary coated with SP-2330, and was temperature-programmed from 190°C to 250°C at 20°C/min, then was held isothermally at 250°C. Splitless injection was used with hydrogen as the carrier gas. A few only of the major peaks are identified here for illustrative purposes. (Reproduced by kind permission of the authors and of the *Canadian Journal of Biochemistry and Cell Biology*, and redrawn from the original paper.)

containing non-polar thermostable stationary phases, which separate broad fractions on the basis of molecular weight only and not with regard to degree of unsaturation. However, it has proved possible to obtain data on both these parameters by deuterating the double bonds in diacylglycerol derivatives for separation and identification by GLC-MS [186]. Recently, it has been shown that molecular species of diacylglycerol derivatives and even of triacylglycerols could be resolved both according to the combined chain-lengths and number of double bonds in the fatty acyl constituents by high-temperature GLC on capillary columns of glass or fused silica coated with thermostable polymers of medium polarity. Thus Myher and Kuksis [562, 563] separated diacylglycerols, prepared from the phospholipids of rat heart and liver by phospholipase C hydrolysis, on a 10 m glass open-tubular column coated with a polar cyanopropylsilicone (Supelco SP-2330). The nature of the separation achieved is illustrated in Fig. 8.11. In this instance, the presence of double bonds in the fatty acyl moieties increased the retention time relative to the corresponding saturated component. This technique used in conjunction with chemical-ionization MS has proved to be an extremely powerful one [453, 455–458].

In spite of the greater molecular weight of triacylglycerols, surprisingly good resolution of some natural fats, both by chain-length and degree of unsaturation, has been achieved on fused-silica capillaries coated with phases of moderate polarity, such as phenylmethylsilicones [236, 561]. The authors of a systematic study of the quantitative recovery of triacylglycerols from a fused-silica column with a chemically-bonded stationary phase (non-polar) concluded that excellent reproducibility was possible, and that small response factors only, in comparison to packed columns, were necessary [513].

CHAPTER 9

HPLC Separation of Sphingolipids

A. Introduction

It is increasingly being recognized that the glycosphingolipids are essential participants in a number of vital process in living tissues. For example, they perform a definitive function in the immunogenicity and antigenicity of cells, and they are involved in cellular interactions, differentiation and oncogenesis. Methods for the isolation and analysis of such compounds have therefore become of great importance. As they tend to be present in cells at very low levels, some care and skill is necessary on the part of the analyst. TLC has been used most often for the purpose, and HP-TLC especially could be considered to have set the standard. However, HPLC procedures for the analysis of sphingolipids, which have also been reviewed elsewhere [503], have played their part. Their speed, resolution and convenience have brought them to the fore in a number of laboratories, a trend that is likely to develop rapidly. In fact, HPLC methods are certainly superior to all others for the separation of molecular species of these compounds. The glycosphingolipids contain a complex range of neutral and acidic carbohydrate components, which have profound effects on their chromatographic properties. The first step in their analysis consists in extracting them quantitatively from the tissues, and then in isolating them as a class, apart from the phospholipids. It may be then be advisable to separate the neutral and acidic (ganglioside) glycolipid classes from each other before proceeding further. Techniques for accomplishing these tasks are described in Chapter 4, and they have been reviewed comprehensively and authoritatively in greater detail than is possible here by Hakomori [268] for glycosphingolipids in general, and by Ledeen and Yu [481] for gangliosides specifically.

Trace amounts of glycosyldiacylglycerols are sometimes isolated from animal tissues together with the glycolipids. For convenience, the analysis of such compounds is discussed here rather than in earlier chapters (the same lipids from plant tissues are discussed in Chapter 6). The separation and analysis of sphingomyelin, together with other phospholipids, are discussed in Chapters 5 and 6 above. As methods for the separation of molecular species of sphingomyelin can have much in common with those used for the glycosphingolipids, this aspect is discussed below.

The distinctive structural feature of sphingolipids, including both the glycosphingolipids and sphingomyelin, is the presence of long-chain aliphatic bases (sphingoid bases) linked by amide bonds to long-chain fatty acids, which are rather different in composition from those found in the glycerolipids. HPLC methods for the analyses of some of these constituent parts have also been described. Similarly, the carbohydrate moieties can be analysed by HPLC, but this aspect is outwith the scope of this book.

There have been two general approaches to the separation and analysis of glycosphingolipids by means of HPLC, each having its devotees and being suited to particular purposes, i.e. to subject them to chromatography in the native form or to convert them to non-polar UV-absorbing derivatives prior to fractionation.

B. Separation of Glycosphingolipid Classes in the Native Form

Improved methods are increasingly being sought for the chromatographic isolation of glycosphingolipids in the native state, in part to reduce the number of steps and to minimize possible losses or alterations during derivatization, but mainly to permit direct investigations of the antigenicity of specific components, or for physical chemical studies of their interactions with cellular membranes. Adsorption chromatography with silica gel has generally been favoured, but bonded phases have found some applications. As with other aspects of HPLC analysis of lipids, the choice of solvents for the mobile phase has frequently been dependent upon the availability of a particular detector in a laboratory.

In the first published separation of underivatized neutral glycolipids by Tjaden *et al.*, standard mixtures of ceramide mono- to tetrasaccharides were resolved on a column (2.8 × 250 mm) of LiChrosorb™ SI 60 silica gel, with chloroform–methanol (3:1 by volume) as the mobile phase [828]. A transport-flame ionization detection system was employed, and it gave linear response over a range of 2 to 200 μg for quantification. The amounts of lipid material required for detection was perhaps rather greater than was desirable for many purposes, and this together with the limited availability of the detector has meant that the technique has been little used.

A procedure described by Watanabe and Arao [864] has attracted rather more attention. They used a column (4 × 500 mm) of 10 μm spherical porous silica Iatrobeads™ 6RS-8010, a form of silica gel that appears to be particularly well suited to the separation of glycolipids. As mobile phase, a linear gradient of isopropanol–hexane–water was generated from 55:44:1 to 55:35:10 (by volume) over 30 min, although some minor modifications to this were made to improve the resolution of specific components. Conventional detection was not possible, but fractions were detected and quantified by means of a post-column reaction of the carbohydrate moieties with an anthrone–sulphuric acid reagent. The nature of the separation

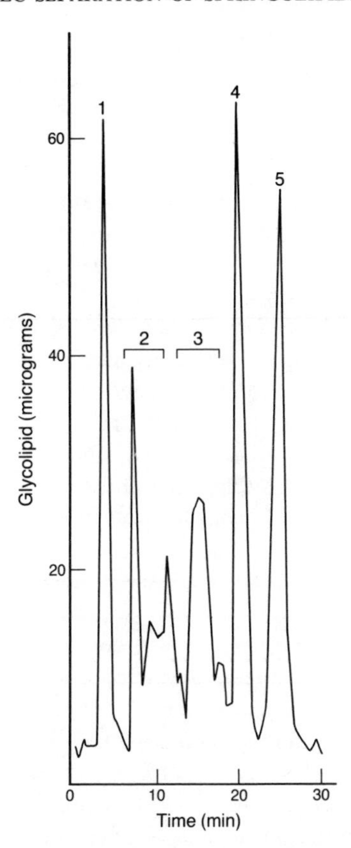

FIG. 9.1. Separation of a standard mixture of mono- to pentaglycosylceramides by HPLC on a column of silica gel (see text for chromatographic conditions) [864]. Abbreviations: 1, glucosylceramide; 2, lactosylceramide; 3, globotriaosylceramide; 4, globotetraosylceramide; 5, Forssman glycolipid. (Reproduced by kind permission of the authors and of the *Journal of Lipid Research*, and redrawn from the original paper.)

achieved is shown in Fig. 9.1. A standard mixture of glycolipids with one to five sugar residues was clearly resolved, with some partial separation according to the nature of the ceramide moeity also being achieved (cf. the separations of sphingomyelin described in Chapter 6). The pattern was highly consistent and reproducible. As an alternative to chemical detection, a post-column bio-assay procedure was used in separations of the glycolipids prepared from erythrocytes (by chromatography on Florisil and DEAE-Sephadex™ A25).

This HPLC procedure was utilized by others [225] for the isolation of branched blood group A-active glycolipids from human erythrocytes, and with some modification of the elution scheme, to the isolation of ceramide hepta- and decasaccharides from human group O erythrocyte membranes

[398], of complex fucolipids from human adenocarcinoma [269], and of related glycolipids and gangliosides from erythrocyte membranes of umbilical cord and from other souces [599]. When the technique was used to isolate a trifucosylceramide from human adenocarcinoma, some overlap of fractions occurred because of heterogeneities in the composition of the acyl moieties, i.e. according to whether they contained medium- or long-chain fatty acids and/or hydroxy or non-hydroxy fatty acids [595]. A final purification step involving acetylation and HP-TLC was therefore necessary. In similar work, neutral and acidic glycolipid standards were resolved on a column of Nucleosil™ silica gel eluted with various hexane-isopropanol-ethanol mixtures, and with UV detection at 206 nm [280, 281]. The fractions obtained from the column were subjected without derivatization to field-desorption mass spectrometry for structural analysis.

Preparative-scale HPLC on a column of silica gel, eluted with chloroform–methanol–water (80:10:0.8 by volume) and with refractive index detection, was used in the final step of a purification of cerebroside from commercial soyabean lecithin [611].

In a different approach to the analysis of glycolipids by means of HPLC, Watanabe and Tomono [865] utilized a column (2.6 × 250 mm) packed with DEAE-derivatized controlled-pore glass, connected serially to two columns (each of 2.6 × 500 mm) of underivatized controlled-pore glass. Mixtures of neutral glycolipids with one to six carbohydrate residues, and of gangliosides with one to four sialic acid groups, were resolved by a complex binary gradient scheme; chloroform–methanol–water, with increasing concentrations of methanol and water, was used to elute the neutral glycolipids over 80 min; then increasing concentrations of lithium acetate (0.015 to 0.1 M), which appeared to be a particularly effective ion suppressant, were incorporated into the mobile phase in order to bring off the gangliosides over a further 60 min. Discreet fractions were collected, and they were examined by HP-TLC with orcinol–sulphuric acid or resorcinol as the detection reagents. The procedure was subsequently used to study the effect of the age and sex of the animals on the pattern of glycolipids in rat kidney [830].

Tjaden *et al.* [828] had earlier utilized the column of silica gel and the transport-flame ionization detector described above with a mobile phase of chloroform–methanol–0.01 M aqueous hydrochloric acid (60:35:4 by volume) to separate a mixture of six gangliosides. They showed that by using a stepwise change in solvents, both the neutral and acidic glycolipids in a sample could be separated sequentially. When the technique was used as a small-scale preparative procedure, the methanol in the mobile phase was replaced by ethanol to minimize the elution of colloidal silica, which otherwise reduced the recoveries.

Most analysts have preferred to isolate a purified ganglioside fractions, free of neutral sphingolipids, or even to prepare ganglioside sub-fractions before proceeding to a more comprehensive separation by means of HPLC. For example, ganglioside extracts from human erythrocytes and bovine brain were first separated into mono- to tetrasialo species by column chromatography on DEAE-silica gel, and these fractions were each then further resolved by means of HPLC on a conventional silica gel column (Zorbax™ Sil) [464]. A linear gradient of isopropanol–hexane–water of 55:42:3 to 55:25:20 respectively by volume over a period of 2 hr. was the mobile phase for the latter step. It was not possible to monitor the eluent continuously, but fractions were collected once more and the elution profile was monitored by TLC.

A new type of strong anion-exchange resin, Mono Q™ (Pharmacia, Sweden), was used to separate fractions containing mono- to tetrasialo species from natural ganglioside extracts, prepared by chromatography on Iatrobeads™ [511]. A stepwise gradient of zero to 0.225 M potassium acetate in methanol was the mobile phase utilized to elute each fraction, aliquots of which were taken for identification by HP-TLC and for quantification with a resorcinol reagent.

Perhaps the most impressive HPLC separations of gangliosides have been achieved on columns containing a stationary phase with bonded-amine groups. For example, Gazzotti Sonnino and Ghidoni [233] separated natural gangliosides on a column (4 × 250 mm) of LiChrosorb™ NH_2, eluted with a complex gradient of acetonitrile–phosphate buffer (pH 5.6) in which both the relative proportions of the solvents and the ionic strength of the mixture were varied. The eluent was monitored spectrophotometrically at 215 nm. With calf brain gangliosides, a sample of 1 to 50 nmoles could successfully be separated as shown in Fig. 9.2. Components were resolved reproducibly not only according to the number of sialic acid residues, but also according to the nature of the carbohydrate moieties. The UV absorbance of equimolar amounts of different gangliosides increased with an increasing content of carbohydrate and sialic acid. However, the response to each was rectilinear and by careful calibration with authentic standards, it could be used for quantification purposes. In a method from another laboratory, in which a similar principle was employed, a column of *micro*Bondapak™ NH_2 was adjusted to the correct pH (5.4) by pre-elution with a sodium acetate buffer, before the gangliosides were eluted with a gradient of methanol and geometrically increasing proportions of 1 M aqueous sodium chloride [874]. UV detection at 210 nm was used. The method was applied to a crude sphingolipid extract from bovine pineal gland, and neutral glycosphingolipids and sphingomyelin eluted almost immediately, enabling gangliosides such as G_{D3} to be isolated later for further analysis.

C. Separation of Glycosphingolipid Classes as Perbenzoyl and Other Derivatives

One of the more successful early applications of HPLC in the field of lipid analysis was to the separation and quantification of glycolipids after they had been converted to the perbenzoyl derivatives. In this form, the lipids are

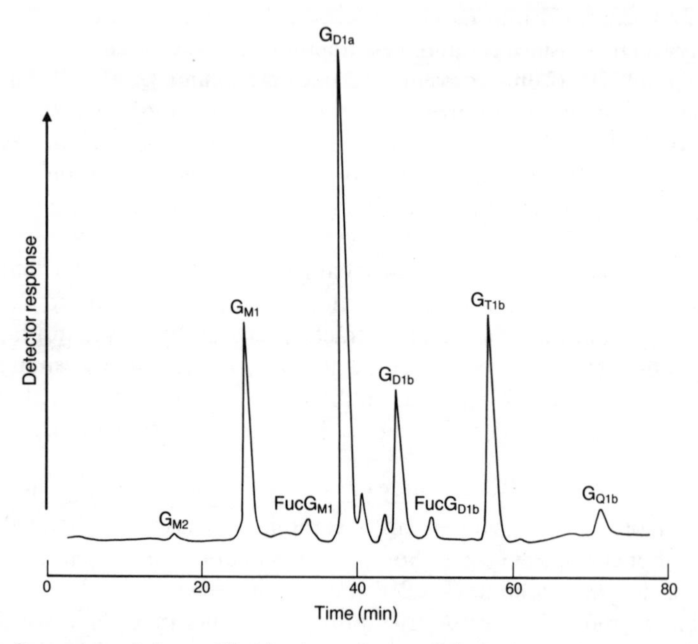

FIG. 9.2. Separation of the ganglioside classes from calf brain by HPLC on a column of LiChrosorb™ NH_2, eluted with a complex gradient of acetonitrile–phosphate buffer (pH 5.6) in which the relative proportions of the solvents and their ionic strength was varied [233]. UV detection at 215 nm was used. (Reproduced by kind permission of the authors and of the *Journal of Chromatography*, and redrawn from the original paper.)

non-polar and can be eluted with non-aqueous mobile phases, but more importantly they can be detected spectrophotometrically with high sensitivity and specificity by their absorbance at 230 nm. Such methods were developed principally in the laboratories of McCluer and Jungalwala in the USA, but they have been taken up or adapted widely elsewhere. Initially, benzoylation of glycolipid samples was accomplished by reaction with benzoyl chloride in pyridine at 60°C for 1 hr [208, 498]. (Some valuable practical advice was given in a later review [503].) The free hydroxyl groups of the carbohydrate moieties and of any 2-hydroxy fatty acids were *O*-benzoylated, and amide groups in sphingolipids that contained only non-hydroxy fatty acids were *N*-benzoylated by this procedure; the latter could not be restored to their native state by alkali-catalyzed transesterification. *N*-benzoylation did not occur with benzoic anhydride as the derivatization reagent. The following is now the recommended benzoylation method, in which benzoyl chloride is used [503]. All reagents should be fresh and free from moisture.

> "The glycolipids (up to 150 nmoles) are heated with 10% benzoyl chloride in pyridine (0.5 ml) for 16 hr at 37°C in a well-stoppered tube. The excess reagents are blown off in a stream of nitrogen, and hexane

(3 ml) is added to the residue. The hexane layer is washed four times with methanol saturated with sodium carbonate (1.8 ml portions), and finally the hexane is removed in a stream of nitrogen and the sample is taken up in carbon tetrachloride for injection onto the HPLC column".

A column of silica gel and a gradient of 0.2 to 0.75% methanol in hexane was employed in the first application of the benzoylation procedure for the separation of glucosylceramide, lactosylceramide, galactosyllactosylceramide and *N*-acetylgalactosaminyllactosylceramide from human plasma [208, 498].

Free ceramides containing hydroxy and non-hydroxy fatty acids in tissues of a patient with Farber's disease were determined by HPLC of the benzoylated (with benzoic anhydride) derivatives [361, 802]. The mobile phase for elution of the required compounds from a silica gel column was either 0.05% methanol in pentane or 2.5% ethyl acetate in hexane, while detection was at 254 nm. Improved benzoylation and chromatographic conditions were developed for the analysis and quantification of free ceramides, containing non-hydroxy and 2-hydroxy fatty acids and phytosphingosine, from various rat tissues [360]. Later, it was suggested that a gradient of isopropanol in hexane with monitoring at 230 nm might have given better sensitivity and resolution [503]. (Ceramides and other sphingolipids containing 4-sphingenine were also determined by HPLC following oxidation to the 3-keto derivative [362].)

Hexane–ethyl acetate (93:7 by volume) was the mobile phase for the separation and quantification of benzoylated (benzoyl chloride and pyridine) cerebrosides, containing either non-hydroxy or hydroxy fatty acids [499]. These lipids from relatively crude extracts of various tissues were analysed on a scale of 10 to 150 nmoles. In this instance, detection was at 280 nm, since both derivatives appeared to exhibit a similar molar response at this wavelength. It also proved possible to separate glucosyl- and galactosylceramides containing non-hydroxy fatty acids on a column of MicroPak™ NH_2 with cyclopentane–isopropanol (98.8:1.2 by volume) as the mobile phase. (Bonded-amine stationary phases have proved of great value in the separation of carbohydrates *per se*, and it appears suprising that relatively little use has been made of them for glycosphingolipid separations.) Sulphatides did not interfere with the analysis, and indeed they could be determined as cerebrosides after desulphation by means of mild acidic methanolysis. Subsequently, the same lipid classes were detected and quantified in the range of 0.5 to 10 nmoles by employing UV detection at 230 nm and gradient elution with 2.8 to 5.5% dioxane in hexane, from a column of silica gel [382]. The response to benzoylated fractions containing hydroxy and non-hydroxy fatty acids was nearly the same.

A method for the analysis of intact cerebroside sulphate in benzoylated form was also described (reported in reference [503]). The glycolipids, benzoylated with benzoic anhydride and 4-dimethylaminopyridine (see below), were separated on a column of silica gel with hexane–isopropanol–propionic acid (100:35:2 by volume) as the mobile phase. Subsequently, improved separations and quantifications of benzoylated cerebroside sulphate from mouse glia cells were attained on a column of silica gel, eluted with hexane–isopropanol–water (100:20:0.5 by volume) [384].

In later work, rather more comprehensive separations of concentrates of neutral glycolipids (and free ceramides) from plasma and other tissues were achieved on a column (2.1 × 500 mm) containing Zipax™, a pellicular silica gel, and elution with a gradient of 2 to 17% ethyl acetate in hexane at a flow-rate of 2 ml/min with UV detection at 280 nm [840], or by elution with a gradient of 1 to 20% dioxane in hexane with higher sensitivity detection at 230 nm (the absorption maximum) [127, 384, 841]. With the latter mobile phase, improved base-line stability was attained by directing the solvent gradient through a high-pressure reference cell ahead of the column and injector to negate the residual absorption of the dioxane. Glycolipids with one to four carbohydrate residues were clearly resolved and detected, in addition to some unidentified minor components, in under 20 minutes. The limit of detection was 2 to 20 pmoles, depending on the lipid class. To assist with the quantification, *N*-acetylpsychosine was added as an internal standard. The procedure was applied in studies of glycolipid synthesis and excretion by mouse kidney [500, 504], and of kidney glycolipids in a lizard [205]. Dioxane–hexane mixtures as the mobile phase are still reported to give the best resolution of minor glycolipids, since they do not simultaneously cause a partial separation on the basis of the chain-lengths of the fatty acid constituents [503].

Monosialogangliosides, after benzoylation with benzoyl chloride and pyridine in toluene and in amounts as little as 3 nmole, were successfully resolved into a number of defined fractions in the same laboratory on a column of LiChrosphere™ SI 4000 silica gel, with a linear gradient of 7 to 23% dioxane in hexane over 18 min at flow-rate of 2 ml/min, and with detection at 230 nm [91]. Later, monosialoganglioside fractions, obtained by a rapid mini-column procedure, were separated following perbenzoylation in a more comprehensive manner [128, 842]. The sample was applied to a column (4.6 × 150 mm) of Rainin™ 3 μm silica gel, maintained at 90°C, and it was eluted with a linear gradient of 1.8 to 12% isopropanol in hexane as the mobile phase. Although the detector response for each benzoylated ganglioside was slightly different, it was always related linearly to mass, and could be used for quantification purposes. The method was applied routinely to the analysis of the gangliosides from as little as 25 mg of human grey matter, and for subcellular fractions of rat nerve tissues.

By using a high column temperature in this last work, there was no requirement for the incorporation of ionic species into the mobile phase to overcome the polarity of the sialic acid residues, which were not derivatized, and to ensure that they migrated in a single ionic form. A higher column temperature (65°C) also enabled a separation of perbenzoylated glucocerebroside from galactocerebroside under similar elution conditions [411].

A major disadvantage of perbenzoylation with benzoyl chloride in pyridine was that the *N*-benzoylated lipids could not be converted back to the parent glycolipids. To circumvent this difficulty, it was shown that no amide benzoylation occurred when the reaction was carried out in benzoic anhydride with 5% *N*,*N*-dimethyl-4-aminopyridine in pyridine as a catalyst [254]. Complete *O*-benzoylation with no *N*-benzoylation was achieved in 4 hr at 37°C. If need be, the parent glycolipids could be recovered later by mild alkaline methanolysis. The glycosphingolipids, derivatized in this way, were separated under similar conditions, i.e. on the Zipax silica gel column with a gradient of 2.5 to 25% dioxane in hexane as the mobile phase and with UV detection at 230 nm. The nature of the separation obtained is shown in Fig. 9.3. Mono- to tetraglycosylceramides were cleanly resolved, and components containing hydroxy fatty acids eluted slightly ahead of, rather than

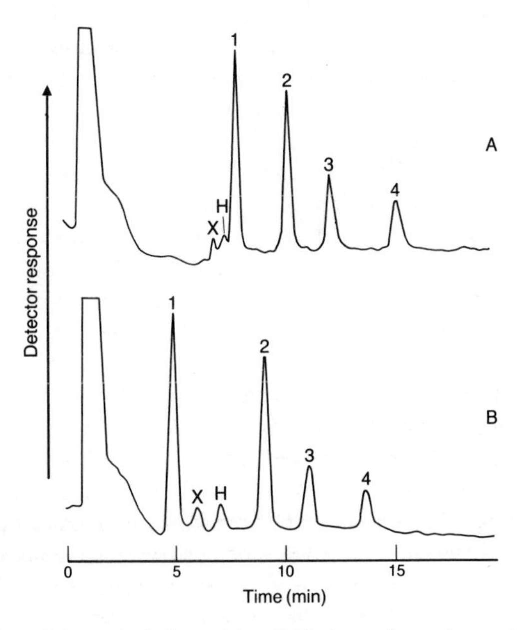

FIG. 9.3. Separation of the neutral glycosphingolipid classes from plasma, in the form of the perbenzoyl derivatives, by HPLC on a column (2.1 × 500 mm) of Zipax™ silica gel, eluted with a gradient of 2.5 to 25% dioxane in hexane over 13 min with UV detection at 230 nm [254]. A, benzoylation with benzoyl chloride and pyridine. B, benzoylation with benzoic anhydride and 4-dimethylaminopyridine. Abbreviations: 1, glucosylceramide; 2, lactosylceramide; 3, globotriaosylceramide; 4, globotetraosylceramide; X, unknown; H, galactosylceramide containing hydroxy fatty acids. (Reproduced by kind permission of the authors and of *Analytical Biochemistry*, and redrawn from the original paper.)

after, the corresponding components with non-hydroxy fatty acids. In subsequent publications, the benzoyl chloride procedure was utilized for work that was simply analytical, since this gave better resolution of those components containing hydroxy fatty acids or phytosphingosine. The benzoic anhydride procedure was preferred for sulphatides and when components were isolated for specific purposes.

The above methods developed in the laboratories of McCluer and Jungalwala have been used by many others, sometimes, with minor adaptations. For example, they were used to study the neutral glycosphingolipids from human lymphocyte subpopulations [742]. By extending the gradient in the mobile phase up to 31% dioxane in hexane, it proved possible to separate benzoylated neutral glycolipids with as many as ten carbohydrate residues together with gangliosides in one step [484]. Further information on the ganglioside composition was obtained by analysing the samples before and after neuraminidase treatments. Similar methodology, except that the column (4.6 × 250 mm plus pre-column) was of 5 μm. Ultrasphere™ silica gel and the mobile phase was a gradient of 0.85 to 15% isopropanol in hexane (to eliminate the need for a reference cell, since these solvents are UV-transparent), was used for the analysis of plasma glycolipids as an aid to the diagnosis of lysosomal storage disease [799]. An additional novel feature of this work was that benzoylated monogalactosyldiacylglycerol of plant origin was added as an internal standard to assist with the quantification. On the other hand, it has been reported that isopropanol–hexane combinations in the mobile phase bring about some peak broadening, because of partial chain-length separations, with a consequent loss of resolution [503].

Benzoylation was also used in the separation and determination of 1-*O*-alkyl-2-*O*-acyl-*beta*-3′-sulphogalactosylglycerol ("seminolipid") in boar spermatozoa by means of HPLC [804]. A fraction enriched in benzoylated seminolipid was prepared by Florisil™ and Sephadex™ LH-20 chromatography, and was applied to an HPLC column (2.1 × 250 mm) of Zorbax™ SIL silica gel; the mobile phase was chloroform–methanol (93:7 by volume) and UV detection was at 254 nm. After careful calibration, quantification in the range of 1 to 15 nmoles was possible.

An important method for the isolation of a glycosphingolipid concentrate involves the preparation of *O*-acetate derivatives. In one laboratory, therefore, *N*-*p*-nitrobenzoyl derivatives of such *O*-acetylated sphingoglycolipids were prepared and separated on a column (2.1 × 250 mm) of Zorabax™ SIL silica gel, with a gradient of 0.5 to 7% isopropanol in hexane–chloroform (2:1 by volume), and with spectrophotometric detection at 254 nm [903]. When applied to the analysis of the glycolipids from erythrocytes, the lower practical limit of detection was about 50 pmoles. Glycolipids containing amino sugars were resolved by this method, and it was noted that the order of elution of individual components was similar but not identical to the order of migration of the underivatized components on TLC. In addition, the same components were separated on a column containing bonded-phenyl groups, eluted isocratically with hexane–isopropanol–chloroform (94:3:3 by volume), but the resolution was not as good as on silica gel. One advantage of this procedure was that a single UV-absorbing moiety was introduced into each molecule of sphingolipid regardless of the number of hexose units; only if an *N*-acetylhexosamine unit was present was there a second *p*-nitrobenzoyl moiety introduced. Thus for the common range of glycosphingolipids, a single calibration curve sufficed for all components without amino sugars, while the response to sphingolipids containing one *N*-acetylhexosamine group was exactly doubled. An important disadvantage of the method was that components containing 2-hydroxy fatty acids did not form UV-absorbing derivatives, although this was apparently not a problem with erythrocyte glycolipids. This information was confirmed later, and the elution conditions were improved by utilizing a gradient of 1 to 5% isopropanol into hexane–dichloroethane (2:1 by volume) [807]. The method was subsequently applied by others to the analysis of erythrocyte glycolipids in patients with congenital dyserythropoietic anaemia [89].

Glycolipids, which were acetylated only, had earlier been separated preparatively by HPLC on a column of silica gel eluted with various isopropanol–chloroform mixtures, depending on the polarity of the lipid of interest [544]. Fractions were collected for identification by TLC. In particular, hexosamine-containing glycolipids, such as the Forssman-active compound, could be isolated with relative ease, and three pure glycolipids with ten to twelve carbohydrate residues were obtained and characterized. They were eventually deacetylated by sodium methoxide-catalyzed transesterification.

Glycosphingolipids containing sulphate residues were not easily separated in intact form by any of the above methods until recently, but their analysis has been accomplished in a manner suited to routine assay by desulphation of benzoylated glycolipids with 0.2 M perchloric acid in acetonitrile, prior to chromatography [593]. For the separation, a column of Spherisorb™ silica gel was used, and it was eluted with a gradient of 0.5 to 10% isopropanol in hexane. With a sample of brain lipids, glucosylceramide, galactosylceramide, monogalactosyldiacylglycerol and sulphoglycosylceramide, including forms containing hydroxy and non-hydroxy fatty acids, were clearly separated. Subsequently, it was shown that improved desulphation was achieved by solvolysis with trifluoroacetic acid in ethyl acetate (followed by rebenzoylation), and that better resolution could be attained by incorporating 0.05% of concentrated ammonia into the mobile phase [759]. The procedure was used to study the changes that occurred in the concentrations of these lipids in the brain of the rat as it matured [760], in the metabolism of these compounds in subcellular fractions of brain tissue [761, 897], and in degenerating nervous tissue [895]. Unfortunately, the resolution of galactosyl- and glucosylceramides deteriorated if the ratio of the two departed appreciably from unity. In addition to the methods described above, sulphatides had earlier been determined by an oxidation procedure followed by HPLC [362].

Several alternatives to benzoylation have been developed for the separation and determination of gangliosides. For example, a ganglioside fraction from human brain was derivatized by subjecting the double bond of the ceramide group to ozonolysis, cleavage and reaction of the nascent aldehyde with *p*-nitrobenzyloxyamine [832]. The products were purified by chromatography on DEAE-Sephadex™, before being subjected to HPLC in the reversed-phase mode on a column containing an ODS phase, eluted with a gradient of methanol–water from 5:5 to 7:3 by volume, and with UV detection at 254 nm. Although good quantification was reported, the procedure appears excessively complicated, and molecules containing dihydrosphingosine would not react while those containing unsaturated fatty acids would be further ozonized.

A more convincing technique involved the conversion of the sialic acid residues of gangliosides to the *p*-bromophenacyl derivatives; the products were applied without purification to an HPLC column (4.6 × 250 mm) of Zorbax™ SIL [567], As mobile phase, a gradient of isopropanol–hexane–water from 50:49:1 to 55:35:10 by volume was used, with UV detection at 261 nm. The order of elution of individual components was broadly similar to that found with TLC procedures. The recoveries from the column were quantitative, the detector response was rectilinear, and the analytical results (with a lower limit of detection of 10 ng of sialic acid) corresponded well those obtained by an established TLC procedure.

Rather similar in its sensitivity and specificity was a method in which gangliosides in total lipid extracts, after removal of water-soluble compounds, were converted to the 2,4-dinitrophenylhydrazide derivatives of the sialic acid residues by reaction with 2,4-dinitrophenylhydrazine hydrochloride and dicyclohexylcarbodiimide in dimethylformamide at 0°C [541]. The derivatives were purified on a mini-column of silicic acid, before being subjected to HPLC on a column of silica gel. Various isocratic mixtures of chloroform–methanol–water–acetic acid were used as mobile phases to achieve particular separations, while $CaCl_2$ was incorporated to effect a separation of G_{M3} ganglioside containing *N*-acetylneuraminic acid from that containing *N*-glycollylneuraminic acid. UV detection was employed at 342 nm. As an example, one separation achieved is shown in Fig. 9.4. The response of the detector was shown to be linear in the range of 0.02 to 1.6 nmoles for the gangliosides G_{M3}, G_{M2}, G_{M1}, G_{D1a}, G_{D1b}, G_{T1b} and L_{M1}; gangliosides G_{M4}, G_{D3}, G_{T1a} and G_{Q1b} also gave distinct peaks. Both of these last procedures appear to have much to offer in comparison to alternative methods for ganglioside analysis in terms of speed, convenience and sensitivity.

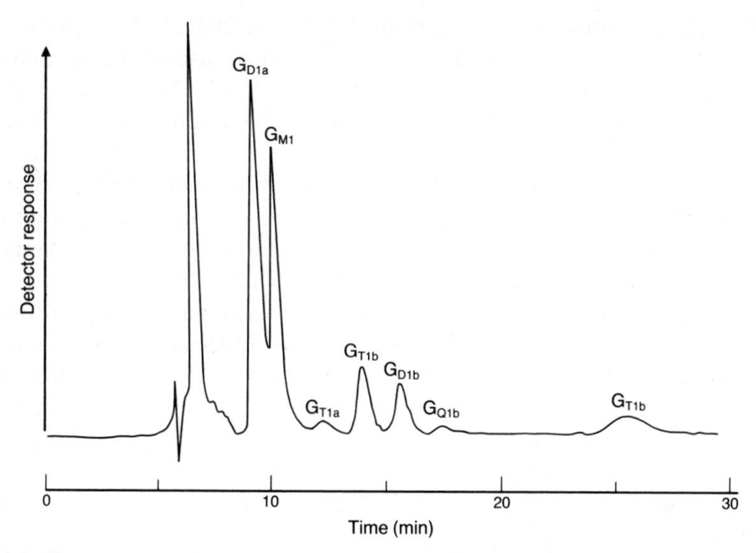

FIG. 9.4. Chromatogram of the 2,4-dinitrophenylhydrazide derivatives of gangliosides prepared from a total lipid extract of rat brain and separated by HPLC [541]. A column (8 × 100 mm) of Resolve™ Si was eluted with a mobile phase of chloroform–methanol–water–acetic acid (63.5:33.2:2.8:0.5 by volume) at a flow-rate of 0.5 ml/min. UV detection was at 342 nm. (Reproduced by kind permission of the authors and of the *Biochemical Journal*, redrawn from the orginal paper.)

D. Separation of Molecular Species of Sphingolipids

1. Glycosphingolipids in underivatized form

As with other lipids, each glycolipid class exists in the form of innumerable molecular species. In this instance, the different fatty acid constitutents are linked to each of the long-chain (sphingoid) bases which may vary in chain-length, degree of unsaturation and the number of hydroxyl groups. It should be noted that the fatty acid constituents are saturated and monoenoic with odd and even numbers of carbon atoms from C_{16} up to about C_{26}, but with essentially no polyunsaturated components. Species containing non-hydroxy and 2-hydroxy fatty acids also exist and are frequently separated from each other during isolation of the lipid classes by the methods described in the previous section. As in the isolation of sphingolipid classes, there have been two approaches to the separation of molecular species, i.e. in the native form or as UV-absorbing derivatives. HPLC in the reversed-phase mode has been almost exclusively the technique of choice.

Several underivatized glycosphingolipid classes, from various tissues and isolated by HPLC on a silica gel column, were each fractionated on a column of FinepackSIL™ C18, with UV detection at 206 nm [280, 281]. For the neutral glycolipids, the mobile phase was methanol–water (99:1 by volume), and for the gangliosides, it was methanol–aqueous 0.01% $CaCl_2 \cdot 2H_2O$ (96:4). As anticipated, the retention times of individual components increased with the total number of carbon atoms and decreased with the number of double bonds in the ceramide part of the molecule; the retention times of glycolipids with the same ceramide moiety also decreased progressively with the number of carbohydrate residues. Each component was characterized by field-desorption mass spectrometry, and this aspect of the analysis was expanded upon later both for the neutral [468] and for the acidic [469] glycolipids.

Monoglycosylceramides obtained from the intestines of Japanese quail were subjected to HPLC on a column of Nucleosil™ C_{18}, eluted with methanol at 1 ml/min and with UV detection at 210 nm [589]. As illustrated in Fig. 9.5, a large number of components were resolved and most were

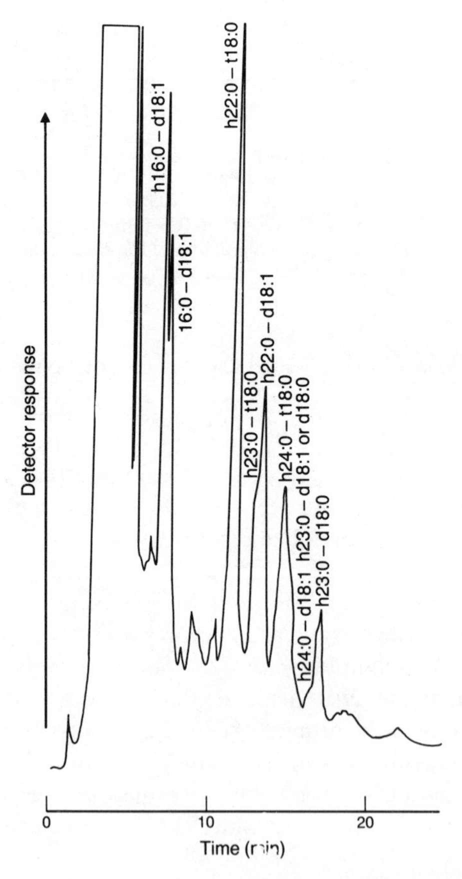

FIG. 9.5. Reversed-phase HPLC separation of molecular species of underivatized monoglycosylceramides from the intestines of Japanese quail [589]. The experimental details are given in the text. The first part of the peak designations represents the fatty acid constituent and the second part the long-chain base. Abbreviations: h, 2-hydroxy; d, dihydroxy; t, trihydroxy. (Reproduced by kind permission of the authors and of *Biochimica Biophysica Acta*, and redrawn from the original paper.)

provisionally identified by their relative retention times as containing homologous series of saturated hydroxy and non-hydroxy fatty acids linked to sphingosine (d18:1). These predictions were confirmed for the most abundant components by fast atom bombardment MS, a technique that is extremely useful for the structural identification of glycolipids. A further series of peaks was identified as containing phytosphingosine (t18:0) linked to non-hydroxy saturated fatty acids. Very recently [318], even better separations were obtained under similar conditions, but with a stationary phase, ERC-ODS-1272™, comprising a very high proportion of ODS groups relative to the inert support (20% by weight). In this instance, critical pairs were completely separated and 49 distinct molecular species were obtained of the monoglycosylceramide from the intestines of Japanese quail.

Native and radioactive semi-synthetic gangliosides G_{M1} and G_{D1a} were separated into molecular species with a homogeneous long-chain base composition by small-scale preparative HPLC in the reversed-phase mode [784, 785]. Amounts of up to 5 mg or as little as 25 μg of each ganglioside were applied to a column (10 × 250 mm) of Spherisorb™ 5S ODS2; acetonitrile–5 mM sodium phosphate buffer (pH 7.0), in the ratio by volume of 3:2 for ganglioside G_{M1} and of 1:1 for G_{D1a}, was the eluent at a flow-rate of 7.5 ml/min, with UV detection at 195 nm. (An adaptation to an analytical scale was also described.) The main fractions obtained in this way contained C_{18} or C_{20} sphingosine and C_{18} or C_{20} sphinganine. Subsequently, the procedure was applied to similar isotopically-labelled gangliosides [234]. In an elution scheme more suited to analysis *per se*, underivatized gangliosides G_{M4}, G_{M3}, G_{M2} and G_{M1}, isolated by column chromatography on Iatrobead™ 6RS-8060, were fractionated into molecular species by HPLC on a column (4.6 × 250 mm) of Ultrasphere™ ODS [391]. The isocratic mobile phase consisted of various methanol–water mixtures, depending on the polarity of each ganglioside, and UV detection at 205 nm was employed; fractions were collected and purified, then they were quantified by HPLC of an aliquot of the perbenzoylated derivative on a column of silica gel. The nature of the separation achieved with ganglioside G_{M4} is illustrated in Fig. 9.6 as an example. Up to twenty components were resolved, and each was identified by direct chemical-ionization MS with ammonia as the reagent gas. With this technique, characteristic fragments were obtained for the ceramide moiety and for the fatty acid and sphingoid base constituents. It was then possible to draw up a graphical relationship to show the "effective carbon number" of a molecular species containing a given fatty acid when linked to a particular base.

2. *Glycosphingolipids in derivatized form*

Among the first applications of derivatization of sphingolipids for separations of molecular species was one of perbenzoylated glucosylceramide from a patient with Gaucher's disease. This was fractionated on a column of *micro*Bondapack™ C_{18}, with methanol as the mobile phase and UV detection

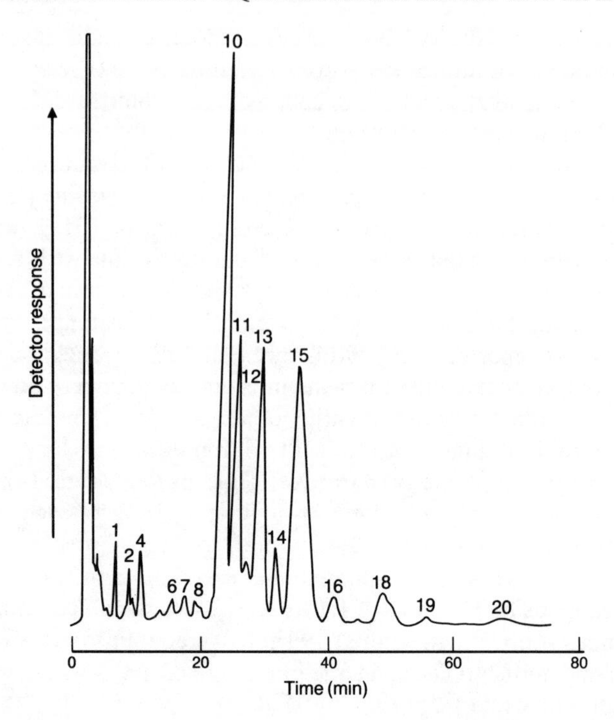

FIG. 9.6. HPLC separation of molecular species of ganglioside G_{M4} from human brain on a column (4.7 × 250 mm) containing Ultrasphere™ ODS, with methanol–water (89:11 by volume) as the mobile phase at a flow-rate of 1.3 ml/min at 26°C, and with UV detection at 205 nm [391]. Peak identifications: 1, 16:0–d18:0; 2, 18:0–d18:1; 4, 18:0–d18:0; 6, h20:0–d18:0; 7, 20:0–d18:0; 8, 21:0–d18:1;10, h22:0–d18:1 & h24:1–d18:1; 11, 22:0–d18:1; 12, 24:1–d18:1; 13, h23:0–d18:1 & h25:1–d18:1; 14, 25:1–d18:1; 15, h24:0–d18:1 & h26:1–d18:1; 16, 24:0–d18:1 & 26:1–d18:1; 18, h25:0–d18:1; 19, 25:0–d18:1; 20, h26:0–d18:1. An explanation of the nomenclature is given in the legend to Fig. 9.5. (Reproduced by kind permission of the authors and of the *Journal of Lipid Research*, and redrawn from the original paper.)

at 254 nm [805]. Nine fractions were obtained, differing in their fatty acid compositions, but apparently not in the compositions of the long-chain bases. Analogous separations were obtained, in another laboratory with similar elution conditions [442]. In this instance, cerebrosides from rat myelin were perbenzoylated and separated by adsorption chromatography into hydroxy and non-hydroxy fatty acid fractions by adsorption chromatography, prior to the resolution of each form into molecular species by reversed-phase HPLC.

O-Acetyl-*N*-*p*-nitrobenzoyl derivatives of a number of glycosphingolipids, isolated as described above, were separated by HPLC on a column of *micro*Bondapak™ C_{18}, with acetonitrile as the mobile phase [806]. The elution patterns were similar to those found with the perbenzoyl derivatives, and fractions were again collected for fatty acid analysis. Similar results were obtained with the glycosphingolipids from erythrocytes [903].

The benzoylation–desulphation products, derived from sulphogalactocerebroside and containing either non-hydroxy or hydroxy fatty acids (see previous section), were further resolved into 7 to 10 species on a column packed with Spherisorb™ 5ODS, and eluted with acetonitrile–methanol (2:3 by volume) with UV detection at 230 nm [759]. Better resolution was attained subsequently on the same glycolipid fractions and on cerebrosides extracted from the peripheral nerves of patients with adrenoleukodystrophy [896]. The relative proportions of the two solvents in the mobile phase were altered to optimize the separation for each lipid class. In this study, it was noted that resolution was accomplished largely on the basis of the nature of the fatty acid component, since the 4-*trans* double bond in the sphingosine base had virtually no effect on the retention time. Indeed with these particular samples, the fatty acid composition of the glycolipids could be determined in this way without hydrolysis. The procedure was also applied to the analysis of molecular species of glycolipids from nervous tissue undergoing Wallerian degeneration [895].

The *p*-bromophenacyl derivatives of gangliosides (see Sections C) were separated essentially into two fractions, containing either sphingosine or eicosasphingosine, by reversed-phase HPLC [567]. A column (4.6 × 150 mm) packed with Zorbax™ C8 was eluted with acetonitrile–methanol (4:1 by volume) as the mobile phase. Much better resolution was attained by others with the perbenzoyl derivatives of the monosialogangliosides G_{M3}, G_{M1} and L_{M1} [128]. In this instance, the fractions were eluted from a short column (4.6 × 100 mm) of Accupak™ ODS phase (3 μm particles) with either methanol–dichloromethane (4:1 by volume) or methanol–acetonitrile–dichloromethane (7:13:5 by volume) as the mobile phase; the latter gave better resolution of critical pairs, but peaks for later-running components were broader. With these samples, the separation was accomplished according to the combined properties of the fatty acid and sphingoid base constituents.

3. *Ceramides and sphingomyelin*

Sphingomyelin contains a similar ceramide backbone to the glycosphingolipids. By means of hydrolysis with phospholipase C, it is possible to obtain these ceramides for analysis. (Representative ceramides are not readily obtainable from glycosphingolipids). However, most analysts have preferred to separate sphingomyelin into molecular species in a form that retained the phosphorus moiety.

Naturally-occurring ceramides (from calf brain) were first resolved into molecular species in the form of the benzoylated derivatives on a column of Spherisorb™ ODS, with methanol–acetonitrile (3:2 by volume) as the mobile phase [896]. A large number of components were separated, largely according to the nature of the fatty acid constituents, since all the main peaks contained both sphingosine and sphinganine.

Sphingomyelin from bovine brain, and thus having only non-hydroxy fatty acid constituents, was converted to a benzoylated ceramide derivative by benzoylation, followed by reaction with hydrofluoric acid to remove the phosphocholine moiety [386, 779]. The product was first resolved by silver ion chromatography on a "Chrompak™ silver column" (apparently consisting of silver ions on a sulphuric acid-based ion-exchange medium), with hexane–isopropanol (9:1 or 9.5:0.5 by volume) as the mobile phase and with UV detection at 230 nm. Ceramides containing hydroxy fatty acids (from a commercial source) were treated similarly, both in derivatized and underivatized forms. Two main components were clearly separated, with the first containing only saturated fatty acids and the second containing monounsaturated fatty acids, in addition to minor constituents that probably contained sphinganine or sphingosine in combination with these fatty acids. The silver ion column was used repeatedly over a 6 month period, and any loss of resolution was corrected by washing with polar solvents, such as methanol, at 55°C to remove impurities. The major fractions from this column were each subjected to HPLC in the reversed-phase mode, i.e. on a Nucleosil™ C_{18} column with methanol as the mobile phase, to separate the homologues. By this combination of methods, critical pairs were resolved. Similarly, ceramides prepared from egg yolk sphingomyelin and non-hydroxy fatty acid-containing ceramides prepared from bovine brain, after conversion to the benzoyl or *p*-nitrobenzoyl derivatives, were separated on a column (4.6 × 250 mm) LiChrosorb™ RP18 with various chloroform–methanol mixtures as the mobile phase [192]. Again, the separation was mainly according to the nature of the fatty acid constituents with a partial resolution only of species containing sphingosine and sphinganine.

Underivatized sphingomyelin from erythrocytes was successfully fractionated on columns (4 × 300 mm) containing *micro*Bondapak™ or Nucleosil™ ODS packing materials, and isocratic elution with methanol–5 mM potassium phosphate buffer (pH 7.4) in proportions which were varied from 9:1 to 9.8:0.2 by volume, according to the resolution required for particular species [383]. The detection was at 203–205 nm, and each of the ten to twelve peaks observed were collected for identification, by analysis of the fatty acid and sphingoid base constituents. By constructing semilogarithmic plots of retention times against "effective carbon numbers" for each component, it was possible to predict retention times for other species, which could potentially be present. Again it was observed that a 4-*trans* double bond in the long-chain base had much less effect on the hydrophobicity of the molecule than had a *cis* double bond in a fatty acyl chain. With the system described, up to 3 mg of sample could be separated with good resolution, and by moving to a full-scale preparative system, up to 2 g was resolved. Later, an improved resolution was obtained by others [821] on a column (4.6 × 100 mm) of Microsorb™ C_{18} (3 μm particles) eluted with a complex linear gradient of acetonitrile–methanol–5 mM phosphate buffer containing 20 mM choline chloride, and with UV detection at 205 nm. In this instance, as many as 25 species were resolved, and the main components were isolated and identified by MS in the chemical-ionization mode.

As an alternative, sphingomyelin was converted to the 3-*O*-benzoyl derivatives by reaction with benzoic anhydride prior to fractionation [386, 779]. In this form, it was separated into species containing either saturated or monoenoic fatty acids on the silver ion column described earlier in this section, with methanol–isopropanol (4:1 by volume) as the mobile phase, and with UV detection at 230 nm. Further resolution of each of these fractions into species containing a single fatty acid constituent was accomplished on an ODS column with methanol–acetonitrile–5 mM potassium phosphate buffer (pH 7.4) (100:20:1 by volume) as the mobile phase. A similar procedure, except that methanol–15 M ammonia (99:1 by volume) was the mobile phase, was used in a study of the turnover of molecular species of sphingomyelin in rat brain [479].

E. Isolation and Separation of Sphingoid Bases

Before the sphingoid or long-chain bases of sphingolipids can be analysed, it is first necessary to hydrolyse any glycosidic linkage or phosphate bond as well as the amide bond to the fatty acyl group. This should ideally be accomplished in such a manner that no degradation or rearrangement of the bases occurs. The perfect method has yet to be devised. Acid-catalysed hydrolysis has been use in most laboratories, although rearrangement and substitution at C-3 and C-5 inevitably occurs to a certain extent, altering the configuration of the bases to the *threo* form and giving *O*-methoxy artefacts in the presence of methanol, for example. A recently described procedure [390], in which aqueous hydrochloric acid in the aprotic solvent, acetonitrile, was employed for hydrolysis, reportedly produced fewer artefacts than earlier methods, and gave particularly good yields with gangliosides. The method is as follows.

> "The hydrolysis reagent (0.3 ml), consisting of 0.5 M HCl and 4 M water in acetonitrile, was added to the glycolipids (up to 200 μg) in a teflon-lined screw-capped tube. This was flushed with nitrogen, sealed and heated at 75°C for 2 hr. The samples were taken to dryness in a stream of nitrogen, chloroform (5 ml) was added followed by 0.05 M sodium hydroxide in methanol–0.9% saline solution–chloroform (48:47:3 by volume) (1 ml), and the mixture was throughly shaken before being centrifuged at 3000*g*. The lower phase was collected and washed with three further portions of the NaOH solution, and then with two portions of the same solvents without NaOH. The lower phase was finally taken to dryness in a stream of nitrogen to recover the required bases."

Somewhat milder hydrolysis conditions were preferred for other sphingolipids, i.e. sphingomyelins for 1 hr in 1 M HCl in water–methanol (2:1 by volume) at 70°C for 16 hr, and cerebrosides in 3 M HCl in water–methanol (1:1 by volume) at 60°C for 1.5 hr [503]. An alternative method in which hydrolysis was carried out under basic conditions, i.e. with barium hydroxide in aqueous dioxane, also reportedly gave little artefact formation, but surprisingly it has not been taken up by other analysts [553]. (A procedure derived from the above, but for the isolation of the fatty acid constituents of sphingolipids for further analysis, has been described [39].)

Most work on the separation of long-chain bases by means of HPLC again appears to have been carried out in the laboratories of Jungalwala and McCluer, and they have favoured the preparation of the biphenylcarbonyl derivatives, which absorb strongly in the UV region at 280 nm. These compounds are prepared by reacting the sphingoid bases (as little as 0.1 μg or 0.33 nmoles) with 1% biphenylcarbonylchloride in tetrahydrofuran (0.05 ml) and saturated sodium acetate in water (0.1 ml); the products are partitioned in a "Folch" mixture, and washed carefully prior to analysis. For the separation, a Waters "Fatty Acid Analysis" column (4 × 300 mm) was used initially, and it was eluted with tetrahydrofuran–methanol–water (5:8:8 by volume) as the mobile phase [383]. This procedure was used to separate and quantify the sphingoid bases from bovine brain, and sheep and pig erythrocytes [383], and in studies of the metabolism of mouse kidney glycosphingolipids [504]. Later, a more rigorous study of the method was carried out; the derivatives were separated on a column (4.6 × 250 mm) of Ultrasphere™ 5ODS, eluted with a mobile phase containing the same solvents in the ratios 5:8:7 or 5:8:4 respectively [378]. The nature of the separation achieved with the long-chain bases of cerebrosides from bovine brain is shown in Fig. 9.7. The main component was *erythro*-C_{18}-sphingenine, and it was accompanied by only 2% of artefacts, i.e the *threo*, 3-methoxy and 5-methoxy compounds; C_{18}-sphinganine and C_{20}-sphingenine were also detected. Mass spectra of each of the bases were obtained to support the structures assigned. In addition, the detector response was shown to be rectilinear over the range 0.33 to 3.3 nmoles. When a measure of the total sphingoid bases was required, the biphenylcarbonyl derivatives were eluted as a single easily-measurable peak from a column (50 × 2.1 mm) of MicroPak™ SI-10 silica gel, with acetonitrile–dichloromethane (3:1 by volume) as the mobile phase, in only 2 min. The procedure has been applied to the bases from brain gangliosides [391]. In a further extension of the method, the HPLC separation was combined with chemical-ionization MS (with ammonia or methane as the reagent gas) by means of a moving belt interface [379]. It was thus possible to characterize the sphingoid bases directly, together with the side-products, and to quantify them by means of a selective ion-monitoring technique. A short column containing a 3 μm

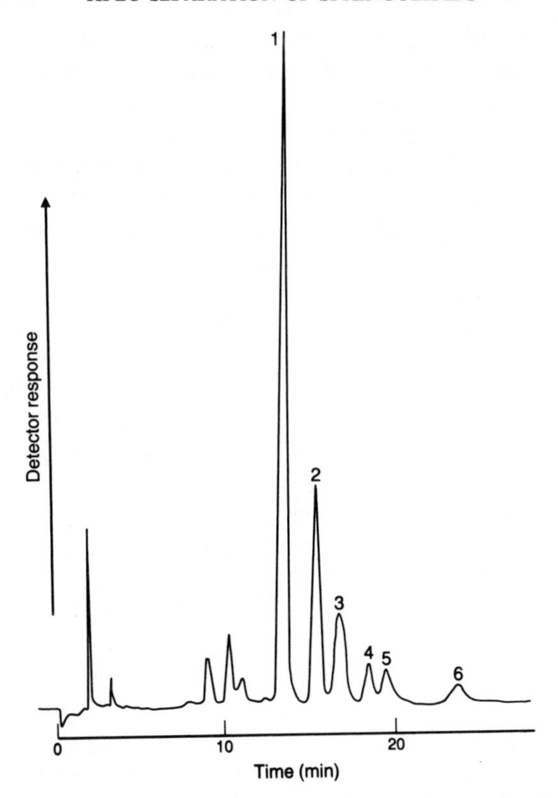

FIG. 9.7. Reversed-phase HPLC separation of sphingoid bases from bovine brain cerebrosides in the form of the biphenylcarbonyl derivatives [378]. A column of Ultrasphere™-ODS was used, with tetrahydrofuran–methanol–water (5:8:4 by volume) as the mobile phase at a flow-rate of 1.5 ml/min, and with UV detection at 280 nm. Peak identifications: 1, *erythro*-C_{18}-sphingenine; 2, *threo*-C_{18}-sphingenine; 3, 5-*O*-methyl-C_{18}-sphingenine; 4, *erythro*-C_{18}-sphinganine; 5, 3-*O*-methyl-C_{18}-sphingenine; 6, *erythro-C_{20}*- sphingenine. (Reproduced by kind permission of the authors and of the *Journal of Lipid Research*, and redrawn from the original paper.)

ODS phase was employed, with methanol–water (94:6 by volume) as the mobile phase.

Others workers [706] converted the sphingoid bases to the 4-dimethylaminoazobenzenesulphonyl ("dabsyl") derivatives, which could be detected spectrophotometrically with great sensitivity (as little as 5 pmoles) at 430 nm. In place of an ODS phase, a column (4.6 × 250 mm) of Zorbax™ CN was employed, with acetonitrile–0.0175 M aqueous sodium acetate (pH 6.0) (3:2 by volume) as the mobile phase. However, saturated and unsaturated isomers, or those of the *erthro* and *threo* configuration, were not resolved. Sphingoid bases have also been successfully separated and quantified in the form of the *p-N*-nitrophenylacetyl derivatives [590]. In this instance, the elution conditions were similar to those described above for the biphenylcarbonyl derivatives. The method was applied to the complex range of bases, including both di- and trihydroxy isomers, in the glycolipids from the intestines of the Japanese quail. A brief report of the separation of the *erythro* and *threo* forms of dihydrosphingosines as their *N*-acetyl derivatives by HPLC in the reversed-phase mode has been published [322].

Gas chromatography methods have hitherto been widely used for the analysis of sphingoid bases, but the sensitivity and resolution possible are not as good as with HPLC.

CHAPTER 10

Some Miscellaneous HPLC Separations

A. Lipoproteins

The lipids in plasma do not occur simply in a free state, but rather exist in the form of complexes with proteins, known as ‘lipoproteins”. As these have much of the solubility properties of the protein constituents, they are able to transport lipids, which are essentially hydrophobic, through an aqueous environment from the liver or intestines to the peripheral tissues. The linkage between the lipids and the proteins does not involve covalent binding, but rather London or Van der Waal's forces between hydrophobic regions of the proteins and the aliphatic moeities of the lipids. Lipid–lipid interaction are also important. As a consequence of variation in the nature and composition of the protein subunits (“apolipoproteins”) and of the lipids classes, the lipoproteins consist of heterogenous families or classes of compounds. These have been defined in terms of the separatory techniques first used to effect some simplification, i.e. differential flotation by means of ultracentrifugation and electrophoresis. Thus, the *high density lipoproteins* (HDL) are that fraction which distributes in the solution density band of 1.063 to 1.21 g/ml upon ultracentrifugation. Subfractions of these are designated HDL_1, HDL_2 and HDL_3, for example. Similarly, *low density lipoproteins* (LDL) are found in the density band 1.006 to 1.063 g/ml, *very low density liproteins* (VLDL) occur between 0.93 and 1.006 g/ml, while *chylomicrons* are large particles of enteric origin with a density of less than 0.95 g/ml. In an alternative nomenclature, HDL, LDL and VLDL have been termed *alpha-*, *beta-* and *pre-beta*-lipoproteins respectively, because of their relative mobilities on electrophoresis.

Electrophoretic methods of separation and analysis are rapid and convenient, and they are usually preferred when information on the relative proportions of the different subclasses only is required. Ultracentrifugation procedures require a heavy capital expenditure on equipment and tend to be slow and tedious; they have been reported to give rise to some limited decomposition or denaturation of the lipoprotein fractions. However, such methods do afford excellent separations on a scale that is suitable for detailed analyses of the individual lipid and protein constituents of the lipoprotein classes, and they now provide the standard against which other procedures are judged. They have been described in some detail in two recent monographs [531, 751]. A simple guide to the more widely-used techniques has also been published [133]. Only the blinkered would consider such methods to be the province of the protein expert rather than of the lipid analyst. On the other hand, a number of applications of HPLC in the gel-permeation mode to the separation of lipoprotein classes have been described in recent years, and many lipid analysts, including the author, would feel more at home with these than with the traditional methods. Potentially, HPLC offers rapid separations both for analytical and small-scale preparative purposes, while utilizing conventional chromatographic equipment, already available in many lipid laboratories. Whether this potential has been fully realized is a matter for debate and is discussed below.

Note that all lipoprotein separation methods require *fresh* not frozen plasma.

Much of the pioneering work on HPLC of lipoproteins has been carried out in the laboratories of Okazaki and Hara in Japan, and they have briefly reviewed progress [284]. In virtually all of the published work, the gel-permeation media used were made by the Toyo Soda Manufacturing Company of Japan, and consisted of rigid polystyrene beads, 10 to 17 μm in diameter, with controlled pore-sizes and with a surface layer of hydroxyl groups to render them hydrophilic. The more important of these gels for lipoprotein work are designated TSK 3000 SW, used for proteins in the molecular weight range 1000 to 3×10^5, TSK 4000 SW, used for proteins in the molecular weight range 5000 to 1×10^6, TSK 5000 PW, with an exclusion limit for dextran of 7×10^6, and TSK 6000 PW, with an exclusion limit for dextran of 4×10^7. They are supplied in stainless steel columns of 7.5 mm internal diameter and lengths of either 300 or 600 mm.

Prior to analysis, plasma samples were adjusted to a density of 1.21 with solid sodium bromide and diluted five fold with an aqueous solution of sodium bromide of density 1.21 [607]. Aliquots of 5 ml of this solution were centrifuged at 105,000 g for 24 hr at 8°C, the top 0.5 to 1 ml, which contained the lipoproteins, was collected and 20 to 40 μl was injected onto the HPLC column.

In a systematic study of the properties of each of the individual types of column, it was shown that the 3000 SW gel gave three sharp peaks with a standard mixture of VLDL, LDL, HDL and albumin, with VLDL and LDL eluting together [607]. With a 4000 SW column, four partially resolved peaks were obtained, while the 5000 PW and 6000 PW columns gave good resolutions of VLDL and LDL, and divided the HDL fraction into two peaks. It was concluded that for the complete separation of all fractions, a combination of columns was required, and the optimum results initially appeared to be obtained with a 500 PW column (600 mm) and two 3000 SW columns (each 600 mm) in series. In this work, the mobile phase was 0.1 M Tris-HCl buffer (pH 7.4) at a flow-rate of 1 ml/min, and the separation

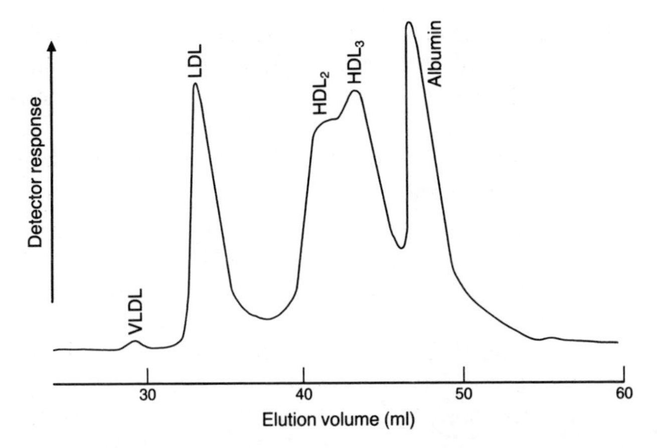

FIG. 10.1. Separation of human serum lipoproteins by HPLC in the gel-permeation mode [607]. One Toyo Soda 5000 PW and two 3000 SW columns (each 7.5 × 600 mm) in series were eluted with an 0.1 M Tris-HCl buffer (pH 7.4) at a flow-rate of 1 ml/min. UV detection at 280 nm was employed. Chapter 1 contains a list of abbreviations. (Reproduced by kind permission of the authors and of the *Journal of Chromatography*, and redrawn from the original paper.)

was monitored spectrophotometrically by the absorbance of the protein constituents at 280 nm. The nature of the separation that could be achieved is shown in Fig. 10.1. VLDL and LDL were clearly separated from a twin peak comprising HDL_2 and HDL_3, and these were distinguishable in turn from an albumin peak. When the salt content of the mobile phase was increased or the pH was decreased, small increases in the elution volumes of the HDL fractions and albumin were observed, but there was virtually no effect on the mobilities of the LDL and VLDL. A plot of the logarithms of the molecular weights of each lipoprotein component versus their retention volumes showed an inverse rectilinear relationship for the column combination 5000 PW + 3000 SW + 6000 PW + 3000 SW, so this could be used to determine the molecular weights of unknown lipoproteins.

In a distinctive and novel extension of the work, the more important lipid components of the lipoprotein fractions, i.e. cholesterol, triacylglycerols and the choline-containing phospholipids, were monitored continuously as they eluted from the columns by specific reactions in a post-column reaction chamber. For example, the total cholesterol was detected and quantified by using a commercial enzyme kit containing cholesterol esterase to hydrolyze the cholesterol esters and cholesterol oxidase to oxidize the free cholesterol to cholest-4-en-3-one; this was in turn reacted with specific dyes and the absorbance of the products at 550 nm was measured [285, 603, 608, 609]. A commercial enzyme kit containing lipoprotein lipase, glycerol oxidase and a peroxidase was used in the same way to continuously monitor the triacylglycerols [286], while one containing phospholipase D, choline oxidase and a peroxidase was utilized for the assay of the choline-containing

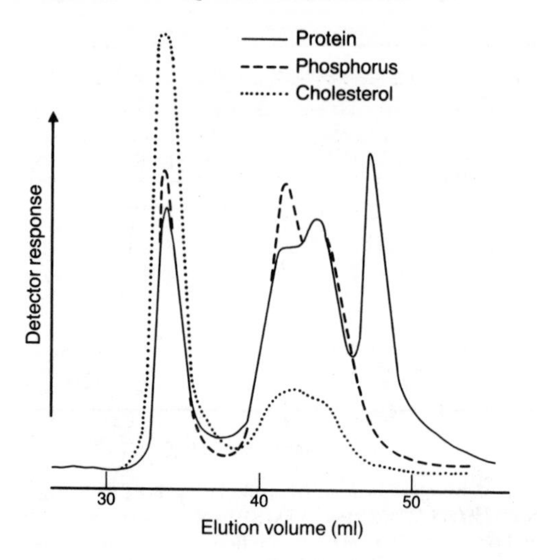

FIG. 10.2. The relative compositions of protein, phosphorus and cholesterol in lipoprotein fractions separated by HPLC in the gel-permeation mode [598]. Plasma samples were taken from normal young women. The column configuration is shown in the legend to Fig. 10.1, while the mobile phase was 0.15 M aqueous sodium chloride solution at a flow-rate of 1 ml/min. The protein trace was the absorbance at 280 nm, phosphorus was determined by a chemical procedure (about 2 μg/ml at full-scale deflection), and cholesterol was determined by an enzymatic procedure (about 90 μg/ml at full-scale deflection); triacylglycerols were also determined but are not shown here. (Reproduced by kind permission of the authors and of the *Journal of Biochemistry* (*Tokyo*), and redrawn from the original paper.)

phospholipids [267, 600, 601]. The last technique was found to be of particular value for detecting heterogeneities due to particle size differences [602] and any abnormalities in the HDL fraction [600] of plasma, for example, while the triacylglycerol assay was important for the recognition of the LDL and VLDL fractions. The cholesterol procedure was used to detect a remarkable decrease of HDL_3 cholesterol levels in cirrhosis of the liver [604]. Used in concert, these methods gave a great deal of information on the relative proportions and compositions of the lipoprotein fractions in human serum [598]. Figure 10.2 illustrates the nature of the chromatograms obtained.

In one comparison of lipoprotein quantification by gel-permeation HPLC with agarose-gel electrophoresis, cholesterol determination was accomplished more easily by means of the former technique, with greater precision and resolution [610]. Similarly in a comparison of the HPLC procedure with differential ultracentrifugal flotation, excellent agreement was obtained for the composition of most of the fractions from the sera of normal and hyperlipidemic patients [605]. The agreement was found to be less good for patients with liver diseases or with a deficiency in the enzyme lecithin:cholesterol acyltransferase, because of heterogeneities in the lipoprotein subfractions. In particular, the fractions designated LDL, HDL_2 and HDL_3 from the the gel-permeation column tended to correspond well with the analogous fractions obtained by ultracentrifugation. On the other hand, the fraction designated as VLDL/chylomicrons from the HPLC column appeared to contain some bilirubin, which affected the quantification procedure, together with some phospholipid-containing material of high density; the HDL fraction produced by ultracentrifugation appeared to have lost some portion of its material. Nevertheless, the general conclusion was that the HPLC method had a number of advantages as an analytical tool in terms of simplicity, resolution, reproducibility and speed, in addition to having small sample requirements.

The optimum configuration of columns in this and related work was now considered to be one TSK 4000 SW plus one 3000 SW (each of 600 mm length) in series [600, 601, 605]. The preferred mobile phase was 0.15 M aqueous sodium chloride at a flow-rate of 0.6 ml/min,. With this system, it was also shown that there was an inverse rectilinear relationship between the logarithms of the diameters of the lipoprotein particles and their retention volumes.

Other workers in Japan and elsewhere have evaluated gel-permeation HPLC of lipoproteins and have found much to commend. For example, Busbee *et al.* [100] prestained the lipoproteins with diformazin dye, which permitted the column effluent to be monitored spectrophotometrically at 580 nm, and which appeared to give peak area measurements that were proportional to the relative amounts of the fractions. (Unfortunately, others reported that this prestaining procedure shortened the life span of the column, decreased recoveries and prevented a subsequent immunological assay for the apolipoprotein constituents [849].) Two TSK 4000 SW columns (each of 300 mm length) and a guard column were used, while the mobile phase was an 0.2 M phosphate buffer (pH 6.8) containing sodium azide (0.5 g/l), as an antibacterial agent, and at a flow-rate of 1 ml/min. This procedure gave two fractions from human serum, one comprising VLDL/LDL and the other HDL.

A more thorough study of the effect of column type on the nature of the separation was made by Carroll and Rudel [110], who tried various configurations and combinations of 3000 SW, 4000 SW and 5000 PW columns. While accepting that a 3000 SW and a 4000 SW column in series gave the optimum separation, they concluded that a single 5000 PW column was to be preferred, since it gave acceptable separations and it appeared to be much more stable in use than the other types of column; economy and simplicity in use were further considerations. The recovery of lipoprotein cholesterol averaged 91%. By means of immunological, chemical and electrophoretic examination of the fractions, it was concluded that they were intact, and that there was very little cross-contamination. After calibrating the apparent particle size distributions against those obtained by a standard agarose-gel procedure, it proved possible to use the HPLC technique to determine the molecular weights of lipoprotein fractions from the serum of non-human primates. Yet others made use of a single TSK 4000 SW column and obtained good separations of VLDL/chylomicrons, an intermediate density lipoprotein fraction, LDL and HDL; these corresponded well with analogous fractions obtained by a standard ultracentrifugation procedure [849].

Gel-permeation HPLC has been employed for the isolation of an abnormal lipoprotein fraction in the serum of a colony of baboons [877], to detect and characterize an abnormal HDL fraction in patients suffering from primary biliary cirrhosis [825], and to subfractionate and study native and incubated HDL [331]. In addition, the technique has been utilized to study the turnover of radioactively-labelled (^{125}I) LDL in fasting rabbits; fractions were monitored with an on-line gamma counter [878].

The general conclusion of a number of users has therefore been that gel-permeation HPLC with the Toyo Soda columns affords a rapid, mild and sensitive alternative to ultracentrifugation for the isolation of lipoprotein fractions. Generally, it has been treated as an analytical technique, but in one study up to 5 mg of protein was reportedly fractionated without loss of resolution [110]. The technique makes use of standard HPLC equipment, and as isocratic elution only is required, a single relatively-inexpensive pump suffices. The columns are certainly exceedingly costly, currently five to ten times as much as for more conventional HPLC columns. A few authors only have reported on the life span of the columns, and in one instance, it was stated that "the lack of stability of the 3000 SW and 4000 SW columns was evidenced by adsorption of proteins to the columns after 1 to 3 months of use" [110]. The author has heard similar anecdotal accounts. On the other hand, proteins bound to columns were removable by elution with a detergent in a buffer, and a number of commercial companies offer a (not inexpensive) column regeneration service. The polystyrene gel in Toyo Soda columns is not nearly as rigid as stationary phases based on silica gel, and it can be compacted if treated roughly. Some problems are certainly avoidable, if columns are handled with care. Sudden movements, shocks or pressure fluctuations to the columns must be prevented, slow flow rates and low inlet pressures are essential, and the columns should be filled with distilled water, ideally containing the minimum amount of a bacteriostatic agent, when not in use. In addition, a TSK guard column can be a useful investment by damping pulsations in the mobile phase, and by removing adsorptive and insoluble substances from samples; it can be regenerated for re-use relatively easily.

One other form of gel-permeation chromatography is worthy of note. Conventional column chromatography on agarose gels has been available for the separation of lipoprotein fractions for a number of years, but a lengthy elution time has tended to be required for a complete separation. On the other hand, a new form of cross-linked agarose manufactured by Pharmacia of Sweden, i.e. Sepharose™ 6B, is intended for HPLC use, with the advantages in terms of speed and convenience that usually result. One application to lipoproteins has been described, with 0.15 M sodium chloride containing Na_2EDTA (1 g/l) and sodium azide (0.2 g/l) of pH 7.2 as the mobile phase, and with detection at 280 nm [265]. The procedure was applied to the plasma lipoproteins of various species of animals, and the quality of the separation achieved varied with the species. In general, an acceptable separation of VLDL, LDL and HDL was achieved in about 2 hr. The cholesterol and triacylglycerol profiles were monitored by discontinuous assay procedures. Although the resolution did not match that obtained with the Toyo Soda columns, further improvements may be possible.

HPLC in the gel-permeation, reversed-phase and ion-exchange modes has been applied with considerable success towards the analysis and isolation of the apolipoprotein constituents of lipoproteins. For example, it has been used for the total apolipoproteins of human serum [363, 614, 867], those of VLDL [278, 868] and HDL [422, 423, 606, 666], and the individual apolipoproteins (and their subfractions) A [279, 740, 908], C [705, 741, 869] and E [647, 848]. The list is not necessarily complete, and a comprehensive review has appeared elsewhere [200]. Further discussion of this aspect is not appropriate here.

B. Sterol, Wax, Retinol and Xanthophyll Esters

1. Sterol esters

Procedures for the isolation and analysis of cholesterol esters as a class are described in Chapter 5. As with other lipid classes, it is often desirable to distinguish individual molecular species, and reversed-phase HPLC has been much used for the purpose. Only one fatty acid constituent is present in each molecule, so the basis of the separation is identical to that for simple fatty acid derivatives described in Chapter 7. On the other hand, if esters of sterols other than cholesterol are also present, the elution pattern is more complex. ODS stationary phases have been favoured, with acetonitrile as the mobile phase, although it has been necessary to add a modifier component to increase the solubility of the cholesterol esters. For example, Duncan *et al.* [197] employed a column packed with an ODS phase and acetonitrile–isopropanol (1:1 by volume) as the mobile phase, with spectrophotometric detection at 200 nm, to obtain acceptable separations. When the method was applied to serum extracts, triacylglycerol species overlapped with cholesterol esters to a certain extent, so it was desirable to isolate the pure lipid class before commencing the analysis. The same method in essence was used to study plasma cholesterol esters in diabetic rats [781], in normal human plasma [67], and in patients with lipoprotein abnormalities [668]. A column packed with a Zorbax™ ODS phase at 60°C, and acetonitrile–isopropanol (3:2 by volume) as the mobile phase at 2 ml/min, were used to study the elution characteristics of sterol esters prepared biosynthetically [820] or by chemical synthesis from 10 different fatty acids and 14 sterols [74]. It was noted that the presence of a double bond in a fatty acyl chain had a greater effect on the retention volume than a similar feature in a sterol moiety.

Other mobile phases used for the separation of molecular species of cholesterol esters have included a gradient of zero to 3% water in acetonitrile–tetrahydrofuran (13:7 by volume) [109], acetonitrile–chloroform–methanol (1:1:1 by volume) [639, 640], a gradient of 30 to 90% propionitrile in acetonitrile [453], and methanol–acetone (17:3 by volume) [310]. In terms of resolution, none of these appeared to have any particular advantage, and the choice may be governed by their suitability to the detection systems available to the analyst.

The plasma sterol esters in patients suffering from phytosterolemia have been found to contain appreciable amounts of such plant sterols as campesterol, stigmasterol, avenasterol and *beta*-sitosterol, in addition to cholesterol. The different forms could not be distinguished by adsorption chromatography, and were eluted together in one band on a TLC plate [460]. On further separation by HPLC in the reversed-phase mode on a Supercosil™ LC-18 column with a gradient of 30 to 90% propionitrile in acetonitrile as the mobile phase, a large number of fractions were resolved. The main components, differing in the nature of both the sterol and fatty acid moieties, were then identified directly by means of coupled mass spectrometry (with chemical ionization).

2. *Wax esters*

It might be expected that wax esters could be separated into molecular fractions with relative ease by means of reversed-phase HPLC, but only one report of an analysis of a natural sample appears to have been published to date [789]. In this instance, jojoba oil was separated into components differing simply in their partition numbers on a column of *micro*Bondapak™ C_{18} with acetonitrile–acetone (2:1 by volume) as the mobile phase, and with refractive index detection. Elution conditions similar to those for sterol esters were employed for two synthetic wax ester species [74]. Further separations will no doubt be described in due course.

3. *Retinol and retinol esters*

Although retinol or vitamin A itself would not be considered a lipid if the definition of Chapter 3 were applied strictly, retinol esters are certainly lipids by any definition. HPLC is a particularly appropriate technique for the analysis of these compounds, as they exhibit a relatively specific absorbance at 330 nm with a high extinction coefficient. Cholesterol esters, with which they tend to co-chromatograph, do not interfere with the analysis at this wavelength.

Both adsorption and reversed-phase partition chromatography have been used for the separation of retinol esters. With the latter technique, for example, a column packed with an ODS phase (10 μm particles) and methanol as the mobile phase were used to separate and quantify standard mixtures and serum extracts containing retinol and retinol esters of various fatty acids [183]. The detector response was rectilinear for each component, although for reasons that were not immediately obvious, the lines for the palmitoyl and stearoyl derivatives diverged rather more than might have been anticipated. Unfortunately, critical pairs such as the palmitoyl and oleoyl derivatives were not resolved, and while this was accomplished by adding silver ions to the mobile phase [184], most analysts would prefer to avoid using such a corrosive reagent (see Chapter 7 and 8). Much better separation, including base-line resolution of critical pairs, was obtained by Ross [707] with a column (4.6 × 250 mm) packed with Supelcosil™ LC-8 or a bonded-phenyl phase, and with acetonitrile–water (88:12 later changed to 92:8 by volume) as the mobile phase. It was also determined that equimolar amounts of different retinol esters had essentially the same extinction coefficient under these conditions. The procedure was applied to the analysis of retinol esters in rat lymph chylomicrons, and with some modification to the mobile phase to the analysis of retinol, retinaldehyde and retinyl acetate. Others obtained equally satisfactory results with retinol esters on ODS stationary phases with either methanol–water (98:2 by volume) [70] or acetonitrile–methanol–chloroform (47:47:6 by volume) [97] as the mobile phase.

With adsorption chromatography, little resolution of molecular species of retinol esters would be expected. However, it was possible to separate and quantify the total retinol esters and free retinol in horse serum and milk on a column containing *micro*Porasil™ with hexane–chloroform (3:2 by volume) as the mobile phase [798]. Similarly, retinol and retinol esters were determined in as little as 100 microlitres of human serum [52]. In this instance, retinyl acetate was added as an internal standard; a column (4 × 125 mm) of LiChrosorb™ Si60 (5 μm particles) was eluted with a gradient of 0.5 to 20% dioxane into hexane.

4. Xanthophyll esters

Diesters of xanthophyll and fatty acids, predominantly 14:0, 16:0 and 18:0, are known to occur in the petals of the marigold, *Tagetes erecta*. These were separated into the different molecular species by reversed-phase HPLC [231]. For analytical purposes, a Merck Hibar™ RP-18 column (4 × 125 mm) was used with dichloromethane–acetonitrile (1:3 by volume) at a flow-rate of 3 ml/min as the mobile phase; the required compounds were detected spectrophotometrically at 471 nm. The dipalmitoyl derivative was the major component, but all other possible fatty acid combinations were present. In addition, it proved possible to isolate the main fractions on a preparative scale in a similar manner.

C. Acyl-coenzyme A Esters

Coenzyme A (CoA) esters tend to occur at trace levels only in tissues, but they are of very great metabolic importance as intermediates in lipid biosynthesis and as moderators of a number of enzymatic processes. They are polar molecules with a strong amphipathic character, but they have been successfully analysed by HPLC. Detection has been facilitated by a relatively strong molecular absorbance at 254 nm. Short-chain CoA esters tend to behave somewhat differently from those of longer chain-length on chromatography with aqueous mobile phases, as the latter tend to form micelles rather readily.

CoA itself has been separated and quantified by ion-exchange HPLC on a microparticulate strong anion-exchange resin; a column of Partisil™ 10 SAX with 196 mM potassium phosphate buffer (pH 3.9), containing 2% isopropanol and 0.05% thiodiglycol, as the mobile phase [351]. By measuring CoA concentrations before and after alkaline hydrolysis, the amount of CoA present in the esterified form could also be calculated.

Most analysts have approached the problem of the separation of individual acyl-CoA esters by adopting reversed-phase HPLC with a mobile phase containing ion-pairing reagents or of high ionic strength. Thus, Baker and Schooley [46, 47] successfully resolved CoA itself and a number of short-chain acyl-CoA derivatives on either an ODS or an octyl-bonded phase with various methanol–water mixtures, containing tetrabutylammonium phosphate (pH 5.5), as the mobile phase. Both the concentration of the counterion and of the sample had an effect on the elution volume. In preparative-scale applications, the tetrabutylammonium ions were subsequently removed from the fractions by ion-exchange chromatography on an Amberlite™ IR-120 resin. Others employed this method with minor modifications to separate and quantify the medium-chain acyl-CoA esters formed biosynthetically by the fatty acid synthase of ruminant mammary gland *in vitro* [432]. In similar work, an ODS stationary phase, and methanol–220 mM potassium phosphate buffer (pH 4.0) [64, 65, 176, 350] or methanol–50 mM potassium phosphate buffer (pH 5.3) [160] in various proportions as the mobile phase, were used to separate and quantify CoA and different acyl-CoA esters in extracts of mitochondria and liver tissue.

A wider range of short-chain acyl-CoA esters (17 different components) from freeze-clamped liver were separated on a column (4.6 × 75 mm) containing Ultrasphere™ ODS (3 μm particles) with a complex gradient of a phosphate buffer into acetonitrile [420]. Under these conditions, non-CoA UV-absorbing material in the extract eluted ahead of the CoA esters. The sample prepration involved extraction with perchloric acid, centrifugation to remove the proteins, and adjustment of the pH to 3.0, where the acyl-CoA esters were more stable. Other workers concentrated samples prepared in a similar way on a Sep-PAK™ C_{18} cartridge, from which the required compounds were removed by elution with ethanol–water (65:35 by volume) containing 0.1 M ammonium acetate, prior to detailed analysis [336]. The individual short-chain acyl-CoA esters were then separated by HPLC on a column (4.6 × 400 mm) of Develosil™ ODS, with a gradient of 1.75 to 10% acetonitrile into 0.2 M ammonium acetate as the mobile phase. The lower limit of detection for malonyl-CoA, for example, was 50 pmole. In one further study, individual long-chain acyl-CoA esters (C_{12} to C_{18}, including unsaturated components) from freeze-clamped livers were separated on a column packed with an ODS phase, with a complex gradient of acetonitrile and a phosphate buffer (pH 5.3) as the mobile phase [884].

D. Acyl Carnitines

Acyl carnitine derivatives are important intermediates in the *beta*-oxidation of fatty acids in animal tissues, and procedures for their analysis have been reviewed [73]. They have proved difficult to analyse by HPLC as they do not contain a distinctive chromophore. In addition, the author (unpublished work) has found that they tend to co-chromatograph with the choline-containing phospholipids, presumably because both contain zwitter ion moieties.

A rather distinctive approach to the analysis of the short-chain fatty acid derivatives has therefore been required, and Kerner and Bieber [413] found it necessary to convert the acyl carnitines to an isotopically-labelled form, i.e. with a ^{3}H-label in the carnitine moiety, by an enzyme-catalyzed exchange reaction. The products were separated on a column (4.6 × 250 mm) packed with Partisil™ 10 ODS-3, with 5 mM butanesulphonic acid (pH 3.4) and a complex gradient of methanol into acetic acid as the mobile phase; the component emerging from the column were detected and quantified by a means of liquid scintillation counting with a continuous-flow detector. This procedure was applied to the analysis of the short-chain acyl carnitines in sows' colostrum and milk and in the tissues of the piglet; isovaleroyl carnitine was found to be a major component [414].

A brief description of a method in which *p*-bromophenacyl derivatives of acylcarnitines were prepared prior to separation by HPLC in the reversed-phase mode has appeared [831]. This should have wider applicability, and further details are awaited with interest.

Carnitine itself, together with the metabolically-related butyrobetaine and betaine, have also been determined by HPLC in the form of the *p*-bromophenacyl derivatives [532].

References

1. ABE, K. and KOGURE, K., *J. Neurochem.*, **47**, 577–582 (1986).
2. ACKMAN, R.G., *Methods in Enzymology*, **72**, 205–252 (1981).
3. ACKMAN, R.G., in *Analysis of Oils and Fats*, pp. 137–206 (1986) (edited by R.J. Hamilton and B.J. Rossell, Elsevier Applied Science Publishers, London).
4. ACKMAN, R.G. and BURGHER, R.D., *J. Am. Oil Chem. Soc.*, **42**, 38–42 (1965).
5. ACKMAN, R.G. and EATON, C.A., *Fette Seifen Anstrichm.*, **80**, 21–37 (1978).
6. ADLOF, R.O. and EMKEN, E.A., *J. Am. Oil Chem. Soc.*, **57**, 276–278 (1980).
7. ADLOF, R.O. and EMKEN, E.A., *J. Am. Oil Chem. Soc.*, **58**, 99–101 (1981).
8. ADLOF, R.O. and EMKEN, E.A., *J. Am. Oil Chem. Soc.*, **62**, 1592–1595. (1985).
9. ADLOF, R.O., RAKOFF, H. and EMKEN, E.A., *J. Am. Oil Chem. Soc.*, **57**, 273–275 (1980).
10. AGRAWAL, V.P. and SCHULTE, E., *Anal. Biochem.*, **131**, 356–359 (1983).
11. AIGNER, R., SPITZY, H. and FREI, R.W., *Anal. Chem.*, **48**, 2–7 (1976).
12. AITZETMULLER, K., *J. Chromatogr.*, **71**, 355–360 (1972).
13. AITZETMULLER, K., *J. Chromatogr.*, **113**, 231–266 (1975).
14. AITZETMULLER, K., *J. Chromatogr.*, **139**, 61–68 (1977).
15. AITZETMULLER, K., *Prog. Lipid Res.*, **21**, 171–193 (1982).
16. AITZETMULLER, K., *Fette Seifen Anstrichm.*, **86**, 318–322 (1984).
17. AITZETMULLER, K. and GUHR, G., *Fette Seifen Anstrichm.*, **78**, 83–88 (1976).
18. AITZETMULLER, K. and HANDT, D., *Fette Seifen Anstrichm.*, **86**, 322–325 (1984).
19. AITZETMULLER, K. and KOCH, J., *J. Chromatogr.*, **145**, 195–202 (1978).
20. ALAM, I., SMITH, J.B. and SILVER, M.J., *Lipids*, **18**, 534–538 (1983).
21. ALAM, I., SMITH, J.B., SILVER, M.J. and AHERN, D., *J. Chromatogr.*, **234**, 218–221 (1982).
22. ALBERGHINA, M., FIUMARA, A., PAVONE, L. and GUIFFRIDA, A.M., *Neurochem. Res.*, **9**, 1719–1727 (1984).
23. ALLAN, D. and COCKCROFT, S., *J. Lipid Res.*, **23**, 1373–1374 (1982).
24. ALLEN, C.F., GOOD, P., DAVIS, H.F., CHISUM, P. and FOWLER, S.D., *J. Am. Oil. Chem. Soc.*, **43**, 223–231 (1966).
25. ALLEN, C.F., GOOD, P., DAVIS, H.F. and FOWLER, S.D., *Biochem. Biophys. Res. Commun.*, **15**, 424–430 (1964).
26. ANDERSSON, B.A., *Prog. Chem. Fats other Lipids*, **16**, 279–308 (1978).
27. ANDERSSON, B.A., DINGER, F. and DINH-NGUYEN, D., *Chem. Scripta*, **19**, 118–121 (1982).
28. ANDO, N., ANDO, S. and YAMAKAWA, T., *J. Biochem.* (*Tokyo*), **70**, 341–348 (1971).
29. ANDREWS, A.G., *J. Chromatogr.*, **336**, 139–150 (1984).
30. AOSHIMA,H., *J. Biochem.* (*Tokyo*), **82**, 569–573 (1977).
31. AOSHIMA, H., *Anal. Biochem.*, **87**, 49–55 (1978).
32. AOSHIMA, H., KAJIWARA, T. and HATANAKA, A., *Agric. Biol. Chem.*, **45**, 2245–2251 (1981).
33. ARGOUDELIS, C.J. and PERKINS, E.G., *Lipids*, **3**, 379–381 (1968).
34. ARMSTRONG, D.W., *J. Liq. Chromatogr.*, **7**, 353–376 (1984).
35. ARTHUR, G. and SHELTAWY, A., *Biochem. J.*, **191**, 523–532 (1980).
36. ARVIDSON, G.A.E., *J. Chromatogr.*, **103**, 201–204 (1975).
37. ASHWORTH, E.N., St. JOHN, J.B., CHRISTIANSEN, M.N. and PATTERSON, G.W., *J. Agric. Food Chem.*, **29**, 881–884 (1981).
38. ATKIN, D.S.J., HAMILTON, R.J., MITCHELL, S.F. and SEWELL, P.A., *Chromatographia*, **15**, 97–100 (1982).
39. AVELDANO, M.I. and HORROCKS, L.A., *J. Lipid Res.*, **24**, 1101–1105 (1983).
40. AVELDANO, M.I., VANROLLINS, M. and HORROCKS, L.A., *J. Lipid Res.*, **24**, 83–93 (1983).

41. AWL, R.A., NEFF, W.E., FRANKEL, E.N., PLATTNER, R.D. and WEISLEDER, D., *Chem. Phys. Lipids*, **39**, 1–17 (1986).
42. BAILLIE, A.G., WILSON, T.D., O'BRIEN, R.K., BEEBE, J.M., STUART, J.D., MCCOSH-LILIE, E.J. and HILL, D.W., *J. Chromatogr. Sci.*, **20**, 466–470 (1982).
43. BAKALYAR, S.R., BRADLEY, M.P.T. and HONGANEN, R., *J. Chromatogr.*, **158**, 277–293 (1978).
44. BAKALYAR, S.R. and HENRY, R.A., *J. Chromatogr.*, **126**, 327–345 (1976).
45. BAKER, C.J. and MELHUISH, J.H., *J. Chromatogr.*, **284**, 251–256 (1984).
46. BAKER, F.C. and SCHOOLEY, D.A., *Anal. Biochem.*, **94**, 417–424 (1979).
47. BAKER, F.C. and SCHOOLEY, D.A., *Methods in Enzymology*, **72**, 41–52 (1981).
48. BANDI, Z.L. and ANSARI, G.A.S., *J. Chromatogr.*, **363**, 402–406 (1986).
49. BANDI, Z.L., MOSLEN, M.T. and REYNOLDS, E.S., *J. Chromatogr.*, **269**, 93–101 (1982).
50. BANDI, Z.L. and REYNOLDS, E.S., *J. Chromatogr.*, **329**, 57–63 (1985).
51. BANERJEE, A.K., RATNAYAKE, W.M.N. and ACKMAN, R.G., *Lipids*, **20**, 121–125 (1985).
52. BANKSON, D.D., RUSSELL, R.M. and SADOWSKI, J.A., *Clin. Chem.*, **32**, 35–40 (1986).
53. BARANSKA, J., *Adv. Lipid Res.*, **19**, 163–184 (1982).
54. BARKER, S.A., MONTI, J.A., CHRISTIAN, S.T., BENINGTON, F., and MORIN, R.D., *Anal. Biochem.*, **107**, 116–123 (1980).
55. BARTLETT, G.R., *J. Biol. Chem.*, **234**, 466–468 (1959).
56. BASCETTA, E., GUNSTONE, F.D. and SCRIMGEOUR, C.M., *Lipids*, **19**, 801–803 (1984).
57. BATLEY, M., PACKER, N.H. and REDMOND, J.W., *J. Chromatogr.*, **198**, 520–525 (1980).
58. BATLEY, M., PACKER, N.H. and REDMOND, J.W., *Biochim. Biophys. Acta*, **710**, 400–405 (1982).
59. BATTA, A.K., DAYAL, V., COLMAN, R.W., SINHA, A.K., SHEFER, S. and SALEN, G., *J. Chromatogr.*, **284**, 257–260 (1984).
60. BATTAGLIA, R. and FROHLICH, D., *Chromatographia*, **13**, 428–431 (1980).
61. BATY, J.D. and PAZOUKI, C., *Chromatogr. Int.*, issue **20**, 28–30 (1986).
62. BATY, J.D., WILLIS, R.G. and TAVENDALE, R., *Biomed. Mass Spectrom.*, **12**, 565–569 (1985).
63. BATY, J.D., WILLIS, R.G. and TAVENDALE, R., *J. Chromatogr.*, **353**, 319–328 (1986).
64. BERGE, R.K., AARSLAND, A., BAKKE, O.M. and FARSTAD, M., *Int. J. Biochem.*, **15**, 191–204 (1983).
65. BERGE, R.K., OSMUNDSEN, H., AARSLAND A. and FARSTAD, M., *Int. J. Biochem.*, **15**, 205–209 (1983).
66. BERGLUND, R. and THENTE, K., *Int. Lab.*, (Nov. issue), 34–37 (1983).
67. BERNERT, J.T., AKINS, J.R. and MILLER, D.T., *Clin. Chem.*, **28**, 676–680 (1982).
68. BERNERT, J.T., MEREDITH, N.K., AKINS, J.R. and HANNON, W.H., *J. Liq. Chromatogr.*, **8**, 1573–1591 (1985).
69. BEZARD, J.A. and OUEDRAOGO, M.A., *J. Chromatogr.*, **196**, 279–293 (1980).
70. BHAT, B.V. and LACROIX, A., *J. Chromatogr.*, **272**, 269–278 (1983).
71. BHATI, A., BENBOUZID, M., HAMILTON, R.J. and SEWELL, P.A., *Chem. Ind.* (*Lond.*), 70–71 (1986).
72. BIANCHINI, J-P., RALAIMANARIVO, A. and GAYDOU, E.M., *J. High Res. Chromatogr., Chromatogr. Commun.*, **5**, 199–204 (1982).
73. BIEBER, L.L. and LEWIN. L.M., *Methods in Enzymology*, **72**, 276–287 (1981).
74. BILLHEIMER, J.T., AVART, S. and MILANI, B., *J. Lipid Res.*, **24**, 1646–1651 (1983).
75. BINDER, H., WEBER, P.C. and SIESS, W., *Anal. Biochem.*, **148**, 220–227 (1985).
76. BITMAN, J., WOOD, D.L., HAMOSH, M., HAMOSH, P. and MEHTA, N.R., *Am. J. Clin. Nutr.*, **38**, 300–312 (1983).
77. BITMAN, J., WOOD, D.L., MEHTA, N.R., HAMOSH, P. and HAMOSH, M., *Am. J. Clin. Nutr.*, **40**, 1103–1119 (1984).
78. BJERVE, K.S., DAAE, L.N.W. and BREMER, J., *Anal. Biochem.*, **58**, 238–245 (1974).
79. BLANK, M.L., CRESS, E.A., LEE, T-C., STEPHENS, N., PIANTADOSI, C. and SNYDER, F., *Anal. Biochem.*, **133**, 430–436 (1983).
80. BLANK, M.L., CRESS, E.A., ROBINSON, M. and SNYDER, F., *Biochim. Biophys. Acta*, **833**, 366–371 (1985).
81. BLANK, M.L., ROBINSON, M., FITZGERALD, V. and SNYDER, F., *J. Chromatogr.*, **298**, 473–482 (1984).
82. BLANK, M.L. and SNYDER, F., *J. Chromatogr.*, **273**, 415–420 (1983).
83. BLIGH, E.G. and DYER, W.J., *Can. J. Biochem. Physiol.*, **37**, 911–917 (1959).
84. BLOM, C.P., DEIERKAUF, F.A. and RIEMERSMA, J.C., *J. Chromatogr.*, **171**, 331–338 (1979).

85. Bloor, W.R., *Proc. Soc. Exp. Biol. Med.*, **17**, 138–140 (1920).
86. Bocckino, S.B., Blackmore, P.F. and Exton, J.H., *J. Biol. Chem.*, **260**, 14201–14207 (1985).
87. Boeynaems, J.M., Brash, A.R., Oates, J.A. and Hubbard, W.C., *Anal. Biochem.*, **104**, 259–267 (1980).
88. Borch, R.F., *Anal. Chem.*, **47**, 2437–2439 (1975).
89. Bouhours, J-F., Bouhours, D. and Delaunay, J., *J. Lipid Res.*, **26**, 435–441 (1985).
90. Breckenridge, W.C., in *Handbook of Lipid Research. Vol. 1, Fatty Acids and Glycerides*, pp. 197–232 (1978) (edited by A. Kuksis, Plenum Press, New York).
91. Bremer, E.G., Gross, S.K. and McCluer, R.H., *J. Lipid Res.*, **20**, 1028–1035 (1979).
92. Briand, R.L., Harold, S. and Blass, K.G., *J. Chromatogr.*, **223**, 277–284 (1981).
93. Brockerhoff, H., *Comp. Biochem. Physiol.*, **19**, 1–12 (1966).
94. Brockerhoff, H. and Hoyle, R.J., *Archs Biochem. Biophys.*, **102**, 452–455 (1963).
95. Brockerhoff, H., Hoyle, R.J. and Wolmark, N., *Biochim. Biophys. Acta*, **116**, 67–72 (1966).
96. Brockerhoff, H. and Jensen, R.G., *Lipolytic Enzymes* (1974) (Academic Press, New York).
97. Broich, C.R., Gerber, L.E. and Erdman, J.W., *Lipids*, **18**, 253–258 (1983).
98. Bruins, A.P., *J. Chromatogr.*, **323**, 99–112 (1985).
99. Brunngraber, E.G., Tettamanti, G. and Berra, B., in *Glycolipid Methodology*, pp. 159–186 (1976) (edited by L.A. Witting, American Oil Chemists' Society, Champaign, Ill.).
100. Busbee, D.L., Payne, D.M., Jasheway, D.W., Carlisle, S. and Lacko, A.G., *Clin. Chem.*, **27**, 2052–2058 (1981).
101. Bussell, N.E., Gross, A. and Miller, R.A., *J. Liq. Chromatogr.*, **2**, 1337–1365 (1979).
102. Bussell, N.E., and Miller, R.A., *J. Liq. Chromatogr.*, **2**, 697–718 (1979).
103. Byrne, M.C., Sbaschnig-Agler, M., Aquino, D.A., Sclafani, J.R. and Ledeen, R.W., *Anal. Biochem.*, **148**, 163–173 (1985).
104. Caboni, M.F., Lercker, G. and Ghe, A.M., *J. Chromatogr.*, **315**, 223–231 (1984).
105. Cantafora, A., Di Biase, A., Alvaro, D., Angelico, M., Marin, M. and Attili, A.F., *Clin. Chim. Acta*, **134**, 281–295 (1983).
106. Capella, P. and Zorzut, C.M., *Anal. Chem.*, **40**, 1458–1463 (1968).
107. Careaga, M.M. and Sprecher, H., *J. Biol. Chem.*, **259**, 14413–14417 (1984).
108. Carroll, K.K., in *Lipid Chromatographic Analysis* (2nd edition), Vol. 1, pp. 173–214 (1976) (edited by G.V. Marinetti, Marcel Dekker, New York).
109. Carroll, R.M. and Rudel, L.L., *J. Lipid Res.*, **22**, 359–363 (1981).
110. Carroll, R.M. and Rudel, L.L., *J. Lipid Res.*, **24**, 200–207 (1983).
111. Carter, T.P. and Kanfer, J.N., *Lipids*, **8**, 537–548 (1973).
112. Carunchio, V., Nicoletti, I., Frezza, L. and Sinibaldi, M., *Ann. Chim.*, **74**, 331–339 (1984).
113. Chan, H.W-S. and Levett, G., *Lipids*, **12**, 99–104 (1977).
114. Chan, H.W-S. and Levett, G., *Chem. Ind. (Lond.)*, 692–693 (1977).
115. Chan, H.W-S. and Levett, G., *Chem. Ind. (Lond.)*, 578–579 (1978).
116. Chan, H.W-S. and Prescott, F.A.A., *Biochim. Biophys. Acta*, **380**, 141–144 (1975).
117. Chan, H.W-S., Prescott, F.A.A. and Swoboda, P.A.T., *J. Am. Oil Chem. Soc.*, **53**, 572–576 (1976).
118. Charlesworth, J.M., *Anal. Chem.*, **50**, 1414–1420 (1978).
119. Chen, S.F. and Chan, P.H., *J. Chromatogr.*, **344**, 297–303 (1985).
120. Chen, S.S-H. and Kou, A.Y., *J. Chromatogr.*, **227**, 25–31 (1982).
121. Chen, S.S-H. and Kou, A.Y., *J. Chromatogr.*, **232**, 237–249 (1982).
122. Chen, S.S-H. and Kou, A.Y., *J. Chromatogr.*, **307**, 261–269 (1984).
123. Chen, S.S-H., Kou, A.Y. and Chen, H-H.Y., *J. Chromatogr.*, **208**, 339–346 (1981).
124. Chen, S.S.-H., Kou, A.Y. and Chen, H-H.Y., *J. Chromatogr.*, **276**, 37–44 (1983).
125. Chilton, F.H., O'Flaherty, J.T., Ellis, J.M., Swendsen, C.L. and Wykle, R.L., *J. Biol. Chem.*, **258**, 7268–7271 (1983).
126. Chobanov, D. Tarandjiska, R. and Chobanova, R., *J. Am. Oil Chem. Soc.*, **53**, 48–51 (1976).
127. Chou, K-H. and Jungalwala, F.B., *J. Neurochem.*, **36**, 394–401 (1981).
128. Chou, K-H., Nolan, C.E. and Jungalwala, F.B., *J. Neurochem.*, **44**, 1898–1912 (1985).
129. Christie, W.W., *J. Chromatogr.*, **34**, 405–406 (1968).
130. Christie, W.W., in *Topics in Lipid Chemistry*, Vol. 1, pp. 1–49 (1970) (edited by F.D. Gunstone, Logos Press, London).
131. Christie, W.W., *Biochim. Biophys. Acta*, **316**, 204–211 (1973).

132. CHRISTIE, W.W., in *Lipid Metabolism in Ruminant Animals*, pp. 95–191 (1981) (edited by W.W. Christie, Pergamon Press, Oxford).
133. CHRISTIE, W.W., *Lipid Analysis* (2nd edition) (1982) (Pergamon Press, Oxford).
134. CHRISTIE, W.W., *J. Lipid Res.*, **23**, 1072–1075 (1982).
135. CHRISTIE, W.W., in *Developments in Dairy Chemistry. 2. Lipids*, pp. 1–35 (1983) (edited by P.F. Fox, Applied Science Publishers, London).
136. CHRISTIE, W.W., *J. Lipid Res.*, **26**, 507–512 (1985).
137. CHRISTIE, W.W., *Z. Lebensm. Unters. Forsch.*, **181**, 171–182 (1985).
138. CHRISTIE, W.W., in *Analysis of Oils and Fats*, pp. 313–339 (1986) (edited by R.J. Hamilton and J.B. Rossell, Elsevier Applied Science Publishers, London).
139. CHRISTIE, W.W., *J. Chromatogr.*, **361**, 396–399 (1986).
140. CHRISTIE, W.W., BRECHANY, E.Y., JOHNSON, S.B. and HOLMAN, R.T., *Lipids*, **21**, 657–661 (1986).
141. CHRISTIE, W.W., CALVERT, D.T., SHAND, J.H. and NOBLE, R.C., *Comp. Biochem. Biophys.*, **80B**, 617–621 (1985).
142. CHRISTIE, W.W., CONNOR, K. and NOBLE, R.C., *J. Chromatogr.*, **298**, 513–515 (1984).
143. CHRISTIE, W.W. and HUNTER, M.L., *J. Chromatogr.*, **171**, 517–518 (1979).
144. CHRISTIE, W.W. and HUNTER, M.L., *J. Chromatogr.*, **294**, 489–493 (1984).
145. CHRISTIE, W.W. and HUNTER, M.L., *J. Chromatogr.*, **325**, 473–476 (1985).
146. CHRISTIE, W.W. and HUNTER, M.L., *Biochem. J.*, **235**, 833–838 (1986).
147. CHRISTIE, W.W. and MOORE, J.H., *Biochim. Biophys. Acta*, **210**, 46–56 (1970).
148. CHRISTIE, W.W. and MOORE, J.H., *J. Sci. Fd. Agric.*, **22**, 120–124 (1971).
149. CHRISTIE, W.W., NOBLE, R.C. and DAVIES, G., *J. Soc. Dairy Technol.*, **40**, 10–12 (1987).
150. CHRISTIE, W.W., NOBLE, R.C. and MOORE, *Analyst* (*Lond.*), **95**, 940–944 (1970).
151. CHRISTOPOULOU, C.N. and PERKINS, E.G., *J. Am. Oil Chem. Soc.*, **63**, 679–684 (1986).
152. CLARKE, M.J., HAWKE, R.L. and WELCH, R.M., *J. Liq. Chrom.*, **9**, 1711–1725 (1986).
153. COLBEAU, A., NACHBAUR, J. and VIGNAIS, P.M., *Biochim. Biophys. Acta*, **249**, 462–492 (1971).
154. COLBORNE, A.J. and LAIDMAN, D.L., *Phytochemistry* **14**, 2639–2645 (1975).
155. COLSON, C.E. and LOWENSTEIN, J.M., *Methods in Enzymology*, **72**, 53–56 (1981).
156. COMFURIUS, P. and ZWAAL, R.F.A., *Biochim. Biophys. Acta*, **488**, 36–42 (1977).
157. CONNELLAN, J.M. and MASTERS, C.J., *Biochem. J.*, **94**, 81–84 (1965).
158. CONNOLLY, T.M., WILSON, D.B., BROSS, T.E. and MAJERUS, P.W., *J. Biol. Chem.*, **261**, 122–126 (1986).
159. COOPER, M.J. and ANDERS, M.W., *Anal. Chem.*, **46**, 1849–1852 (1974).
160. CORKEY, B.E., BRANDT, M., WILLIAMS, R.J. and WILLIAMSON, J.R., *Anal. Biochem.*, **118**, 30–41 (1981).
161. COXON, D.T. and PRICE, K.R., *J. Chromatogr.*, **285**, 392–394 (1984).
162. COXON, D.T., PRICE, K.R. and CHAN, H.W-S., *Chem. Phys. Lipids*, **28**, 365–378 (1981).
163. CRANE, R.T., GOHEEN, S.C., LARKIN, E.C. and RAO, G.A., *Lipids*, **18**, 74–80 (1983).
164. CRAWFORD, C.G., PLATTNER, R.D., SESSA, D.J. and RACKIS, J.J., *Lipids*, **15**, 91–94 (1980).
165. CREEK, K.E., RIMOLDI, D., CLIFFORD, A.J., SILVERMAN-JONES, C.S. and DE LUCA, L.M., *J. Biol. Chem.*, **261**, 3490–3500 (1986).
166. CREER, M.H. and GROSS, R.W., *Lipids*, **20**, 922–928 (1985).
167. CREER, M.H. and GROSS, R.W., *J. Chromatogr.*, **338**, 61–69 (1985).
168. CREER, M.H., PASTOR, C., CORR, P.B., GROSS, R.W. and SOBEL, B.E., *Anal. Biochem.*, **144**, 65–74 (1985).
169. CURSTEDT, T., *Biochim. Biophys. Acta*, **489**, 79–88 (1977).
170. CURSTEDT, T., in *Ether Lipids; Biochemical and Biomedical Aspects*, pp. 1–15 (1983) (edited by H.K. Mangold and F. Paltauf, Academic Press, New York).
171. D'AMBOISE, M. and GENDREAU, M., *Anal. Lett.*, **12**, 381–395 (1979).
172. DAPPEN, R., ARM, H. and MEYER, V.R., *J. Chromatogr.*, **373**, 1–20 (1986).
173. DARBRE, A., in *Handbook of Derivatives for Chromatography*, pp. 36–103 (1978) (edited by K. Blau and G.S. King, Heyden & Son, London).
174. DAWSON, R.M.C. and EICHBERG, J., *Biochem. J.*, **96**, 634–643 (1965).
175. D'COSTA, M., DASSIN, R., BRYAN, H. and JOUTSI, P., *Clin. Biochem.*, **18**, 27–31 (1985).
176. DEBUYSERE, M.S. and OLSON, M.S., *Anal. Biochem.*, **133**, 373–379 (1983).
177. DEFFENSE, E., *Rev. Franc. Corps Gras*, **31**, 123–129 (1984).
178. DEFFENSE, E., *J. Am. Oil Chem. Soc.*, **62**, 376–385 (1985).

179. DeJarlais, W.J., Adlof, R.O. and Emken, E.A., *J. Am. Oil Chem. Soc.*, **60**, 975–978 (1983).
180. DeJarlais, W.J. and Emken, E.A., *Lipids*, **21**, 662–665 (1986).
181. Demandre, C., Tremolieres, A., Justin, A-M. and Mazliak, P., *Phytochem.*, **24**, 481–485 (1985).
182. Demandre, C., Tremolieres, A., Justin, A-M. and Mazliak, P., *Biochim. Biophys. Acta*, **877**, 380–386 (1986).
183. DeRuyter, M.G.M. and de Leenheer, A.P., *Clin. Chem.*, **24**, 1920–1923 (1978).
184. DeRuyter, M.G.M. and de Leenheer, A.P., *Anal. Chem.*, **51**, 43–46 (1979).
185. Dethloff, L.A., Gilmore, L.B. and Hook, G.E.R., *J. Chromatogr.*, **382**, 79–87 (1986).
186. Dickens, B.F., Ramesha, C.S. and Thompson, G.A., *Anal. Biochem.*, **127**, 37–48 (1982).
187. Distler, W., *J. Chromatogr.*, **192**, 240–246 (1980).
188. Dittmar, K.E.J., Heckers, H. and Melcher, F.W., *Fette Seifen Anstrichm.*, **80**, 279–303 (1978).
189. Dittmer, J.C. and Wells, M.A., *Methods in Enzymology*, **14**, 482–530 (1969).
190. Dix, T.A. and Marnett, L.J., *J. Biol. Chem.*, **260**, 5351–5357 (1985).
191. Dixon, J.B., *Chimia*, **38**, 82–86 (1984).
192. Do, U.H., Pei, P.T. and Minard, R.D., *Lipids*, **16**, 855–862 (1981).
193. Dong, M.W. and DiCesare, J.L., *J. Am. Oil Chem. Soc.*, **60**, 788–791 (1983).
194. Duck-Chong, C.G. and Baker, G.J., *Lipids*, **18**, 387–396 (1983).
195. Dudley, P.A. and Anderson, R.E., *Lipids*, **10**, 113–115 (1975).
196. Dugan, L.L., Demediuk, P., Pendley, C.E. and Horrocks, L.A., *J. Chromatogr.*, **378**, 317–327 (1986).
197. Duncan, I.W., Culbreth, P.H. and Burtis, C.A., *J. Chromatogr.*, **162**, 281–292 (1979).
198. Durst, H.D., Milano, M., Kikta, E.J., Connelly, S.A. and Grushka, E., *Anal. Chem.*, **47**, 1797–1801 (1975).
199. Eberendu, A.R.N., Venables, B.J. and Daugherty, K.E., *Liq. Chromatogr. Mag.*, **3**, 424–432 (1985).
200. Edelstein, C. and Scanu, A.M., *Methods in Enzymology*, **128**, 339–353 (1986).
201. El-Hamdy, A.H. and Perkins, E.G., *J. Am. Oil Chem. Soc.*, **58**, 49–53 (1981).
202. El-Hamdy, A.H. and Perkins, E.G., *J. Am. Oil Chem. Soc.*, **58**, 867–872 (1981).
203. Emken, E.A., Hartman, J.C. and Turner, C.R., *J. Am. Oil Chem. Soc.*, **55**, 561–563 (1978).
204. Engelhardt, H. and Elgass, H., *J. Chromatogr.*, **158**, 249–259 (1978).
205. Erdahl, W.L. and Privett, O.S., *J. Am. Oil Chem. Soc.*, **62**, 786–792 (1985).
206. Erdahl, W.L., Stolyhwo, A. and Privett, O.S., *J. Am. Oil Chem. Soc.*, **50**, 513–515 (1973).
207. Evans, J.C., Rao, K.R.N., Jackson, S.K., Rowlands, C.C. and Barratt, M.D., *J. High Res. Chromatogr. Chromatogr. Commun.*, **8**, 829–830 (1985).
208. Evans, J.E. and McCluer, R.H., *Biochim. Biophys. Acta*, **270**, 565–569 (1972).
209. Fager, R.S., Shapiro, S. and Litman, B.J., *J. Lipid Res.*, **18**, 704–709 (1977).
210. Farines, M., Soulier, J., Charrouf, M. and Soulier, R., *Rev. Franc. Corps Gras*, **31**, 283–286 (1984).
211. Farinotti, R., Siard, P., Bourson, J., Kirkiacharian, S., Valeur, B. and Mahuzier, G., *J. Chromatogr.*, **269**, 81–90 (1983).
212. Farnworth, E.R., Thompson, B.K. and Kramer, J.K.G., *J. Chromatogr.*, **240**, 463–474 (1982).
213. Farooqui, A.A., *Adv. Lipid Res.*, **18**, 159–227 (1981).
214. Fex, G., *Biochim. Biophys. Acta*, **231**, 161–169 (1971).
215. Fiebig, H-J., *Fette Seifen Anstrichm.*, **87**, 53–57 (1985).
216. Figlewicz, D.A., Nolan, C.E., Singh, I.N., and Jungalwala, F.B., *J. Lipid Res.*, **26**, 140–144 (1985).
217. Fine, J.B. and Sprecher, H., *J. Lipid Res.*, **23**, 660–663 (1982).
218. Folch, J., Lees, M. and Stanley, G.H.S., *J. Biol. Chem.*, **226**, 497–509 (1957).
219. Foot, M. and Clandinin, M.T., *J. Chromatogr.*, **241**, 428–431 (1982).
220. Ford, G.L., *J. Chromatogr.*, **346**, 431–434 (1985).
221. Fowlis, I.A. and Scott, R.P.W., *J. Chromatogr.*, **11**, 1–10 (1963).
222. Frankel, E.N., Neff, W.E. and Plattner, R.D., *Lipids*, **21**, 333–337 (1986).
223. Frey, B.M. and Frey, F.J., *Clin. Chem.*, **28**, 689–692 (1982).
224. Fricke, H.S.G. and Oehlenschlager, J., *J. Chromatogr.*, **252**, 331–334 (1982).
225. Fukuda, M.N. and Hakomori, S-I., *J. Biol. Chem.*, **257**, 446–455 (1982).

226. Funk, M.O., Isaac, R. and Porter, N.A., *Lipids*, **11**, 113–117 (1976).
227. Galanos, C., Luderitz, O., Rietschel, E.T. and Westphal, O., in *Biochemistry of Lipids II*, pp. 239–335 (1977) (edited by T.W. Goodwin, University Park Press, London).
228. Galanos, D.S. and Kapoulas, V.M., *J. Lipid Res.*, **3**, 134–136 (1962).
229. Galliard, T., *Phytochem.*, **7**, 1907–1914 (1968).
230. Galliard, T., *Phytochem.*, **7**, 1915–1922 (1968).
231. Gau, W., Ploschke, H-J. and Wunsche, C., *J. Chromatogr.*, **262**, 277–284 (1983).
232. Gaydou, E.M., Bianchini, J.P. and Ralaimanarivao, A., *Anal. Chem.*, **55**, 2313–2317 (1983).
233. Gazzotti, G., Sonnino, S. and Ghidoni, R., *J. Chromatogr.*, **348**, 371–378 (1985).
234. Gazzotti, G., Sonnino, S., Ghidoni, R., Orlando, P. and Tettamanti, G., *Glycoconjugate J.*, **1**, 111–121 (1984).
235. Geeraert, E. and De Schepper, D., *J. High Res. Chromatogr. Chromatogr. Commun.*, **6**, 123–132 (1983).
236. Geeraert, E. and Sandra, P., *J. High Res. Chromatogr. Chromatogr. Commun.*, **8**, 415–422 (1985).
237. Geurts van Kessel, W.S.M., Hax, W.M.A., Demel, R.A. and De Gier, J., *Biochim. Biophys. Acta*, **486**, 524–530 (1977).
238. Geurts van Kessel, W.S.M., Tieman, M. and Demel, R.A., *Lipids*, **16**, 58–63 (1981).
239. Geyer, K.G. and Goodman, H.M., *Proc. Soc. Exp. Biol. Med.*, **133**, 404–406 (1970).
240. Ghiggeri, G.M., Candiano, G., Delfino, G., Queirolo, C., Ginevri, F., Perfumo, F. and Gusmano, R., *J. Chromatogr.*, **381**, 411–418 (1986).
241. Gidez, L.I., *J. Lipid Res.*, **25**, 1430–1436 (1984).
242. Gildenberg. L. and Firestone, D., *J. Assoc. Off. Anal. Chem.*, 68, 46–51 (1985).
243. Gillan, F.T. and Johns, R.B., *J. Chromatogr. Sci.*, **21**, 34–38 (1983).
244. Glod, B.K. and Kemula, W., *J. Chromatogr.*, **321**, 433–439 (1985).
245. Goiffon, J-P., Reminiac, C. and Furon, D., *Rev. Franc. Corps Gras*, **28**, 199–207 (1981).
246. Goiffon, J-P., Reminiac, C. and Olle, M., *Rev. Franc. Corps Gras*, **28**, 167–170 (1981).
247. Goldfine, H., *Curr. Top. Membr. Transport*, **17**, 1–43 (1982).
248. Gray, G.M., *Biochim. Biophys. Acta*, **144**, 511–518 (1967).
249. Greenspan, M.D. and Schroeder, E.A., *Anal. Biochem.*, **127**, 441–448 (1982).
250. Griffin, H.D. and Hawthorne, J.N., *Biochem. J.*, **176**, 541–552 (1978).
251. Grogan, W.M., *Lipids*, **19**, 341–346 (1984).
252. Gross, R.W., *Biochemistry*, **24**, 1662–1668 (1985).
253. Gross, R.W. and Sobel, B.E., *J. Chromatogr.*, **197**, 79–85 (1980).
254. Gross, S.K. and McCluer, R.H., *Anal. Biochem.*, **102**, 429–433 (1980).
255. Gross, S.K., McCluer, R.H. and Irwin, L.N., *Arch. Biochem. Biophys.*, **247**, 446–449 (1986).
256. Grove, R.I., Fitzpatrick, D. and Schimmel, S.D., *Lipids*, **16**, 691–693 (1981).
257. Gubitz, G., *J. Chromatogr.*, **187**, 208–211 (1980).
258. Guha, O.K. and Janak, J., *J. Chromatogr.*, **68**, 325–343 (1972).
259. Guichardant, M. and Lagarde, M., *J. Chromatogr.*, **275**, 400–406 (1983).
260. Gunstone, F.D., Harwood, J.L. and Padley, F.B., *The Lipid Handbook* (1986) (Chapman & Hall, London).
261. Gunstone, F.D., Ismail, I.A. and Lie Ken Jie, M.S.F., *Chem. Phys. Lipids*, **1**, 376–385 (1967).
262. Gunstone, F.D., Kilcast, D., Powell, R.G. and Taylor, G.M., *Chem. Commun.*, 295 (1967).
263. Gurr, M.I., *Role of Fats in Food and Nutrition* (1984) (Elsevier Applied Science, London).
264. Gurr, M.I., Prottey, C. and Hawthorne, J.N., *Biochim. Biophys. Acta*, **106**, 357–370 (1965).
265. Ha, Y.C. and Barter, P.J., *J. Chromatogr.*, **341**, 154–159 (1985).
266. Hadley, N.F., *The Adaptive Role of Lipids in Biological Systems* (1985) (J. Wiley & Sons, New York).
267. Hagiwara, N., Okazaki, M. and Hara, I., *Yukagaku*, **31**, 262–267 (1982).
268. Hakomori, S-I., in *Sphingolipid Biochemistry, Handbook of Lipid Research*, **3**, pp. 1–165 (1983) (edited by J.N. Kanfer and S-I. Hakomori, Plenum Press, New York).
269. Hakomori, S-I., Nudelman, E., Levery, S.B. and Kannagi, R., *J. Biol. Chem.*, **259**, 4672–4680 (1984).
270. Halgunset, J., Lund, E.W. and Sunde, A., *J. Chromatogr.*, **237**, 496–499 (1982).
271. Hamilton, J.G. and Comai, K., *J. Lipid Res.*, **25**, 1142–1148 (1984).
272. Hamilton, J.G. and Karol, R.J., *Prog. Lipid Res.*, **21**, 155–170 (1982).

273. HAMILTON, R.J., in *Analysis of Oils and Fats*, pp. 243–311 (1986) (edited by R.J. Hamilton and B.J. Rossell, Elsevier Applied Science Publishers, London).
274. HAMILTON, R.J. and SEWEL, P.A., *Introduction to High Performance Liquid Chromatography* (2nd edition) (1982) (Chapman & Hall, London).
275. HAMMOND, E.W., *J. Chromatogr.*, **203**, 397–403 (1981).
276. HAMMOND, E.W., in *HPLC in Food Analysis*, pp. 167–185 (1982) (edited by R. Macrae, Academic Press, London).
277. HANAHAN, D.J., *Ann. Rev. Biochem.*, **55**, 483–509 (1986).
278. HANCOCK, W.S., BISHOP, C.A., GOTTO, A.M., HARDING, D.R.K., LAMPLUGH, S.M. and SPARROW, J.T., *Lipids*, **16**, 250–259 (1981).
279. HANCOCK, W.S., POWNALL, H.J., GOTTO, A.M. and SPARROW, J.T., *J. Chromatogr.*, **216**, 285–293 (1981).
280. HANDA, S. and KUSHI, Y., *Adv. Exp. Med. Biol.*, **152**, 23–31 (1982).
281. HANDA, S. and KUSHI, Y., in *New Vistas in Glycolipid Research*, pp. 23–31 (1983) (edited by A. Makita, T. Taketomi, S. Handa and Y. Nagai, Plenum Press, New York).
282. HANSON, V.L., PARK, J.Y., OSBORN, T.W. and KIRAL, R.M., *J. Chromatogr.*, **205**, 393–400 (1981).
283. HARA, A., and RADIN, N.S., *Anal. Biochem.*, **90**, 420–426 (1978).
284. HARA, I. and OKAZAKI, M., *Prog. HPLC*, **1**, 95–103 (1985).
285. HARA, I., OKAZAKI, M. and OHNO, Y., *J. Biochem. (Tokyo)*, **87**, 1863–1865 (1980).
286. HARA, I., SHIRAISHI, K. and OKAZAKI, M., *J. Chromatogr.*, **239**, 549–557 (1982).
287. HARVEY, D.J., *Biomed. Mass Spectrom.*, **9**, 33–38 (1982).
288. HARVEY, D.J., *Biomed. Mass Spectrom.*, **11**, 187–192 (1984).
289. HARVEY, D.J., *Biomed. Mass Spectrom.*, **11**, 340–347 (1984).
290. HARWOOD, J.L., *Phytochemistry*, **14**, 1985–1990 (1975).
291. HARWOOD, J.L., in *The Biochemistry of Plants. Vol. 4. Lipids: Structure and Function*, pp. 1–55 (1980) (edited by P.K. Stumpf, Academic Press, New York).
292. HARWOOD, J.L. and RUSSELL, N.J., *Lipids in Plants and Microbes* (1984) (Allen & Unwin, London).
293. HASLBECH, F. and GROSCH, W., *Lipids*, **18**, 706–713 (1983).
294. HASLBECH, F. and GROSCH, W., *Fette Seifen Anstrichm.*, **86**, 408–412 (1984).
295. HASLBECH, F., GROSCH, W. and FIRL, J., *Biochim. Biophys. Acta*, **750**, 185–193 (1983).
296. HATSUMI, M., KIMATA, S-I. and HIROSAWA, K., *J. Chromatogr.*, **380**, 247–255 (1986).
297. HAUSER, G. and EICHBERG, J., *Biochim. Biophys. Acta*, **326**, 201–209 (1973).
298. HAVERKATE, F. and VAN DEENEN, L.L.M., *Biochim. Biophys. Acta*, **106**, 78–92 (1965).
299. HAWTHORNE, J.N. and ANSELL, G.B. (editors), *Phospholipids* (1982) (Elsevier, Amsterdam).
300. HAX, W.M.A. and GEURTS VAN KESSEL, W.S.M., *J. Chromatogr.*, **142**, 735–741 (1977).
301. HAYASHI, K., KAWASE, J., YOSHIMURA, K., ARA, K. and TSUJI, K., *Anal. Biochem.*, **136**, 314–320 (1984).
302. HAZLEWOOD, G.P. and DAWSON, R.M.C., *Biochem. J.*, **153**, 49–53 (1976).
303. HEATH, R.R., TUMLINSON, J.H. and DOOLITTLE, R.E., *J. Chromatogr. Sci.*, **15**, 10–13 (1977).
304. HEATH, R.R., TUMLINSON, J.H., DOOLITTLE, R.E. and PROVEAUX, A.T., *J. Chromatogr. Sci.*, **13**, 380–382 (1975).
305. HECKERS, H., MELCHER, F.W. and SCHLOEDER, U., *J. Chromatogr.*, **136**, 311–317 (1977).
306. HEEMSKERK, J.W.M., BOGEMANN, G., SCHEIJEN, M.A.M. and WINTERMANS, J.F.G.M., *Anal. Biochem.*, **154**, 85–91 (1986).
307. HEFTMANN, E. and LIN, J.T., *J. Liq. Chromatogr.*, **5**, 121–173 (1982).
308. HEINZ, E. and TULLOCH, A.P., *Hoppe-Seyler's Z. Physiol. Chem.*, **350**, 493–498 (1969).
309. HEINZE, F.J., LINSCHEID, M. and HEINZ, E., *Anal. Biochem.*, **139**, 126–133 (1984).
310. HELMICH, O., SPACEK, P. and HRADEC, J., *J. Chromatogr.*, **213**, 105–111 (1981).
311. HELMY, F.M. and HACK, M.H., *Lipids*, **1**, 279–281 (1966).
312. HENKE, H. and SCHUBERT, J., *J. High Res. Chromatogr., Chromatogr. Commun.*, **3**, 69–78 (1980).
313. HERNQVIST, L., HERSLOF, B. and HERSLOF, M., *Fette Seifen Anstrichm.*, **86**, 393–397 (1984).
314. HERSLOF, B., *J. High Res. Chromatogr., Chromatogr. Commun.*, **4**, 471–473 (1981).
315. HERSLOF, B., HERSLOF, M. and PODLAHA, O., *Fette Seifen Anstrichm.*, **82**, 460–462 (1980).
316. HERSLOF, B. and KINDMARK, G., *Lipids*, **20**, 783–790 (1985).
317. HERSLOF, B., PODLAHA, O. and TOREGARD, B., *J. Am. Oil. Chem. Soc.*, **56**, 864–866 (1979).

318. HIRABAYASHI, Y., HAMAOKA, A., MATSUMOTO, M. and NISHIMURA, K., *Lipids*, **21**, 710–714 (1986)
319. HIRAMATSU, K. and ARIMORI, S., *J. Chromatogr.*, **227**, 423–431 (1982).
320. HIRATA, H., HIGUCHI, K. and NAKASATO, S., *Yukagaku*, **33**, 11–19 (1984).
321. HIRATA, H., HIGUCHI, K. and NAKASATO, S., *Yukagaku*, **33**, 290–293 (1984).
322. HIRATA, H. and NAKASATO, S., *Yukagaku*, **32**, 754–755 (1983).
323. HIRATA, Y. and SUMIYA, E., *J. Chromatogr.*, **267**, 125–131 (1983).
324. HITCHCOCK, C. and NICHOLS, B.W., *Plant Lipid Biochemistry* (1971) (Academic Press, London).
325. HOFFMAN, N.E. and LIAO, J.C., *Anal. Chem.*, **48**, 1104–1106 (1976).
326. HOKIN, L.E., *Ann. Rev. Biochem.*, **54**, 205–235 (1985).
327. HOLMAN, R.T., *Prog. Chem. Fats Other Lipids*, **9**, 1–12 (1966).
328. HOLMAN, R.T. (editor), *Essential Fatty Acids and Prostaglandins* (*Prog. Lipid Res., Vol. 20*) (1982) (Pergamon Press, Oxford).
329. HOLMAN, R.T. and HOFSTETTER, H.H., *J. Am. Oil Chem. Soc.*, **42**, 540–544 (1965).
330. HOLMAN. R.T. and RAHM, J.J., *Prog. Chem. Fats other Lipids.*, **9**, 13–90 (1966).
331. HOLMQUIST, L. and CARLSON, L.A., *Lipids*, **20**, 378–388 (1985).
332. HOLUB, B.J. and KUKSIS, A., *J. Lipid Res.*, **12**, 699–705 (1971).
333. HOLUB, B.J. and KUKSIS, A., *Adv. Lipid Res.*, **16**, 1–125 (1978).
334. HORI, T. and NOZAWA, Y., in *Phospholipids*, pp. 95–128 (1982) (edited by J.N. Hawthorne and G.B. Ansell, Elsevier, Amsterdam).
335. HORSTMANN, P. and MONTAG, A., *Fette Seifen Anstrichm.*, **88**, 262–264 (1986).
336. HOSOKAWA, Y., SHIMOMURA, Y., HARRIS, R.A. and OZAWA, T., *Anal. Biochem.*, **153**, 45–49 (1986).
337. HOUX, N.W.H. and VOERMAN, S., *J. Chromatogr.*, **129**, 456–459 (1976).
338. HRESKO, R.C., MARKELLO, T.C., BARENHOLZ, Y. and THOMPSON, T.E., *Chem. Phys. Lipids.*, **38**, 263–273 (1985).
339. HSIEH, J.Y-K., WELCH, D.K. and TURCOTTE, J.G., *Lipids*, **16**, 761–763 (1981).
340. HSIEH, J.Y-K., WELCH, D.K. and TURCOTTE, J.G., *J. Chromatogr.*, **208**, 398–403 (1981).
341. HUGHES, H., SMITH, C.V., HORNING, E.C. and MITCHELL, J.R., *Anal. Biochem.*, **130**, 431–436 (1983).
342. HULLET, D.A. and EISENREICH, S.J., *Anal. Biochem.*, **51**, 1953–1960 (1979).
343. HUNDRIESER, K.E., CLARK, R.M. and JENSEN, R.G., *Am. J. Clin. Nutr.*, **41**, 988–993 (1985).
344. HURST, W.J., ALEO, M.D. and MARTIN, R.A., *J. Dairy Sci.*, **66**, 2192–2194 (1983).
345. HURST, W.J., ALEO, M.D. and MARTIN, R.A., *J. Agric. Fd Chem.*, **33**, 820–822 (1985).
346. HURST, W.J. and MARTIN, R.A., *J. Am. Oil Chem. Soc.*, **57**, 307–310 (1980).
347. ICHINOSE, N., NAKAMURA, K., SHIMIZU, C., KUROKURA, H. and OKAMOTO, K., *J. Chromatogr.*, **295**, 463–469 (1984).
348. IKEDA, M., SHIMADA, K. and SAKAGUCHI, T., *J. Chromatogr.*, **272**, 251–259 (1983).
349. IKEDA. M., SHIMADA, K. SAKAGUCHI, T. and MATSUMOTO, U., *J. Chromatogr.*, **305**, 261–270 (1984).
350. INGEBRETSEN, O.C. and FARSTAD, M., *J. Chromatogr.*, **202**, 439–445 (1980).
351. INGEBRETSEN, O.C., NORMANN, P.T. and FLATMARK, T., *Anal. Biochem.*, **96**, 181–188 (1979).
352. INNIS, S.M. and CLANDININ, M.T., *J. Chromatogr.*, **205**, 490–492 (1981).
353. IOANNOU, P.V. and GOLDING, B.T., *Prog. Lipid Res.*, **17**, 279–318 (1979).
354. ISHIZUKA, T., ISHIKAWA, K., MASEKI, M., TOMODA, Y. and TSUDA, T., *J. Chromatogr.*, **380**, 43–53 (1986).
355. ITABASHI, Y. and TAKAGI, T., *Lipids*, **21**, 413–416 (1986).
356. ITASAKA, O., HORI, T., SASAHARA, K., WAKABAYASHI, Y., TAKAHASHI, F. and RHEE, H., *J. Biochem.* (*Tokyo*), **95**, 1671–1675 (1984).
357. ITOH, K., SUZUKI, A., KUROKI, Y. and AKINO, T., *Lipids*, **20**, 611–616 (1985).
358. IUPAC-IUB Commission on Biochemical Nomenclature, *Eur. J. Biochem.*, **2**, 127–131 (1967); *Biochem. J.*, **105**, 897–902 (1967).
359. IUPAC-IUB Commission on Biochemical Nomenclature, *Hoppe-Seyler's Z. Physiol. Chem.*, **358**, 599–616 (1977); *J. Lipid Res.*, 19, 114–125 (1978).
360. IWAMORI, M., COSTELLO, C. and MOSER, H.W., *J. Lipid Res.*, **20**, 86–96 (1979).
361. IWAMORI, M. and MOSER, H.W., *Clin. Chem.*, **21**, 725–729 (1975).
362. IWAMORI, M., MOSER, H.W., MCCLUER, R.H. and KISHIMOTO, Y., *Biochim. Biophys. Acta*, **380**, 308–319 (1975).

363. JABS, H-U., ASSMANN, G., GREIFENDORF, D. and BENNINGHOVEN, A., *J. Lipid Res.*, **27**, 613–621 (1986).
364. JACKSON, E.M., MOTT, G.E., HOPPENS, C., MCMANUS, L.M., WEINTRAUB, S.T., LUDWIG, J.C. and PINCKARD, R.N., *J. Lipid Res.*, **25**, 753–757 (1984).
365. JAEGER, H., KLOER, H.U. and DITSCHUNEIT, H., *J. Lipid Res.*, **17**, 185–190 (1976).
366. JAEGER, H., KLOER, H.U., DITSCHUNEIT, H. and FRANK, H., in *Applications of Glass Capillary Chromatography*, pp. 395–453 (1981) (edited by W.G. Jennings, Marcel Dekker, N.Y.).
367. JAEGER, H., KLOER, H.U. DITSCHUNEIT, H. and FRANK, H., in *Glass Capillary Chromatography in Clinical Medicine and Pharmacology*, pp. 271–314 (1985) (edited by H. Jaeger, Marcel Dekker, N.Y.).
368. JAMES, J.L., CLAWSON, G.A., CHAN, C.H. and SMUCKLER, E.A., *Lipids*, **16**, 541–545 (1981).
369. JAMIESON, G.R., in *Topics in Lipid Chemistry*, Vol. 1, pp. 107–159 (1970) (edited by F.D. Gunstone, Logos Press, London).
370. JANCA, J., *Chromatogr. Sci.*, **25**, 1–51 (1984).
371. JATZEKEWITZ, H. and MEHL, E., *Hoppe-Seyler's Z. Physiol. Chem.*, **320**, 251–257 (1960).
372. JENSEN, G.W., *J. Chromatogr.*, **204**, 407–411 (1981).
373. JENSEN, O.N. and MOLLER, J., *Fette Seifen Anstrichm.*, **88**, 352–357 (1986).
374. JORDI, H.C., *J. Liq. Chromatogr.*, **1**, 215–230 (1978).
375. JUANEDA, P. and ROCQUELIN, G., *Lipids*, **20**, 40–41 (1985).
376. JUANEDA, P. and ROCQUELIN, G., *Lipids*, **21**, 239–240 (1986).
377. JUNGALWALA, F.B., in *Phospholipids in Nervous Tissues*, pp. 1–44 (1985) (edited by J. Eichberg, Wiley, New York).
378. JUNGALWALA, F.B., EVANS, J.E., BREMER, E. and MCCLUER, R.H., *J. Lipid Res.*, **24**, 1380–1388 (1983).
379. JUNGALWALA, F.B., EVANS, J.E., KADOWAKI, H. and MCCLUER, R.H., *J. Lipid Res.*, **25**, 209–216 (1984).
380. JUNGALWALA, F.B., EVANS, J.E. and MCCLUER, R.H., *Biochem. J.*, **155**, 55–60 (1976).
381. JUNGALWALA, F.B., EVANS, J.E. and MCCLUER, R.H., *J. Lipid Res.*, **25**, 738–749 (1984).
382. JUNGALWALA, F.B., HAYES, L. and MCCLUER, R.H., *J. Lipid Res.*, **18**, 285–292 (1977).
383. JUNGALWALA, F.B., HAYSSEN, V., PASQUINI, J.M. and MCCLUER, R.H., *J. Lipid Res.*, **20**, 579–587 (1979).
384. JUNGALWALA, F.B., KOUL, O., STOOLMILLER, A. and SAPIRSTEIN, V.S., *J. Neurochem.*, **45**, 191–198 (1985).
385. JUNGALWALA, F.B. and MILUNSKY, A., *Pediat. Res.*, **12**, 655–659 (1978).
386. JUNGALWALA, F.B., SANYAL, S. and LEBARON, F., in *Phospholipids in the Nervous System, Vol. 1: Metabolism*, pp. 91–103 (1982) (edited by L.A. Horrocks, G.B. Ansell and G. Porcellati, Raven Press, New York).
387. JUNGALWALA, F.B., TUREL, R.J., EVANS, J.E. and MCCLUER, R.H., *Biochem. J.*, **145**, 517–526 (1975).
388. JUPILLE, T., *J. Chromatogr. Sci.*, **17**, 160–167 (1979).
389. JUSTIN, A.M., DEMANDRE, C., TREMOLIERES, A. and MAZLIAK, P., *Biochim. Biophys. Acta*, **836**, 1–7 (1985).
390. KADOWAKI, H., BREMER, E.G., EVANS, J.E., JUNGALWALA, F.B. and MCCLUER, R.H., *J. Lipid Res.*, **24**, 1389–1397 (1983).
391. KADOWAKI, H., EVANS, J.E. and MCCLUER, R.H., *J. Lipid Res.*, **25**, 1132–1139 (1984).
392. KADUCE, T.L., NORTON, K.C. and SPECTOR, A.A., *J. Lipid Res.*, **24**, 1398–1403 (1983).
393. KAITARANTA, J.K. and BESSMAN, S.P., *Anal. Chem.*, **53**, 1232–1235 (1981).
394. KAITARANTA, J.K., GEIGER, P.J. and BESSMAN, S.P., *J. Chromatogr.*, **206**, 327–332 (1981).
395. KALUZNY, M.A., DUNCAN, L.A., MERRITT, M.V. and EPPS, D.E., *J. Lipid Res.*, **26**, 135–140 (1985).
396. KANFER, J.N., *Methods in Enzymology.*, **14**, 660–664 (1969).
397. KANFER, J.N. and HAKOMORI, S-I. (editors), *Sphingolipid Biochemistry* (*Handbook of Lipid Research Vol. 3.*) (1983) (Plenum Press, New York).
398. KANNAGI, R., NUDELMAN, E., LEVERY, S.B. and HAKOMORI, S-I., *J. Biol. Chem.*, **257**, 14865–14874 (1982).
399. KANNER, J. and KINSELLA, J.E., *Lipids*, **18**, 204–210 (1983).
400. KAPLAN, E. and ANSARI, A., *J. Chromatogr.*, **350**, 435–443 (1985).
401. KAPOULAS, V.M. and ANDRIKOPOULOS, *J. Chromatogr.*, **366**, 311–320 (1986).

402. KARGER, B.L. and VOUROS, P., *J. Chromatogr.*, **323**, 13–22 (1985).
403. KARLSSON, K-A., *Chem. Phys. Lipids*, **5**, 6–43 (1970).
404. KARLSSON, K-A., *Lipids*, **5**, 878–891 (1970).
405. KARLSSON, K-A., SAMUELSSON, B.E. and STEEN, G.O., *Biochim. Biophys. Acta*, **316**, 317–335 (1973).
406. KARLSSON, K.A., SAMUELSSON, B.E. and STEEN, G.O., *Biochim. Biophys. Acta*, **316**, 336–362 (1973).
407. KATES, M., *Prog. Chem. Fats other Lipids*, **15**, 301–342 (1978).
408. KAULEN, H.D., *Anal. Biochem.*, **45**, 664–667 (1972).
409. KAWAGUCHI, A., KOBAYASHI, Y., OGAWA, Y. and OKUDA, S., *Chem. Pharm. Bull.*, **31**, 3228–3232 (1983).
410. KAWASAKI, T., KAMBAYASHI, J., MORI, T. and KOSAKI, G., *Thromb. Res.*, **36**, 335–344 (1984).
411. KAYE, E.M. and ULLMAN, M.D., *Anal. Biochem.*, **138**, 380–385 (1984).
412. KELLER, R.K., ADAIR, W.L. and CAFMEYER, N., *Anal. Biochem.*, **155**, 119–122 (1986).
413. KERNER, J. and BIEBER, L.L., *Anal. Biochem.*, **134**, 459–466 (1983).
414. KERNER, J., FROSETH, J.A., MILLER, E.R. and BIEBER, L.L., *J. Nutr.*, **114**, 854–861 (1984).
415. KESSELMEIER, J. and HEINZ, E., *Anal. Biochem.*, **144**, 319–328 (1985).
416. KIHARA, K., ROKUSHIKA, S. and HATANO, H., *Bunseki Kagaku*, **33**, 647–652 (1984).
417. KIM, H-Y. and SALEM, N., *Anal. Chem.*, **58**, 9–14 (1986).
418. KIMMEY, R.L. and PERKINS, E.G., *J. Am. Oil Chem. Soc.*, **61**, 1209–1211 (1984).
419. KING, J.W., ADAMS, E.C. and BIDLINGMEYER, B.A., *J. Liq. Chromatogr.*, **5**, 275–304 (1982).
420. KING, M.T. and REISS, P.D., *Anal Biochem.*, **146**, 173–179 (1985).
421. KINNUNEN, P.M. and LANGE, L.G., *Anal. Biochem.*, **140**, 567–576 (1984).
422. KINOSHITA, M., OKAZAKI, M., KATO, H., TERAMOTO, T., MATSUSHIMA, T., NAITO, C., OKA, H. and HARA, I., *J. Biochem. (Tokyo)*, **94**, 615–618 (1983).
423. KINOSHITA, M., OKAZAKI, M., KATO, H., TERAMOTO, T., MATSUSHIMA, T., NAITO, C., OKA, H. and HARA, I., *J. Biochem. (Tokyo)*, **95**, 1111–1118 (1984).
424. KINSELLA, J.E., BRUCKNER, G., MAI, J. and SHARP, J., *Am. J. Clin, Nutr.*, **34**, 2307–2318 (1981).
425. KITO, M., TAKAMURA, H., NARITA, H. and URADE, R., *J. Biochem. (Tokyo)*, **98**, 327–331 (1985).
426. KIUCHI, K., OHTA, T. and EBINE, H., *J. Chromatogr. Sci.*, **13**, 461–466 (1975).
427. KIUCHI, K., OHTA, T. and EBINE, H., *J. Chromatogr.*, **133**, 226–230 (1977).
428. KIUCHI, K., OHTA, T., ITOH, H., TAKAHASHI, T. and EBINE, H., *J. Agric. Fd Chem.*, **24**, 404–407 (1976).
429. KLEIN, R.A. and SCHMITZ, B., *Biomed. Envir. Mass Spectrom.*, **13**, 429–437 (1986).
430. KLUMP, B., MELCHERT, H.U. and RUBACH, K., *Freserius Z. Anal. Chem.*, **313**, 553–560 (1982).
431. KNAPP, D.R. and KRUEGER, S., *Anal. Lett.*, **8**, 603–610 (1975).
432. KNUDSEN, J. and GRUNNET, I., *Biochem. J.*, **202**, 139–143 (1982).
433. KOBAYASHI, T., KATAYAMA, M., SUZUKI, S., TOMODA, H., GOTO, I. and KUROIWA, Y., *J. Neurol.*, **230**, 209–215 (1983).
434. KODALI, D.R., REDGRAVE, T.G., SMALL, D.M. and ATKINSON, D. *Biochemistry*, **24**, 519–525 (1985).
435. KOLAROVIC, L. and FOURNIER, N.C., *Anal. Biochem.*, **156**, 244–250 (1986).
436. KOLATTUKUDY, P.E. (editor), *The Chemistry and Biochemistry of Natural Waxes* (1976) (Elsevier, Amsterdam).
437. KOLATTUKUDY, P.E., CROTEAU, R. and BUCKNER, J.S., in *The Chemistry and Biochemistry of Natural Waxes*, pp. 289–347 (1976) (edited by P.E. Kolattukudy, Elsevier, Amsterdam).
438. KONDOH, Y. and TAKANO, S., *Anal. Chem.*, **58**, 2380–2383 (1986).
439. KORTE, W.D., *J. Chromatogr.*, **243**, 153–157 (1982).
440. KORTE, K., CHIEN, K.R. and CASE, M.L., *J. Chromatogr.*, **375**, 225–231 (1986).
441. KOSKAS, J.P., CILLARD, J. and CILLARD, P., *J. Chromatogr.*, **258**, 280–283 (1983).
442. KOUL, O. and JUNGALWALA, F.B., *Biochem. J.*, **194**, 633–637 (1981).
443. KRAMER, J.K.G., *Lipids*, **15**, 651–660 (1980).
444. KRAMER, J.K.G., FARNWORTH, E.R. and THOMPSON, B.K., *Lipids*, **20**, 536–541 (1985).
445. KRAMER, J.K.G., FOUCHARD, R.C. and FARNWORTH, E.R., *Lipids*, **18**, 896–899 (1983).
446. KRUEMPELMAN, M. and DANIELSON, N.D., *J. Liq. Chromatogr.*, **5**, 1679–1689 (1982).
447. KRUGER, J., RABE, H., REICHMANN, G. and RUSTOW, B., *J. Chromatogr.*, **307**, 387–392 (1984).

448. KUCERA, P. (editor), *Microcolumn High-performance Liquid Chromatography* (1984) (Elsevier, Amsterdam).
449. KUHNZ, W., ZIMMERMANN, B. and NAU, H., *J. Chromatogr.*, **344**, 309–312 (1985).
450. KUKSIS, A., *Separation Purification Methods*, **6**, 353–395 (1977).
451. KUKSIS, A., in *Handbook of Lipid Research Vol. 1. Fatty Acids and Glycerides*, pp. 381–442 (1978) (edited by A. Kuksis, Plenum Press, New York).
452. KUKSIS, A., in *Lipid Research Methodology*, pp. 78–132 (1984) (edited by J.A. Story, A.R. Liss Inc., New York).
453. KUKSIS, A., MARAI, L. and MYHER, J.J., *J. Chromatogr.*, **273**, 43–66 (1983).
454. KUKSIS, A., MARAI, L., MYHER, J.J., CERBULIS, J. and FARRELL, H.M., *Lipids*, **21**, 183–190 (1986).
455. KUKSIS, A. and MYHER, J.J., *J. Chromatogr.*, **379**, 57–90 (1986).
456. KUKSIS, A., MYHER, J.J. and MARAI, L., *J. Am. Oil Chem. Soc.*, **61**, 1582–1589 (1984).
457. KUKSIS, A., MYHER, J.J. and MARAI, L., *J. Am. Oil Chem. Soc.*, **62**, 762–767 (1985).
458. KUKSIS, A., MYHER, J.J. and MARAI, L., *J. Am. Oil Chem. Soc.*, **62**, 767–773 (1985).
459. KUKSIS, A., MYHER, J.J., MARAI, L. and GEHER, K., *J. Chromatogr. Sci.*, **13**, 423–430 (1975).
460. KUKSIS, A., MYHER, J.J., MARAI, L., LITTLE, J.A., MCARTHUR, R.G. and RONCARI, D.A.K., *Lipids*, **21**, 371–377 (1986).
461. KUKSIS, A., STACHNYK, O. and HOLUB, B.J., *J. Lipid Res.*, **10**, 660–667 (1969).
462. KUNDU, S.K., CHAKRAVARTY, S.K., ROY, S.K. and ROY, A.K., *J. Chromatogr.*, **170**, 65–72 (1979).
463. KUNDU, S.K. and ROY, S.K., *J. Lipid Res.*, **19**, 390–395 (1978).
464. KUNDU, S.K. and SCOTT, D.D., *J. Chromatogr.*, **232**, 19–27 (1982).
465. KUNZE, D., RUSTOW, B., RABE, H. and ULLRICH, K-P., *Clin. Chim. Acta*, **140**, 215–222 (1984).
466. KUPKE, I.R. and ZEUGNER, S., *J. Chromatogr.*, **146**, 261–272 (1978).
467. KUPRANYCZ, D.B., AMER, M.A. and BAKER, B.E., *J. Am. Oil Chem. Soc.*, **63**, 332–337 (1986).
468. KUSHI, Y. and HANDA, S., *J. Biochem.* (*Tokyo*), **91**, 923–931 (1982).
469. KUSHI, Y., HANDA, S., KAMBARA, H. and SHIZUKUISHI, K., *J. Biochem.* (*Tokyo*), **94**, 1841–1850 (1983).
470. LADISCH, S. and GILLARD, B., *Anal. Biochem.*, **146**, 220–231 (1985).
471. LAIRON, D., AMIC, J., LAFONT, H., NALBONE, G., DOMINGO, N. and HAUTON, J., *J. Chromatogr.*, **88**, 183–186 (1974).
472. LAM, S. and GRUSHKA, E., *J. Chromatogr. Sci.*, **15**, 234–238 (1977).
473. LAM, S. and GRUSHKA, E., *J. Chromatogr.*, **158**, 207–214 (1978).
474. LAM, S. and GRUSHKA, E., *Separation Purification Methods*, **14**, 67–96 (1985).
475. LANSER, A.C. and EMKEN, E.A., *J. Chromatogr.*, **256**, 460–464 (1983).
476. LAUDE, D.A. and WILKINS, C.L., *Anal. Chem.*, **56**, 2471–2475 (1984).
477. LAWRENCE, J.F., *J. Chromatogr. Sci.*, **17**, 147–151 (1979).
478. LAWRENCE, J.G., *J. Chromatogr.*, **84**, 299–308 (1973).
479. LEBARON, F.N., SANYAL, S. and JUNGALWALA, F.B., *Neurochem. Res.*, **6**, 1081–1089 (1981).
480. LECHEVALIER, M.P., *CRC Crit. Rev. Microbiol.*, **5**, 109–210 (1977).
481. LEDEEN, R.W. and YU, R.K., *Methods in Enzymology*, **83**, 139–191 (1982).
482. LEDEEN, R.W., YU, R.K. and ENG, L.F., *J. Neurochem.*, **21**, 829–839 (1973).
483 LEE, K.J., *J. Lipid Res.*, **12**, 635–636 (1971).
484. LEE, W.M.F., WESTRICK, M.A. and MACHER, B.A., *Biochem. Biophys. Acta*, **712**, 498–504 (1982).
485. LIE KEN JIE, M.S.F., *Adv. Chromatogr.*, **18**, 1–57 (1980).
486. LIE KEN JIE, M.S.F., *J. Chromatogr.*, **192**, 457–462 (1980).
487. LINDQVIST, B., SJOGREN, I. and NORDIN, R., *J. Lipid Res.*, **15**, 65–73 (1974).
488. LINGEMAN, H., UNDERBERG, W.J.M., TAKADE, A., and HULSHOFF, A., *J. Liq. Chromatogr.*, **8**, 789–874 (1985).
489. LISCOVITCH, M., FREESE, A., BLUSTAJN, J.K. and WURTMAN, R. J. , *Anal. Biochem.*, **151**, 182–187 (1985).
490. LITCHFIELD, C., *Analysis of Triglycerides* (1972) (Academic Press, New York).
491. LLOYD, J.B.F., *J. Chromatogr.*, **189**, 359–373 (1980).
492. LOUGH, A.K., FELINSKI, L. and GARTON, G.A., *J. Lipid Res.*, **3**, 478–480 (1962).
493. LOVELAND, P.M., PAWLOWSKI, N.E., LIBBEY, L.M., BAILEY, G.S. and NIXON, J.E., *J. Am. Oil Chem. Soc.*, **60**, 1786–1788 (1983).

494. LOZANO, Y., *Rev. Franc. Corps Gras*, **30**, 333–346 (1983).
495. LUTHRA, M.G. and SHELTAWY, A., *Biochem. J.*, **126**, 251–253 (1972).
496. LYNCH, D.V., GUNDERSEN, R.E. and THOMPSON, G.A., *Pl. Physiol.*, **72**, 903–905 (1983).
497. MACALA, L.J., YU, R.K. and ANDO, S., *J. Lipid Res.*, **24**, 1243–1250 (1983).
498. MCCLUER, R.H. and EVANS, J.E., *J. Lipid Res.*, **14**, 611–617 (1973).
499. MCCLUER, R.H. and EVANS, J.E., *J. Lipid Res.*, **17**, 412–418 (1976).
500. MCCLUER, R.H. and GROSS, S.K., *J. Lipid Res.*, **26**, 593–599 (1985).
501. MCCLUER, R.H. and JUNGALWALA, F.B., *Adv. Exp. Med. Biol.*, **68**, 533–554 (1976).
502. MCCLUER, R.H. and JUNGALWALA, F.B., *Chromatogr. Sci.*, **10**, 7–30 (1979).
503. MCCLUER, R.H., ULLMAN, M.D. and JUNGALWALA, F.B., *Adv. Chromatogr.*, **25**, 309–353 (1986).
504. MCCLUER, R.H., WILLIAMS, M.A., GROSS, S.K. and MEISLER, M.H., *J. Biol. Chem.*, **256**, 13112–13120 (1981).
505. MACRAE, R. (editor), *HPLC in Food Analysis* (1982) (Academic Press, London).
506. MALLET, A.I., CUNNINGHAM, F.M. and DANIEL, R., *J. Chromatogr.*, **309**, 160–164 (1984).
507. MANCUSO, C.A., NICHOLS, P.D. and WHITE, D.C., *J. Lipid Res.*, **27**, 49–56 (1986).
508. MANET, H.G., GAREIL, P.C. and ROSSET, R.H., *Anal. Chem.*, **56**, 1770–1773 (1984).
509. MANGOLD, H.K. and PALTAUF, F. (editors) *Ether Lipids: Biochemical and Biomedical Aspects* (1983) (Academic Press, New York).
510. MANKU, M.S., *J. Chromatogr. Sci.*, **21**, 367–369 (1983).
511. MANSSON, J-E., ROSENGREN, B. and SVENNERHOLM, L., *J. Chromatogr.*, **322**, 465–472 (1985).
512. MARAI, L., MYHER, J.J. and KUKSIS, A., *Can. J. Biochem. Cell Biol.*, **61**, 840–849 (1983).
513. MARES, P. and HUSEK, P., *J. Chromatogr.*, **350**, 87–103 (1985).
514. MARES, P., RANNY, M., SEDLACEK, J. and SKOREPA, J., *J. Chromatogr.*, **275**, 295–305 (1983).
515. MARION, D., GANDEMER, G. and DOUILLARD, R., in *Structure, Function and Metabolism of Plant Lipids*, pp. 139–143 (1984) (edited by P-A. Siegenthaler and W. Eichenberger, Elsevier, Amsterdam).
516. MARMER, W.N., FOGLIA, T.A. and VAIL, P.D., *Lipids*, **19**, 353–358 (1984).
517. MARUYAMA, K. and YONESE, C., *J Am. Oil Chem. Soc.*, **63**, 902–905 (1986).
518. MATSUOKA, C., NOHTA, H., KURODA, N. and OHKURA, Y., *J. Chromatogr.*, **341**, 432–436 (1985).
519. MATTHEES, D.P., *Proc. S. D. Acad. Sci.*, **59**, 62–64 (1980).
520. MAZLIAK, P., in *Lipids and Lipid Polymers of Higher Plants*, pp. 48–74 (1977) (edited by M. Tevini and H.K. Lichtenthaler, Springer Verlag, Berlin).
521. MAZZOLA, G. and KENT, C., *Anal. Biochem.*, **141**, 137–142 (1984).
522. MECHAM, D.K. and MOHAMMAD, A., *Cereal Chem.*, **32**, 405–415 (1955).
523. MEEK, J.L. and NICOLETTI, F., *J. Chromatogr.*, **351**, 303–311 (1986).
524. MELL, L.D., JOSEPH, S.W. and BUSSELL, N.E., *J. Liq. Chromatogr.*, **2**, 407–416 (1979).
525. MERRITT, M.V. and BRONSON, G.E., *Anal. Chem.*, **48**, 1851–1853 (1976).
526. MERRITT, M.V. and BRONSON, G.E., *Anal. Biochem.*, **80**, 392–400 (1977).
527. MEYER, V.R., *J. Chromatogr.*, **334**, 197–209 (1985).
528. MICHELSEN, P., ARONSSON, E., ODHAM, G. and AKESSON, B., *J. Chromatogr.*, **350**, 417–426 (1985).
529. MIKES, F., SCHURIG, V. and GIL-AV, E., *J. Chromatogr.*, **83**, 91–97 (1973).
530. MILLER, R.A., BUSSELL, N.E. and RICKETTS, C., *J. Liq. Chromatogr.*, **1**, 291–304 (1978).
531. MILLS, G.L., LANE, P.A. and WEECH, P.K., *A Guidebook to Lipoprotein Technique*, Vol. 14, Laboratory Techniques in Biochemistry and Molecular Biology (1984) (Elsevier, Amsterdam).
532. MINKLER, P.E., INGALLS, S.T., KORMOS, L.S., WEIR, D.E. and HOPPEL, C.L., *J. Chromatogr.*, **336**, 271–283 (1984).
533. MINNIKIN, D.E., *Chem. Phys. Lipids*, **21**, 313–347 (1978).
534. MINNIKIN, D.E., MINNIKIN, S.M., O'DONNELL, A.G. and GOODFELLOW, M., *J. Microbiol. Methods*, **2**, 243–249 (1984).
535. MINNIKIN, D.E., O'DONNELL, A.G., GOODFELLOW, M., ALDERSON, G., ATHALYE, M., SCHAAL, A. and PARLETT, J.H., *J. Microbiol. Methods*, **2**, 233–241 (1984).
536. MIWA, H., HIYAMA, C. and YAMAMOTO, M., *J. Chromatogr.*, **321**, 165–174 (1985).
537. MIWA, H. and YAMAMOTO, M., *J. Chromatogr.*, **351**, 165–174 (1986).
538. MIWA, H., YAMAMOTO, M. and NISHIDA, T., *Clin. Chim. Acta*, **155**, 95–102 (1986).

539. MIWA, T.K., MIKOLAJCZAK, K.L., EARLE, F.R. and WOLFF, I.A., *Anal. Chem.*, **32**, 1739–1742 (1960).
540. MIYASHITA, Y., HARA, N., FUJIMOTO, K. and KANEDA, T., *Lipids*, **20**, 578–587 (1985).
541. MIYAZAKI, K., OKAMURA, N., KISHIMOTO, Y. and LEE, Y.C., *Biochem. J.*, **235**, 755–761 (1986).
542. MOCK, T., PELLETIER, M.J.P., MAN, R.Y.K. and CHOY, P.C., *Anal. Biochem.*, **137**, 277–281 (1984).
543. MOGELSON, S., WILSON, G.E. and SOBEL, B.E., *Biochim. Biophys. Acta*, **619**, 680–688 (1980).
544. MOMOI, M. and YAMAKAWA, T., *J. Biochem. (Tokyo)*, **84**, 317–325 (1978).
545. MONSEIGNY, A., VIGNERON, P-Y., LEVACQ, M. and ZWOBADA, F., *Rev. Franc. Corps Gras*, **26**, 107–120 (1979).
546. MOORE, J.P., SMITH, G.A., HESKETH, T.R. and METCALFE, J.C., *J. Biol. Chem.*, **257**, 8183–8189 (1982).
547. MORITA, M., MIHASHI, S., ITOKAWA, H. and HARA, S., *Anal. Chem.*, **55**, 412–414 (1983).
548. MORRIS, L.J., *J. Lipid Res.*, **7**, 717–732 (1966).
549. MORRIS, L.J. and MARSHALL, M.O., *Chem. Ind. (Lond.)*, 460–461 (1966).
550. MORRIS, L.J., WHARRY, D.M. and HAMMOND, E.W., *J. Chromatogr.*, **31**, 69–76 (1967).
551. MORRIS, L.J., WHARRY, D.M. and HAMMOND, E.W., *J. Chromatogr.*, **33**, 471–479 (1968).
552. MORRISON, W.R., in *Topics in Lipid Chemistry*, Vol. 1, pp. 51–106 (1970) (edited by F.D. Gunstone, Logos Press, London).
553. MORRISON, W.R. and HAY, J.D., *Biochim. Biophys. Acta*, **202**, 460–467 (1970).
554. MORRISON, W.R., TAN, S.L. and HARGIN, K.D., *J. Sci. Fd Agric.*, **31**, 329–340 (1980).
555. MOSCHIDIS, M.C., *Prog. Lipid Res.*, **23**, 223–246 (1984).
556. MOUREY, T.H. and OPPENHEIMER, L.E., *Anal. Chem.*, **56**, 2427–2434 (1984).
557. MUELLER, H.W., O'FLAHERTY, F.T. and WYKLE, R.L., *J. Biol. Chem.*, **259**, 14554–14559 (1984).
558. MURATA, T., *Anal. Chem.*, **49**, 2209–2213 (1977).
559. MURATA, T. and TAKAHASHI, S., *Anal. Chem.*, **49**, 728–731 (1977).
560. MURRAY, R.K., LEVINE, M. and KORNBLATT, M.J., in *Glycolipid Methodology*, pp. 305–327 (1976) (edited by L.A. Witting, American Oil Chemists' Soc., Champaign, Ill.).
561. MUUSE, B.G. and VAN DER KAMP, H.J., *Neth. Milk Dairy J.*, **39**, 1–13 (1985).
562. MYHER, J.J. and KUKSIS, A., *Can. J. Biochem.*, **60**, 638–650 (1982).
563. MYHER, J.J. and KUKSIS, A., *Can. J. Biochem. Cell Biol.*, **62**, 352–362 (1984).
564. MYHER, J.J. and KUKSIS, A., *J. Biochem. Biophys. Methods*, **10**, 13–23 (1984).
565. MYHER, J.J., KUKSIS, A., MARAI, L. and MANGANARO, F., *J. Chromatogr.*, **283**, 289–301 (1984).
566. NAITO, H.K. and DAVID, J.A., in *Lipid Research Methodology*, pp. 1–76 (1984) (edited by J.A. Story, A.R. Liss Inc., New York).
567. NAKABAYASHI, H., IWAMORI, M. and NAGAI, Y., *J. Biochem. (Tokyo)*, **96**, 977–984 (1984).
568. NAKAGAWA, Y., FUJISHIMA, K. and WAKU, K., *Anal. Biochem.*, **157**, 172–178 (1986).
569. NAKAGAWA, Y. and HORROCKS, L.A., *J. Lipid Res.*, **24**, 1268–1275 (1983).
570. NAKAGAWA, Y. and HORROCKS, L.A., *J. Lipid Res.*, **27**, 629–636 (1986).
571. NAKAGAWA, Y., SUGIURA, T. and WAKU, K., *Biochim. Biophys. Acta*, **833**, 323–329 (1985).
572. NAKAGAWA, Y. and WAKU, K., *Lipids*, **20**, 482–487 (1985).
573. NAKAGAWA, Y. and WAKU, K., *J. Chromatogr.*, **381**, 225–231 (1986).
574. NAKAGAWA, Y., WAKU, K., HIROSE, A., KAWASHIMA, Y. and KOSUKA, H., *Lipids*, **21**, 634–638 (1986).
575. NASNER, A. and KRAUS, L., *Fette Seifen Anstrichm.*, **83**, 70–73 (1981).
576. NASNER, A. and KRAUS, L., *J. Chromatogr.*, **216**, 389–394 (1981).
577. NEFF, W.E. and FRANKEL, E.N., *Lipids*, **15**, 587–593 (1980).
578. NES, W.R. and NES, W.D., *Lipids in Evolution* (1980) (Plenum Press, New York).
579. NETTING, A.G., *J. Am. Oil Chem. Soc.*, **63**, 1197–1199 (1986).
580. NETTING, A.G. and DUFFIELD, A.M., *J. Chromatogr.*, **257**, 174–179 (1983).
581. NETTING, A.G. and DUFFIELD, A.M., *J. Chromatogr.*, **336**, 115–123 (1984).
582. NEWKIRK, D.R. and SHEPPARD, A.J., *J. Assoc. Off. Anal. Chem.*, **64**, 54–57 (1981).
583. NIBBERING, N.M.M., *J. Chromatogr.*, **251**, 93–104 (1982).
584. NICHOLAS, A.W., KHOURI, L.G., ELLINGTON, J.C. and PORTER, N.A., *Lipids*, **18**, 434–438 (1983).
585. NICHOLS, B.W., *Biochim. Biophys. Acta*, **70**, 417–422 (1963).

586. Nichols, B.W., in *New Biochemical Separations*, pp. 321–337 (1964) (edited by A.T. James and L.J. Morris, Van Nostrand, New York).
587. Nichols, B.W. and James, A.T., *Fette Seifen Anstrichm.*, **66**, 1003–1006 (1964).
588. Nimura, N. and Kinoshita, T., *Anal. Lett.*, **13**, 191–202 (1980).
589. Nishimura, K., Hirabayashi, Y., Hamaoka, A., Matsumoto, M., Nakamura, A. and Miseki, K., *Biochim. Biophys. Acta*, **796**, 269–276 (1984).
590. Nishimura, K. and Nakamura, A., *J. Biochem. (Tokyo)*, **98**, 1247–1254 (1985).
591. Nissen, H.P. and Kreysel, H.W., *J. Chromatogr.*, **276**, 29–35 (1983).
592. Nissen, H.P., Topfer-Petersen, E., Schill, W.B. and Kreysel, H.W., *Fette Seifen Anstrichm.*, **85**, 590–595 (1983).
593. Nonaka, G. and Kishimoto, Y., *Biochim. Biophys. Acta*, **572**, 423–431 (1979).
594. Norman, H.A. and Thompson, G.A., *Archs Biochem. Biophys.*, **242**, 168–175 (1985).
595. Nudelman, E., Levery, S.B., Kaizu, T. and Hakomori, S-I., *J. Biol. Chem.*, **261**, 11247–11253 (1986).
596. O'Brien, J.S. and Benson, A.A., *J. Lipid Res.*, **5**, 432–436 (1964).
597. Ogren, L., Csiky, I., Risinger, L., Nilsson, L.G. and Johansson, G., *Anal. Chim. Acta*, **117**, 71–79 (1980).
598. Ohno, Y., Okazaki, M. and Hara, I., *J. Biochem. (Tokyo)*, **89**, 1675–1680 (1981).
599. Okada, Y., Kannagi, R., Levery, S., and Hakomori, S.-I., *Immunol.*, **133**, 835–842, (1984).
600. Okazaki, M., Hagiwara, N. and Hara, I., *J. Chromatogr.*, **231**, 13–23 (1982).
601. Okazaki, M., Hagiwara, N. and Hara, I., *J. Biochem. (Tokyo)*, **91**, 1381–1389 (1982).
602. Okazaki, M., Hagiwara, N. and Hara, I., *J. Biochem. (Tokyo)*, **92**, 517–524 (1982).
603. Okazaki, M. and Hara, I., *J. Biochem. (Tokyo)*, **88**, 1215–1218 (1980).
604. Okazaki, M., Hara, I., Tanaka, A., Kodama, T. and Yokoyama, S., *New Eng. J. Med.*, **304**, 1608 (1981).
605. Okazaki, M., Itakura, H., Shiraishi, K. and Hara, I., *Clin. Chem.*, **29**, 768–773 (1983).
606. Okazaki, M., Kinoshita, M., Naito, C. and Hara, I., *J. Chromatogr.*, **336**, 151–159 (1984).
607. Okazaki, M., Ohno, Y. and Hara, I., *J. Chromatogr.*, **221**, 257–264 (1980).
608. Okazaki, M., Ohno, Y. and Hara, I., *J. Biochem. (Tokyo)*, **89**, 879–887 (1981).
609. Okazaki, M., Shiraishi, K., Ohno, Y. and Hara, I., *J. Chromatogr.*, **223**, 285–293 (1981).
610. Okazaki, M., Takizawa, A. and Hara, I., *Yukagaku*, **32**, 423–428 (1983).
611. Okuyama, E. and Yamazaki, M., *Chem. Pharm. Bull.*, **33**, 2529–2530 (1985).
612. Oppenheimer, L.E. and Mourey, T.H., *J. Chromatogr.*, **323**, 297–304 (1985).
613. Orr, G.A., Brewer, C.F. and Heney, G., *Biochemistry*, **21**, 3202–3206 (1982).
614. Ott, G.S. and Shore, V.G., *J. Chromatogr.*, **231**, 1–12 (1982).
615. Ottenstein, D.M., Witting., L.A., Silvis, P.H., Hometchko, D.J. and Pelick, N., *J. Am. Oil Chem. Soc.*, **61**, 390–394 (1984).
616. Ozcimder, M. and Hammers, W.E., *J. Chromatogr.*, **187**, 307–317 (1980).
617. Palmer, D.N., Anderson, M.A. and Jolly, R.D., *Anal. Biochem.*, **140**, 315–319 (1984).
618. Park, D.K., Terao, J. and Matsushita, S., *Agric. Biol. Chem.*, **45**, 2443–2448 (1981).
619. Parris, N.A., *J. Chromatogr.*, **149**, 615–624 (1978).
620. Parris, N.A., *J. Chromatogr.*, **157**, 161–170 (1978).
621. Parris, N.A., *Instrumental Liquid Chromatography* (2nd edition) (1984) (Elsevier, Amsterdam).
622. Parrish, C.C. and Ackman, R.G., *J. Chromatogr.*, **262**, 103–112 (1983).
623. Parrish, C.C. and Ackman, R.G., *Lipids*, **20**, 521–530 (1985).
624. Passi, S., Nazzarro-Porro, M., Picardo, M., Mingrone, G. and Fasella, P., *J. Lipid Res.*, **24**, 1140–1147 (1983).
625. Passi, S., Rothschild-Boros, M.C., Fasella, P., Nazzarro-Porro, M. and Whitehouse, D., *J. Lipid Res.*, **22**, 778–784 (1981).
626. Patel, K.M. and Sparrow, J.T., *J. Chromatogr.*, **150**, 542–547 (1978).
627. Paton, R.D., McGillivray, A.I., Speir, T.F., Whittle, M.J., Whitfield, C.R. and Logan, R.W., *Clin. Chim. Acta*, **133**, 97–110 (1983).
628. Pattee, H.E. and Singleton, J.A., *J. Am. Oil Chem. Soc.*, **54**, 183–185 (1977).
629. Patton, G.M., Clark, S.B., Fasulo, J.M. and Robins, S.J., *J. Clin. Invest.*, **73**, 231–240 (1984).
630. Patton, G.M., Fasulo, J.M. and Robins, S.J., *J. Lipid Res.*, **23**, 190–196 (1982).

631. PATTON, G.M., ROBINS, S.J., FASULO, J.M. and CLARK, S.B., *J. Lipid Res.*, **26**, 1285–1293 (1985).
632. PATTON, S. and JENSEN, R.G., *Prog. Chem. Fats Other Lipids*, **14**, 163–277 (1975).
633. PAULS, R.E., *J. Am. Oil Chem. Soc.*, **60**, 819–822 (1983).
634. PAYNE-WAHL, K. and KLEIMAN, R., *J. Am. Oil Chem. Soc.*, **60**, 1011–1012 (1983).
635. PAYNE-WAHL, K. PLATTNER, R.D., SPENCER, G.F. and KLEIMAN, R., *Lipids*, **14**, 601–605 (1979).
636. PAYNE-WAHL, K., Spencer, G.F., PLATTNER, R.D. and BUTTERFIELD, R.O., *J. Chromatogr.*, **209**, 61–66 (1981).
637. PEI, P.T-S., HENLY, R.S. and RAMACHANDRAN, S., *Lipids*, **10**, 152–156 (1975).
638. PEI, P.T-S., KOSSA, W.C., RAMACHANDRAN, S. and HENLY, R.S., *Lipids*, **11**, 814–816 (1976).
639. PERKINS, E.G., BAUER, J.E., PELICK, N. and EL-HAMDY, A., in *Dietary Fats and Health*, pp. 184–208 (1981) (edited by E.G. Perkins and W.J. Wisek, American Oil Chemists' Soc., Champaign).
640. PERKINS, E.G., HENDREN, D.J., BAUER, J.E. and EL-HAMDY, A.H., *Lipids*, **16**, 609–613 (1981).
641. PERKINS, E.G., HENDREN, D.J., PELICK, N. and BAUER, J.E., *Lipids*, **17**, 460–463 (1982).
642. PERRIN, J-L. and NAUDET, M., *Rev. Franc. Corps Gras*, **30**, 279–285 (1983).
643. PERRIN, J-L. and NAUDET, M., *Rev. Franc. Corps Gras*, **32**, 301–303 (1985).
644. PERRIN, J-L., PREVOT, A., STOLYHWO, A. and GUIOCHON, G., *Rev. Franc. Corps Gras*, **31**, 495–510 (1984).
645. PERRIN, J-L., REDERO, F. and PREVOT, A., *Rev. Franc. Corps Gras*, **31**, 131–133 (1984).
646. PETERSSEN, B., PODLAHA, O. and TOREGARD, B., *J. Am. Oil Chem. Soc.*, **58**, 1005–1009 (1981).
647. PFAFFINGER, D., EDELSTEIN, C. and SCANU, A.M., *J. Lipid Res.*, **24**, 796–800 (1983).
648. PHILLIPS, F.C., ERDAHL, W.L., NADENICEK, J.D., NUTTER, L.J., SCHMIT, J.A. and PRIVETT, O.S., *Lipids*, **19**, 142–150 (1984).
649. PHILLIPS, F.C., ERDAHL, W.L. and PRIVETT, O.S., *Lipids*, **17**, 992–997 (1982).
650. PHILLIPS, F.C., ERDAHL, W.L., SCHMIT, J.A. and PRIVETT, O.S., *Lipids*, **19**, 880–887 (1984).
651. PHILLIPS, F.C. and PRIVETT, O.S, *Lipids*, **14**, 590–595 (1979).
652. PHILLIPS, F.C. and PRIVETT, O.S., *Lipids*, **14**, 949–952 (1979).
653. PHILLIPS, F.C. and PRIVETT, O.S, *J. Am. Oil Chem. Soc.*, **58**, 590–594 (1981).
654. PIND, S., KUKSIS, A., MYHER, J.J. and MARAI, L., *Can. J. Biochem. Cell Biol.*, **62**, 301–309 (1984).
655. PIND, S., KUKSIS, A., MYHER, J.J. and MARAI, L., *Can. J. Biochem. Cell Biol.*, **63**, 137–144 (1985).
656. PIRKLE, W.H., HYUN, M.H. and BANK, B., *J. Chromatogr.*, **316**, 585–604 (1984).
657. PISON, U., GONO, E., JOKA, T., OBERTACKE, U. and OBLADEN, M., *J. Chromatogr.*, **377**, 79–89 (1986).
658. PLATTNER, R.D., *J. Am. Oil Chem. Soc.*, **54**, 511–515 (1977).
659. PLATTNER, R.D., *Methods in Enzymology*, **72**, 21–34 (1981).
660. PLATTNER, R.D., *J. Am. Oil Chem. Soc.*, **58**, 638–642 (1981).
661. PLATTNER, R.D. and PAYNE-WAHL, K., *Lipids*, **14**, 152–153 (1979).
662. PLATTNER, R.D., WADE, K. and KLEIMAN, R., *J. Am. Oil Chem. Soc.*, **55**, 381–382 (1978).
663. PODLAHA, O. and TOREGARD, B., *J. High Res. Chromatogr., Chromatogr. Commun.*, **5**, 553–558 (1982).
664. PODLAHA, O. and TOREGARD, B., *Fette Seifen Anstrichm.*, **86**, 243–245 (1984).
665. POHL, P., GLASL, H. and WAGNER, H., *J. Chromatogr.*, **49**, 488–492 (1970).
666. POLACEK, D., EDELSTEIN, C. and SCANU, A.M., *Lipids*, **16**, 927–929 (1981).
667. POLITZER, I.R., GRIFFIN, G.W., DOWTY, B.J. and LASETER, J.L., *Anal. Lett.*, **6**, 539–546 (1973).
668. POLL, D., HARDING, D.R.K., HANCOCK, W.S., NYE, E.R., JANUS, E.D., HANNAN, S.F. and SCOTT, P.J., *J. Chromatogr.*, **343**, 149–154 (1985).
669. POLLET, S., ERMIDOU, S., LE SAUX, F., MONGE, M. and BAUMANN, N., *J. Lipid Res.*, **19**, 916–921 (1978).
670. POORTHUIS, B.J.H.M., YAZAKI, P.J. and HOSTETLER, K.Y., *J. Lipid Res.*, **17**, 433–437 (1976).
671. PORTER, N.A., LOGAN, J. and KONTOYIANNIDOU, V., *J. Org. Chem.*, **44**, 3177–3181 (1979).
672. PORTER, N.A., WEBER, B.A., WEENEN, H. and KHAN, J.A., *J. Am. Chem. Soc.*, **102**, 5597–5601 (1980).
673. PORTER, N.A. and WEENEN, H., *Methods in Enzymology*, **72**, 34–40 (1981).
674. PORTER, N.A., WOLF, R.A. and NIXON, J.R., *Lipids*, **14**, 20–24 (1979).

675. PORTER, N.A., WOLF, R.A. and WEENEN, H., *Lipids*, **15**, 163–167 (1980).
676. POWELL, W.S., *Anal. Biochem.*, **115**, 267–277 (1981).
677. PRESS, K., SHEELEY, R., HURST, W.J. and MARTIN, R.A., *J. Agric. Fd Chem.*, **29**, 1096–1098 (1981)
678. PRIVETT, O.S., *Prog. Chem. Fats Other Lipids*, **9**, 91–117 (1966).
679. PRIVETT, O.S., DOUGHERTY, K.A., ERDAHL, W.L. and STOLYHWO, A., *J. Am. Oil Chem. Soc.*, **50**, 516–520 (1973).
680. PRIVETT, O.S. and ERDAHL, W.L., *Anal. Biochem.*, **84**, 449–461 (1978).
681. PRIVETT, O.S. and ERDAHL, W.L., *Methods in Enzymology*, **72**, 56–108 (1981).
682. QURESHI, N., MASCAGNI, P., RIBI, E. and TAKAYAMA, K., *J. Biol. Chem.*, **260**, 5271–5278 (1985).
683. QURESHI, N., TAKAYAMA, K. and SCHNOES, H.K., *J. Biol. Chem.*, **255**, 182–189 (1980).
684. RADIN, N.S., *Methods in Enzymology*, **72**, 5–7 (1981).
685. RAINEY, M.L. and PURDY, W.C., *Anal. Chim. Acta*, **93**, 211–219 (1977).
686. RAKOFF, H. and EMKEN, E.A., *J. Am. Oil Chem. Soc.*, **55**, 564–566 (1978).
687. REED, A.W., DEETH, H.C. and CLEGG, D.E., *J. Assoc. Off. Anal. Chem.*, **67**, 718–721 (1984).
688. REEVE, D.R. and CROZIER, R., *Laboratory Practice*, **32** (March), 59–60 (1983).
689. RENKONEN, O., *J. Am. Oil Chem. Soc.*, **42**, 298–304 (1965).
690. REZANKA, T. and PODOJIL, M., *J. Chromatogr.*, **346**, 453–455 (1985).
691. RHEE, J.S. and SHIN, M.G., *J. Am. Oil Chem. Soc.*, **59**, 98–99 (1982).
692. RIEDMANN, M. and TEVINI, M., *Hewlett Packard Application Note 232/13* (1980).
693. RITCHIE, A.S., and JEE, M.H., *J. Chromatogr.*, **329**, 273–280 (1985).
694. RIVNAY, B., *J. Chromatogr.*, **294**, 303–315 (1984).
695. RIZZO, A.F. and KORKELA, H., *Biochim. Biophys. Acta*, **792**, 367–370 (1984).
696. ROBINS, S.J. and PATTON, G.M., *J. Lipid Res.*, **27**, 131–139 (1986).
697. ROBINSON, J.L. and MACRAE, R., *J. Chromatogr.*, **303**, 386–390 (1984).
698. ROBINSON, J.L., TSIMIDOU, M. and MACRAE, R., *J. Chromatogr.*, **324**, 35–52 (1985).
699. ROBINSON, M., BLANK, M.L. and SNYDER, F., *J. Biol. Chem.*, **260**, 7889–7895 (1985).
700. ROBINSON, M., BLANK, M.L. and SNYDER, F., *Archs Biochem. Biophys.*, **250**, 271–279 (1986)
701. ROBINSON, P.G., *J. Lipid Res.*, **23**, 1251–1253 (1982).
702. RODRIGUEZ DE TURCO, E.B. and BAZAN, N.G., *J. Chromatogr.*, **137**, 194–197 (1977).
703. ROGERS, V.A., VAN ALLER, R.T., PESSONEY, G.F., WATKINS, E.J. and LEGGETT, H.G., *Lipids*, **19**, 304–306 (1984).
704. ROGGERO, J.P. and COEN, S.V., *J. Liq. Chromatogr.*, **4**, 1817–1829 (1981).
705. RONAN, R., KAY, L.L., MENG, M.S. and BREWER, H.B., *Biochim. Biophys. Acta*, **713**, 657–662 (1983).
706. ROSENFELDER, G., CHANG, J-Y. and BRAUN, D.G., *J. Chromatogr.*, **272**, 21–27 (1983).
707. ROSS, A.C., *Anal. Biochem.*, **115**, 324–330 (1981).
708. ROUGHAN, P.G. and BATT, R.D., *Phytochemistry*, **8**, 363–369 (1969).
709. ROUSER, G., BAUMAN, A.J., KRITCHEVSKY, G., HELLER, D. and O'BRIEN, J., *J. Am. Oil Chem. Soc.*, **38**, 544–555 (1961).
710. ROUSER, G., KRITCHEVSKY, G., SIMON, G. and NELSON, G.J., *Lipids*, **2**, 37–40 (1967).
711. ROUSER, G., KRITCHEVSKY, G. and YAMAMOTO, A., in *Lipid Chromatographic Analysis*, Vol. 1, pp. 99–162 (1967) (edited by G.V. Marinetti, Edward Arnold Ltd, London).
712. RUNSER, D.J., *Maintaining and Trouble Shooting HPLC Systems* (1981) (John Wiley & Sons, New York).
713. RUSTOW, B., KUNZE, D., RABE, H. and REICHMANN, G., *Biochim. Biophys. Acta*, **835**, 465–476 (1985).
714. RYAN, P.J. and HONEYMAN, T.W., *J. Chromatogr.*, **312**, 461–466 (1984).
715. RYAN, P.J. and HONEYMAN, T.W., *J. Chromatogr.*, **331**, 177–182 (1985).
716. RYAN, P.J., MCGOLDRICK, K., STICKNEY, D. and HONEYMAN, T.W., *J. Chromatogr.*, **320**, 421–425 (1985).
717. RYU, E.K. and MACCOSS, M., *J. Lipid Res.*, **20**, 561–563 (1979).
718. SAITO, T. and HAKOMORI, S-I., *J. Lipid Res.*, **12**, 257–259 (1971).
719. SALARI, H., *J. Chromatogr.*, **382**, 89–98 (1986).
720. SANDHU, R.S., *Clin. Chem.*, **22**, 1973–1975 (1976).
721. SASAKI, T. and HASEGAWA-SASAKI, H., *Biochim. Biophys. Acta*, **833**, 316–322 (1985).

722. SASTRY, B.V.R., STATHAM, C.N., AXELROD, J. and HIRATA, F., *Archs Biochem. Biophys.*, **211**, 762–773 (1981).
723. SASTRY, B.V.R., STATHAM, C.N., MEEKS, R.G. and AXELROD, J., *Pharmacology*, **23**, 211–222 (1981).
724. SASTRY, P.S., *Adv. Lipid Res.*, **12**, 251–310 (1974).
725. SAUNDERS, R.D. and HORROCKS, L.A., *Anal. Biochem.*, **143**, 71–75 (1984).
726. SAX, S.M., MOORE, J.J., OLEY, A., AMENTA, J.S. and SILVERMAN, J.A., *Clin. Chem.*, **28**, 2264–2268 (1982).
727. SCHACHT, J., *J. Lipid Res.*, **19**, 1063–1067 (1978).
728. SCHLAGER, S.I. and JORDI, H., *Biochim. Biophys. Acta*, **665**, 355–358 (1981).
729. SCHLAME, M., RUSTOW, B., KUNZE, D., RABE, H. and REICHMANN, G. *Biochem. J.*, **240**, 247–252 (1986).
730. SCHMITZ, B. and KLEIN, R.A., *Chem. Phys. Lipids*, **39**, 285–311 (1986).
731. SCHOLFIELD, C.R., *J. Am. Oil Chem. Soc.*, **52**, 36–37 (1975).
732. SCHOLFIELD, C.R., *Anal. Chem.*, **47**, 1417–1420 (1975).
733. SCHOLFIELD, C.R., *J. Am. Oil Chem Soc.*, **56**, 510–511 (1979).
734. SCHOLFIELD, C.R., in *Geometrical and Positional Fatty Acid Isomers*, pp. 17–52 (1979) (edited by E.A. Emken and H.J. Dutton, AOCS, Champaign, Ill.).
735. SCHOLFIELD, C.R. and MOUNTS, T.L., *J. Am. Oil Chem. Soc.*, **54**, 319–321 (1977).
736. SCHOMBURG, G. and ZEGARSKI, K., *J. Chromatogr.*, **114**, 174–178 (1975).
737. SCHULTE, E., *Fette Seifen Anstrichm.*, **83**, 289–291 (1981).
738. SCHULTE, E., *Fette Seifen Anstrichm.*, **84**, 178–180 (1982).
739. SCHULTE, E., HOHN, M. and RAPP, U., *Frezenius' Z. Anal. Chem.*, **307**, 115–119 (1981).
740. SCHWANDT, P., RICHTER, W.O., HEINEMANN, V. and WEISWEILER, P., *J. Chromatogr.*, **345**, 145–149 (1985).
741. SCHWANDT, P., RICHTER, W.O. and WEISWEILER, P., *J. Chromatogr.*, **225**, 185–188 (1981).
742. SCHWARTING, G.A., *Biochem. J.*, **189**, 407–412 (1980).
743. SCHWARZENBACH, R., *J. Chromatogr.*, **202**, 397–404 (1980).
744. SCOREPA, J., *Molecular Species of Triglycerides in Biological Systems* (1975) (Universita Karlova, Prague).
745. SCOTT, R.P.W., *J. Chromatogr.*, **122**, 35–53 (1976).
746. SCOTT, R.P.W. (editor), *Chemical Analysis. Vol. 72. Small Bore Liquid Chromatography Columns: Their Properties and Uses* (1984) (J. Wiley & Sons, New York).
747. SCOTT, R.P.W. and KUCERA, P., *J. Chromatogr.*, **112**, 425–442 (1975).
748. SCOTT, R.P.W. and LAWRENCE, J.G., *J. Chromatogr. Sci.*, **8**, 65–71 (1970).
749. SEBEDIO, J-L., *Fette Seifen Anstrichm.*, **87**, 267–273 (1985).
750. SEBEDIO, J-L., FARQUHARSON, T.E. and ACKMAN, R.G., *Lipids*, **17**, 469–475 (1982).
751. SEGREST, J.P. and ALBERS, J.J. (editors), *Methods in Enzymology. Vol. 128. Plasma Lipoproteins. Part A* (Academic Press, Orlando) (1986).
752. SEMPORE, G. and BEZARD, J., *J. Chromatogr.*, **366**, 261–282 (1986).
753. SHACKLETON, C.H.L., *J. Chromatogr.*, **379**, 91–156 (1986).
754. SHAND, J.H. and NOBLE, R.C., *Anal. Biochem.*, **101**, 427–434 (1980).
755. SHAW, N., *Biochim. Biophys. Acta*, **164**, 435–436 (1968).
756. SHAW, N., *Bacteriol. Rev.*, **34**, 365–377 (1970).
757. SHEN, C-S.J. and SHEPPARD, A.J., *Chromatographia*, **17**, 469–471 (1983).
758. SHEPPARD, A.J., IVERSON, J.L. and WEIHRAUCH, J.L., in *Handbook of Lipid Research. Vol. 1. Fatty acids and Glycerides*, pp. 341–379 (1978) (edited by A. Kuksis, Plenum Press, New York).
759. SHIMOMURA, K. and KISHIMOTO, Y., *Biochim. Biophys. Acta*, **754**, 93–100 (1983).
760. SHIMOMURA, K. and KISHIMOTO, Y., *Biochim. Biophys. Acta*, **794**, 162–164 (1984).
761. SHIMOMURA, K., YAHARA, S., KISHIMOTO, Y. and BENJAMINS, J.A., *Biochim. Biophys. Acta*, **795**, 265–270 (1984).
762. SHIMOMURA, Y., TANIGUCHI, K., SUGIE, T., MURAKAMI, M., SUGIYAMA, S. and OZAWA, T., *Clin. Chim. Acta*, **143**, 361–366 (1984).
763. SHUKLA, V.K.S., NIELSEN, W.S. and BATSBERG, W., *Fette Seifen Anstrichm.*, **85**, 274–278 (1983).
764. SHUKLA, V.K.S. and SPENER, F., *J. Chromatogr.*, **348**, 441–446 (1985).
765. SIAKOTOS, A.N. and ROUSER, G., *J. Am. Oil Chem. Soc.*, **42**, 913–919 (1965).

766. SINGH, H. and PRIVETT, O.S., *Lipids*, **5**, 692–697 (1970).
767. SINGLETON, J.A. and PATTEE, H.W., *J. Am. Oil Chem. Soc.*, **61**, 761–766 (1984).
768. SINGLETON, W.S., GRAY, M.S., BROWN, M.L. and WHITE, J.L., *J. Am. Oil Chem. Soc.*, **42**, 53–56 (1965).
769. SKIPSKI, V.P. and BARCLAY, M., *Methods in Enzymology.*, **14**, 530–598 (1969).
770. SKIPSKI, V.P., BARCLAY, M., REICHMAN, E.S. and GOOD, J.J., *Biochim. Biophys. Acta*, **137**, 80–89 (1967).
771. SKIPSKI, V.P., PETERSON, R.F. and BARCLAY, M., *Biochem. J.*, **90**, 374–378 (1964).
772. SLOMIANY, A. and SLOMIANY, B.L., *J. Biochem. Biophys. Methods*, **5**, 229–236 (1981).
773. SMAAL, E.B., ROMIJN, D., GEURTS VAN KESSEL, W.S.M., DE KRUIJFF, B. and DE GIER, J., *J. Lipid Res.*, **26**, 634–637 (1985).
774. SMITH, C.R., *Prog. Chem. Fats Other Lipids*, **11**, 137–177 (1970).
775. SMITH, C.R., in *Topics in Lipid Chemistry*, Vol. 3, pp. 89–124 (1972) (edited by F.D. Gunstone, Logos Press, London).
776. SMITH, E.C., JONES, A.D. and HAMMOND, E.W., *J. Chromatogr.*, **188**, 205–212 (1980).
777. SMITH, L.A., NORMAN, H.A., CHO, S.H. and THOMPSON, G.A., *J. Chromatogr.*, **346**, 292–299 (1985).
778. SMITH, M. and JUNGALWALA, F.B., *J. Lipid Res.*, **22**, 697–704 (1981).
779. SMITH, M., MONCHAMP, P. and JUNGALWALA, F.B., *J. Lipid Res.*, **22**, 714–719 (1981).
780. SMITH, S.L., JORGENSON, J.W. and NOVOTNY, M., *J. Chromatogr.*, **187**, 111–118 (1980).
781. SMITH, S.L., NOVOTNY, M., MOORE, S.A. and FELTEN, D.L., *J. Chromatogr.*, **221**, 19–26 (1980).
782. SNYDER, L.R., *J. Chromatogr. Sci.*, **16**, 223–234 (1978).
783. SNYDER, L.R. and KIRKLAND, J.J., *Introduction to Modern Liquid Chromatography* (2nd edition) (1979) (John Wiley & Sons, New York).
784. SONNINO, S., GHIDONI, R., GAZZOTTI, G., KIRSCHNER, G., GALLI, G. and TETTAMANTI, G., *J. Lipid Res.*, **25**, 620–629 (1984).
785. SONNINO, S., KIRSCHNER, G., GHIDONI, R., ACQUOTTI, D. and TETTAMANTI, G., *J. Lipid Res.*, **26**, 248–257 (1985).
786. SOTIRHOS, N., HO, C-T. and CHANG, S.S., *Fette Seifen Anstrichm.*, **88**, 6–8 (1986).
787. SOTIRHOS, N., THORNGREN, C. and HERSLOF, B., *J. Chromatogr.*, **331**, 313–320 (1985).
788. SPARK, A.A. and ZIERVOGEL, M., *J. High Res. Chromatogr., Chromatogr. Commun.*, **5**, 206–207 (1982).
789. SPENCER, G.F., PLATTNER, R.D. and MIWA, T., *J. Am. Oil Chem. Soc.*, **54**, 187–189 (1977).
790. STEWART, M.E. and DOWNING, D.T., *Lipids*, **16**, 355–359 (1981).
791. STILLWAY, L.W. and HARMON, S.J., *J. Lipid Res.*, **21**, 1141–1143 (1980).
792. STODOLA, F.H., DEINEMA, M.H. and SPENCER, J.F.T., *Bacteriol. Rev.*, **31**, 194–213 (1967).
793. STOLYHWO, A., COLIN, H. and GUIOCHON, G., *J. Chromatogr.*, **265**, 1–18 (1983).
794. STOLYHWO, A., COLIN, H. and GUIOCHON, G., *Anal. Chem.*, **57**, 1342–1354 (1985).
795. STOLYHWO, A., COLIN, H., MARTIN, M. and GUIOCHON, G., *J. Chromatogr.*, **288**, 253–275 (1984).
796. STOLYHWO, A. and PRIVETT, O.S., *J. Chromatogr. Sci.*, **11**, 20–25 (1973).
797. STOLYHWO, A., PRIVETT, O.S. and ERDAHL, W.L. *J. Chromatogr. Sci.*, **11**, 263–267 (1973).
798. STOWE, H.D., *J. Animal Sci.*, **54**, 76–81 (1982).
799. STRASBERG, P.M., WARREN, I., SKOMOROWSKI, M.A. and LOWDEN, J.A., *Clin. Chim. Acta*, **132**, 29–41 (1983).
800. SUBBAIAH, P.V., CHEN, C-H., BAGDADE, J.D. and ALBERS, J.J., *J. Biol. Chem.*, **260**, 5308–5314 (1985).
801. SUDRAUD, G., COUSTARD, J.M., RETHO, C., CAUDE, M., ROSSET, R., HAGEMANN, R., GAUDIN, D. and VIRELIZIER, H., *J. Chromatogr.*, **204**, 397–406 (1981).
802. SUGITA, M., IWAMORI, M., EVANS, J., MCCLUER, R.H., DULANEY, J.T. and MOSER, H.W., *J. Lipid Res.*, **15**, 223–226 (1974).
803. SULPICE, J.C. and FEREZOU, J., *Lipids*, **19**, 631–635 (1984).
804. SUZUKI, A., HANDA, S., ISHIZUKA, I. and YAMAKAWA, T., *J. Biochem. (Tokyo)*, **81**, 127–134 (1977).
805. SUZUKI, A., HANDA, S. and YAMAKAWA, T., *J. Biochem. (Tokyo)*, **80**, 1181–1183 (1976).
806. SUZUKI, A., HANDA, S. and YAMAKAWA, T., *J. Biochem. (Tokyo)*, **82**, 1185–1187 (1977).
807. SUZUKI, A., KUNDU, S.K. and MARCUS, D.M., *J. Lipid Res.*, **21**, 473–477 (1980).
808. SVENNERHOLM, L., *J. Neurochem.*, **1**, 42–53 (1956).

809. SVENNERHOLM, L. and FREDMAN, P., *Biochim. Biophys. Acta*, **617**, 97–109 (1980).
810. SVENSSON, L., *Lipids*, **18**, 171–178 (1983).
811. SVENSSON, L., SISFONTES, L., NYBORG, G. and BLOMSTRAND, R., *Lipids*, **17**, 50–59 (1982).
812. SWEELEY, C.C. and SIDDIQUI, B., in *The Glycoconjugates*, Vol. 1, pp. 459–540 (1977) (edited by M. Horowitz and W. Pigman, Academic Press, New York).
813. TAKAGI, T. and ITABASHI, Y., *Yukagaku*, **34**, 962–963 (1985).
814. TAKAGI, T. and ITABASHI, Y., *J. Chromatogr.*, **366**, 451–455 (1986).
815. TAKAHASHI, K., HIRANO, T., EGI, M. and ZAMA, K., *J. Am. Oil Chem. Soc.*, **62**, 1489–1491 (1985).
816. TAKAHASHI, K., HIRANO, T. and ZAMA, K., *J. Am. Oil Chem. Soc.*, **61**, 1226–1229 (1984).
817. TAKAMURA, H., NARITA, H., URADE, R. and KITO, M., *Lipids*, **21**, 356–361 (1986).
818. TAKAYAMA, K., JORDI, H.C. and BENSON, F., *J. Liq. Chromatogr.*, **3**, 61–69 (1980).
819. TAKAYAMA, K., QURESHI, N., HYVER, K., HONOVICH, J., COTTER, R.J., MASCAGNI, P. and SCHNEIDER, H., *J. Biol. Chem.*, **261**, 10624–10631 (1986).
820. TAVANI, D.M., NES, W.R. and BILLHEIMER, J.T., *J. Lipid Res.*, **23**, 774–781 (1982).
821. TENG, J.I. and SMITH, L.L., *J. Chromatogr.*, **322**, 240–245 (1985).
822. TENG, J.I. and SMITH, L.L., *J. Chromatogr.*, **339**, 35–44 (1985).
823. TENG, J.I. and SMITH, L.L., *J. Chromatogr.*, **350**, 445–451 (1985).
824. TERABAYASHI, T., OGAWA, T., KAWANISHI, Y., TANAKA, M., TAKASE, K. and ISHII, J., *J. Chromatogr.*, **367**, 280–285 (1986).
825. TERAMOTO, T., KATO, H., HASHIMOTO, Y., KINOSHITA, M., TODA, G. and OKA, H., *Clin. Chim. Acta*, **149**, 135–148 (1985).
826. TERAO, J., ASANO, I. and MATSUSHITA, S., *Archs Biochem. Biophys.*, **235**, 326–333 (1984).
827. TEVINI, M. and STEINMULLER, D., in *High Performance Liquid Chromatography in Biochemistry*, pp. 349–392 (1985) (edited by A. Henschen, K-P. Hupe, F. Lottspeich and W. Voelter, VCH, Weinheim).
828. TJADEN, U.R., KROL, J.H., VAN HOEVEN, R.P., OOMEN-MEULEMANS, E.P.M. and EMMELOT, P., *J. Chromatogr.*, **136**, 233–243 (1977).
829. TOKITA, M. and MORITA, M., *Agr. Biol. Chem.*, **49**, 3545–3550 (1985).
830. TOMONO, Y., NAITO, I. and WATANABI, K., *Biochim. Biophys. Acta*, **796**, 199–204 (1984).
831. TRACEY, B.M., CHALMERS, R.A., ROSANKIEWICZ, J.R., DE SOUSA, C. and STACEY, T.E., *Biochem. Soc. Trans.*, **14**, 700–701 (1986).
832. TRAYLOR, T.D., KOONTZ, D.A. and HOGAN, E.L., *J. Chromatogr.*, **272**, 9–20 (1983).
833. TSIMIDOU, M. and MACRAE, R., *J. Chromatogr.*, **285**, 178–181 (1984).
834. TSIMIDOU, M. and MACRAE, R., *J. Chromatogr. Sci.*, **23**, 155–160 (1985).
835. TSUCHIYA, H., HAYASHI, T., SATO, M., TATSUMI, M. and TAKAGI, N., *J. Chromatogr.*, **309**, 43–52 (1984).
836. TUCHMAN, M. and KRIVIT, W., *J. Chromatogr.*, **307**, 172–179 (1984).
837. TWEETEN, T.N. and WETZEL, D.L., *Cereal Chem.*, **56**, 398–402 (1979).
838. TWEETEN, T.N., WETZEL, D.L. and CHUNG, O.K., *J. Am. Oil Chem. Soc.*, **58**, 664–672 (1981).
839. TYMAN, J.H.P., TYCHOPOULOS, V. and COLENUTT, B.A., *J. Chromatogr.*, **213**, 287–300 (1981).
840. ULLMAN, M.D. and MCCLUER, R.H., *J. Lipid Res.*, **18**, 371–378 (1977).
841. ULLMAN, M.D. and MCCLUER, R.H., *J. Lipid Res.*, **19**, 910–913 (1978).
842. ULLMAN, M.D. and MCCLUER, R.H., *J. Lipid Res.*, **26**, 501–506 (1985).
843. URSINI, F., BONALDO, L., MAIORINO, M. and GREGOLIN, C., *J. Chromatogr.*, **270**, 301–308 (1983).
844. VANCE, D.E. and VANCE, J.E. (Editors), *Biochemistry of Lipids and Membranes* (1985) (Benjamin/Cummings, Menlo Park).
845. VANDAMME, D., BLATON, V. and PEETERS, H., *J. Chromatogr.*, **145**, 151–154 (1978).
846. VANROLLINS, M. and MURPHY, R.C., *J. Lipid Res.*, **25**, 507–517 (1984).
847. VEAZEY, R.L., *J. Am. Oil Chem. Soc.*, **63**, 1043–1046 (1986).
848. VERCAEMST, R., BURY, J. and ROSSENEU, M., *J. Lipid Res.*, **25**, 876 (1984).
849. VERCAEMST, R., ROSSENEU, M. and VAN BIERVLIET, J.P., *J. Chromatogr.*, **276**, 174–181 (1983).
850. VERHAGEN, J., VELDINK, J.A., EGMOND, M.R., VLIEGENTHART, J.F.G., BOLDINGH, J. and VAN DER STAR, J., *Biochim. Biophys. Acta*, **529**, 369–379 (1978).
851. VERZELE, M., USE, L. and VAN KERREBROECK, M., *J. Chromatogr.*, **289**, 333–337 (1984).
852. VESTAL, M.L., *Science*, **226**, 275–281 (1984).
853. VIOQUE, E., MAZA, M.P. and MILLAN, F., *J. Chromatogr.*, **331**, 187–192 (1985).

854. VISWANATH, C.V., HOEVET, S.P., LUNDBERG, W.O., WHITE, J.M. and MUCCINI, G.A., *J. Chromatogr.*, **40**, 225–234 (1969).
855. VOELTER, W., HUBER, R. and ZECH, K., *J. Chromatogr.*, **217**, 491–507 (1981).
856. VONACH, B. and SCHOMBURG, G., *J. Chromatogr.*, **149**, 417–430 (1978).
857. VORBECK, M.L. and MARINETTI, G.V., *J. Lipid Res.*, **6**, 3–6 (1965).
858. WADA, S., KOIZUMI, C. and NONAKA, J., *Yukagaku*, **26**, 11–15 (1977).
859. WADA, S., KOIZUMI, C., TAKIGUCHI, A. and NONAKA, J., *Bull. Jap. Soc. Sci. Fish.*, **45**, 611–614 (1979).
860. WAGNER, H., HORHAMMER, L. and WOLFF, P., *Biochem. Z.*, **334**, 175–184 (1961).
861. WARDLOW, M.L., *J. Chromatogr.*, **342**, 380–384 (1985).
862. WARTHEN, J.D., *J. Am. Oil Chem. Soc.*, **52**, 151–153 (1975).
863. WARTHEN, J.D., *J. Chromatogr. Sci.*, **14**, 513–515 (1976).
864. WATANABE, K. and ARAO, Y., *J. Lipid Res.*, **22**, 1020–1024 (1981).
865. WATANABE, K. and TOMONO, Y., *Anal. Biochem.*, **139**, 367–372 (1984).
866. WAYS, P. and HANAHAN, D.J., *J. Lipid Res.*, **5**, 318–328 (1964).
867. WEHR, C.T., CUNICO, R.L., OTT, G.S. and SHORE, V.G., *Anal. Biochem.*, **125**, 386–394 (1982).
868. WEISWEILER, P., FRIEDL, C. and SCHWANDT, P., *Biochim. Biophys. Acta*, **875**, 48–51 (1986).
869. WEISWEILER, P., FRIEDL, C. and SCHWANDT, P., *Clin. Chem.*, **32**, 992–994 (1986).
870. WELLS, M.A. and DITTMER, J.C., *Biochemistry*, **2**, 1259–1263 (1963).
871. WELLS, M.A. and DITTMER, J.C., *Biochemistry*, **4**, 2459–2468 (1965).
872. WELLS, M.A. and DITTMER, J.C., *J. Chromatogr.*, **18**, 503–511 (1965).
873. WERTZ, P.W. and DOWNING, D.T., *J. Lipid Res.*, **24**, 759–765 (1983).
874. WHALEN, M.M., WILD, G.C., SPALL, W.D. and SEBRING, R.J., *Lipids*, **21**, 267–270 (1986).
875. WHITE, D.A., in *Form and Function of Phospholipids*, pp. 441–482 (1973) (edited by G.B. Ansell, J.N. Hawthorne and R.M.C. Dawson, Elsevier, Amsterdam).
876. WILLIAMS, M.A. and MCCLUER, R.H., *J. Neurochem.*, **35**, 266–269 (1980).
877. WILLIAMS, M.C., KELLEY, J.L. and KUSHWAHA, R.S., *J. Chromatogr.*, **308**, 101–109 (1984).
878. WILLIAMS, M.C., STENOIEN, C.G. and KUSHWAHA, R.S., *J. Chromatogr.*, **375**, 233–243 (1986).
879. WILLIAMSON, P.K. and ZURIER, R.B., *J. Liq. Chromatogr.*, **7**, 2193–2201 (1984).
880. WILSON, D.B., CONNOLLY, T.M., BROSS, T.E., MAJERUS, P.W., SHERMAN, W.R., TYLER, A.N., RUBIN, L.J. and BROWN, J.E., *J. Biol. Chem.*, **260**, 13496–13501 (1985).
881. WING, D.R., HARVEY, D.J., LA DROITE, P., ROBINSON, K. and BELCHER, S. *J. Chromatogr.*, **368**, 103–111 (1986).
882. WINTERMANS, J.F.G.M., *Biochim. Biophys. Acta*, **44**, 49–54 (1960).
883. WITTING, L.A. (editor), *Glycolipid Methodology* (1976) (American Oil Chemists' Society, Champaign).
884. WOLDEGIORGIS, G., SPENNETTA, T., CORKEY, B.E., WILLIAMSON, J.R. and SHRAGO, E., *Anal. Biochem.*, **150**, 8–12 (1985).
885. WOLF, R.A. and GROSS, R.W., *J. Lipid Res.*, **26**, 629–633 (1985).
886. WOOD, R., *J. Chromatogr.*, **287**, 202–208 (1984).
887. WOOD, R., *Biochem. Archs*, **2**, 63–71 (1986).
888. WOOD, R. and HARLOW, R.D., *Archs Biochem. Biophys.*, **131**, 495–501 (1969).
889. WOOD, R. and LEE, T., *J. Chromatogr.*, **254**, 237–246 (1983).
890. WOODFORD, F.P. and VAN GENT, C.M., *J. Lipid Res.*, **1**, 188–190 (1960).
891. WREN, J.J. and SZCZEPANOWSKA, A.D., *J. Chromatogr.*, **14**, 404–410 (1964).
892. WURZENBURGER, M. and GROSCH, W., *Biochim. Biophys. Acta*, **794**, 25–30 (1984).
893. WUTHIER, R.E., *J. Lipid Res.*, **7**, 544–550 (1966).
894. WUTHIER, R.E., *J. Lipid Res.*, **7**, 558–561 (1966).
895. YAHARA, S., KAWAMURA, N., KISHIMOTO, Y., SAIDA, T. and TOURTELLOTTE, W.W., *J. Neurol. Sci.*, **54**, 303–315 (1982).
896. YAHARA, S., MOSER, H.W., KOLODNY, E.H. and KISHIMOTO, Y., *J. Neurochem.*, **34**, 694–699 (1980).
897. YAHARA, S., SINGH, I and KISHIMOTO, Y., *Biochim. Biophys. Acta*, **619**, 177–185 (1980).
898. YAMAGUCHI, M., HARA, S., MATSUNAGA, R., NAKAMURA, M. and OHKURA, Y., *J. Chromatogr.*, **346**, 227–236 (1985).
899. YAMAGUCHI, M., MATSUNAGA, R., FUKUDA, K., NAKAMURA, M. and OHKURA, Y., *Anal. Biochim.*, **155**, 256–261 (1986).

900. YAMAGUCHI, M., MATSUNAGA, R., HARA, S., NAKAMURA, M. and OHKURA, Y., *J. Chromatogr.*, **375**, 27–35 (1986).
901. YAMAMOTO, A., FUJII, Y., YASUMOTO, K. and MITSUDA, H., *Lipids*, **15**, 1–5 (1980).
902. YAMAUCHI, R., KOJIMA, M., ISOGAI, M., KATO, K. and UENO, Y., *Agric. Biol. Chem.*, **46**, 2847–2849 (1982).
903. YAMAZAKI, T., SUZUKI, A., HANDA, S. and YAMAKAWA, T., *J. Biochem.* (*Tokyo*), **86**, 803–809 (1979).
904. YANAGISAWA, I., YAMANE, M. and URAYAMA, T., *J. Chromatogr.*, **345**, 229–240 (1985).
905. YANDRASITZ, J.R., BERRY, G. and SEGAL, S., *J. Chromatogr.*, **225**, 319–328 (1981).
906. YANDRASITZ, J.R., BERRY, G. and SEGAL, S., *Anal. Biochem.*, **135**, 239–243 (1983).
907. YAO, J.K. and RASTETTER, G.M., *Anal. Biochem.*, **150**, 111–116 (1985).
908. YOUNG, P.M., BOEHM, T.M. and BROWN, J.E., *J. Chromatogr.*, **311**, 79–92 (1984).
909. YU, R.K. and LEDEEN, R.W., *J. Lipid Res.*, **13**, 680–686 (1972).
910. ZABKIEWICZ, J.A. and STEELE, K.D., *Chromatographia*, **16**, 92–97 (1982).
911. ZAHLER, P. and NIGGLI, V., in *Methods in Membrane Biology*, Vol. 8, pp. 1–50 (1970) (edited by E.D. Korn, Plenum Press, New York).
912. ZAKARIA, M., GONNORD, M.F. and GUIOCHON, G., *J. Chromatogr.*, **271**, 127–192 (1983).
913. ZELINSKI, S.G. and HUBER, J.W., *Chromatographia*, **11**, 645–646 (1978).
914. ZEMAN, I., RANNY, M. and WINTEROVA, L., *J. Chromatogr.*, **354**, 283–292 (1986).
915. ZHUKOV, A.V. and VERESHCHAGIN, A.G., *Adv. Lipid Res.*, **18**, 247–282 (1981).
916. ZINKEL, D.F. and ROWE, J.W., *Anal. Chem.*, **36**, 1160–1161 (1964).

Index